ATMOSPHERIC CORROSION

THE ELECTROCHEMICAL SOCIETY SERIES

ECS-The Electrochemical Society
65 South Main Street
Pennington, NJ 08534-2839
http://www.electrochem.org

A complete list of the titles in this series appears at the end of this volume.

ATMOSPHERIC CORROSION

Second Edition

CHRISTOFER LEYGRAF
KTH Royal Institute of Technology, Stockholm, Sweden

INGER ODNEVALL WALLINDER
KTH Royal Institute of Technology, Stockholm, Sweden

JOHAN TIDBLAD
Swerea KIMAB, Stockholm, Sweden

THOMAS GRAEDEL
Yale University, New Haven, CT, USA

Published by John Wiley & Sons, Inc., Hoboken, New Jersey
Published simultaneously in Canada

Library of Congress Cataloging-in-Publication Data

Names: Leygraf, Christofer. | Odnevall Wallinder, Inger. | Tidblad, Johan, 1964– | Graedel, T. E.
Title: Atmospheric corrosion.
Description: Second edition / Christofer Leygraf, Inger Odnevall Wallinder, Johan Tidblad, Thomas Graedel. | Hoboken, New Jersey : John Wiley & Sons, Inc., [2016] | Includes bibliographical references and index.
Identifiers: LCCN 2016015705| ISBN 9781118762271 (cloth) | ISBN 9781118762189 (epub)
Subjects: LCSH: Corrosion and anti-corrosives.
Classification: LCC TA418.74 .L49 2016 | DDC 620.1/1223–dc23 LC record available at https://lccn.loc.gov/2016015705

Printed in the United States of America

10 9 8 7 6 5 4 3 2 1

CONTENTS

PREFACE

Iron was first separated from its ore about 4000 B.C. and promptly began to corrode. In the 6000 years since, a wide variety of pure materials have been isolated and alloyed or composite materials have been created from isolated constituents, materials have been worked in various ways, products of almost infinite variety have resulted, and corrosion is still with us.

The global cost of atmospheric corrosion has been estimated, with more bravery than accuracy, at upward of US$100 million per year. Whatever the correct number, it is clear that atmospheric corrosion extracts an enormous toll—electrical and electronic equipment fails to function, bridges collapse, intricate surfaces of statuary grow smooth or disintegrate, and on and on. What can be done about this situation, and how soon?

Atmospheric corrosion has been a subject of engineering study, largely empirical, for nearly a century. Scientists came to the field rather later on and (partly because of inherent experimental and conceptual difficulties discussed in this book) had considerable difficulty bringing their arsenal of tools to bear on the problem. In the decades of the 1990s and 2000s, it was finally possible to initiate controlled field and laboratory studies, as well as computer model investigations of atmospheric corrosion processes. Even so, atmospheric corrosion was traditionally studied by specialists in corrosion having little knowledge of atmospheric chemistry, history, or prospects. In the first version of this book, the approach was to combine the fields, one of the authors (C.L.) being principally an experimental corrosion scientist and the other (T.E.G.) principally an atmospheric chemist. In the second edition the approach has been extended further, by including also an author that has pioneered the field of environmental aspects of corrosion (I.O.W.) and one with substantial insight into atmospheric corrosion modeling and also in international corrosion exposure programs (J.T.). The combination of specialities provides a more comprehensive view than what results from a single specialist picture. The perspectives emerging from our relatively recent efforts and those of many others begin to tell us what is happening when atmospheric corrosion occurs and how it might best be prevented or minimized. These scientific insights into the corrosion process and its amelioration are the focus of this book.

This book concerns primarily the atmospheric corrosion of metals and is written at a level suitable for advanced undergraduates or beginning graduate students in any of the physical or engineering sciences and is designed to be suitable for a one-semester course. In addition, we anticipate that practicing corrosion scientists, corrosion engineers, conservators, and other relevant specialists may find it valuable as a reference guide. Recent concerns about the input of metals to the environment as a result of dissipative corrosion may make the volume of interest to environmental scientists as well. The book begins with five chapters that introduce the subjects atmospheric corrosion of metals and atmospheric chemistry. Chapters 6–9 present information on corrosion mechanisms in a variety of laboratory, outdoor, and indoor environments; our intent is to present a scientific picture for corrosion of primarily metals under these circumstances without being exhaustive. In Chapters 10–14, we discuss more practical topics: how do metals typically used in architectural and structural applications, electronics, and cultural artifacts degrade or disperse, how might such materials be protected, and what materials choices might be made by the designers of the future. In Chapter 15, a prediction is made on how and where atmospheric corrosion may evolve in the future.

A number of appendices provide more detailed information relating to specific materials, experimental techniques, and other relevant topics. The aim has been to make that material accessible without unnecessarily interrupting the presentation of conceptual material in the body of the book.

We are grateful to many of our colleagues near and far for their help during the preparation of this book, in particular Peter Brimblecombe (United Kingdom), Sara Goidanich (Italy), Beatrice Hannoyer (France), Yolanda Hedberg (Sweden), Gunilla Herting (Sweden), Katerina Kreislova (Czech Republic), Vladimir Kucera (Sweden), Nathalie Le Bozec (France), Manuel Morcillo (Spain), Bo Rendahl (Sweden), Bror Sederholm (Sweden), Pasquale Spezzano (Italy), Costas Varotsos (Greece), Susanna Wold (Sweden), and Tim Yates (United Kingdom). We also thank Xian Zhang (Sweden and China) for assistance with the preparation of figures. One of us (C.L.) is most grateful to Stiftelsen San Michele, Sweden, for a 3-week stay at Villa San Michele on the Island of Capri, Italy, during which part of the writing process took place. Finally, we express our appreciation to our publishers for their interest in this book and for their help in seeing it through to publication.

Christofer Leygraf, Stockholm
Inger Odnevall Wallinder, Stockholm
Johan Tidblad, Stockholm
Thomas Graedel, New Haven

1

THE MANY FACES OF ATMOSPHERIC CORROSION

1.1 DR. VERNON'S LEGACY

Thousands of years ago, humanity wrested materials from beneath the surface of Earth and processed them into spear points, rudimentary tools, and ornamental objects, which immediately began to corrode and have been corroding ever since. As technology has evolved and our atmosphere has come to contain increasing levels of acid gases, the rates of corrosion have increased. Everyday corrosion claims its victims—electronic connectors, towering bridges, and unique statuary. The forces opposing these processes are composed of corrosion scientists and engineers, whose war plan must, of necessity, be based on anticipating, understanding, and overcoming the enemy.

The science of atmospheric corrosion—corrosion that occurs in materials exposed to the ambient air—is less than a century old. Beginning in the 1920s, W.H.J. Vernon in England began systematic experiments in atmospheric corrosion. Except for some increased sophistication in instrumentation, his experiments were very similar to those of today: he cleaned metal samples, exposed them to specific concentrations of gases, such as SO_2 and CO_2, or to natural outdoor environments, and determined corrosion rates and the major corrosion products.

Vernon's work took place some 80 years ago. Werner Heisenberg was just inventing the uncertainty principle of quantum physics, the neutron was not yet discovered, polymer chemistry was barely thought of, continental drift was an unsupported speculation, and the DNA double helix would not be discovered for 30 years. Today, quantum physics is a mature specialty, insight into the atomic nucleus has resulted in the use of nuclear power, polymers are ubiquitous, Earth science has been revolutionized by plate tectonics, and biological scientists have sequenced the human genome. Meanwhile, Vernon's experiments are still cited in the corrosion science literature as

Atmospheric Corrosion, Second Edition. Christofer Leygraf, Inger Odnevall Wallinder, Johan Tidblad and Thomas Graedel.

relevant, at least occasionally. What has caused atmospheric corrosion science to stagnate while other scientific fields were forging ahead in great leaps and bounds?

One answer is that other fields are conceptually more straightforward and more highly specialized, while atmospheric corrosion is enormously complex and interdisciplinary. To understand quantum physics, one only needs the atom, its nucleus, and its electrons, for DNA only the molecule, although characterized by a highly complex structure. For atmospheric corrosion, however, one needs to understand a degraded solid phase, a very thin and transitory liquid phase, and a changing gas phase all at once and all without the ability to monitor everything during the time in which corrosion is actually occurring.

A second answer is that many applied investigations in corrosion science have had as their main emphasis the determination of the corrosion rate of a given metal in a given atmospheric environment. In these investigations, the corrosion products formed are to be removed from the metal by some chemical stripping treatment. However, this procedure not only determines the rate by removing the corrosion products, it also removes all the information hidden in the corrosion products that could tell something of what was going on during the corrosion process.

A third answer is that atmospheric corrosion has not traditionally attracted scientists performing fundamental research. In contrast, during the last three–four decades, a substantial amount of fundamentally orientated work in corrosion science has been devoted to understanding the chemical composition and atomic structure of passive films. Both atmospheric corrosion and passivity are research fields with enormous economic consequences. Yet, the efforts made in passivity have far outnumbered the efforts made in atmospheric corrosion. The main reason is simple: it is easier to set up and perform a well-defined laboratory experiment for fundamental passivity studies than for fundamental atmospheric corrosion studies. The former only needs two phases, the passivating metal and the liquid environment, whereas the latter needs three phases, the solid material, the atmosphere and a thin liquid film in between, and a thorough understanding of an intricate and rapidly changing atmospheric chemical environment.

1.2 CONCEPTS AND CONSEQUENCES

Atmospheric corrosion is the result of interaction between a material—an object made of a metal, a calcareous stone, a glass, or a polymer or covered by paint—and its surrounding atmospheric environment. The mechanisms that govern the corrosion or degradation of these materials differ greatly. The scope of this book has therefore been limited to the atmospheric corrosion of metals and alloys, whereas other types of materials only will be discussed occasionally. As opposed to the situation when the material is immersed in a liquid, atmospheric corrosion occurs during unsheltered exposure to rain or in rain-sheltered exposure indoors or outdoors.

Most frequently, atmospheric corrosion is triggered by atmospheric humidity, which forms a very thin water layer on the object. Depending on the humidity conditions, the water layer exhibits different thicknesses, resulting in various forms of atmospheric corrosion. In dry atmospheric corrosion or dry oxidation, the water layer is virtually absent. A common example of dry oxidation is the tarnishing of copper or silver, which can proceed without any humidity in the presence of reduced sulfur compounds. In damp atmospheric corrosion, humidity and traces of atmospheric pollutants result in a thin, mostly nonvisible, water layer. Wet atmospheric corrosion requires rain or other forms of bulk water together with atmospheric pollutants and results in a relatively thick water layer, often clearly visible to the eye.

The consequences of corrosion on our society are enormous. In the United States, for example, the total costs for all forms of corrosion have been estimated to be around 1000 US$ per capita per year. A substantial part of that amount is due to atmospheric corrosion. To estimate the costs for repair of corrosion-induced failures of our infrastructure, including bridges, elevated highways, railway, or subway systems, is tedious but can be done with a certain accuracy. It is more difficult to estimate the costs of direct or indirect consequences caused by atmospheric corrosion of electronic components or systems and how these can affect the reliability of security systems, aircraft, automobiles, or industrial processes. It is likewise difficult to estimate costs related to the loss of our cultural heritage. International concern has increased over the last decades as it has become evident that acid deposition through rain, snow, fog, or dew has resulted in substantial deterioration of artistic and historic objects, including old buildings and structures of historic value, statues, monuments, and other cultural resources.

1.3 THE EVOLUTION OF A FIELD

Developments in our understanding of atmospheric corrosion have been closely linked with society's need to gain more information about a visibly important process. During the first decades of the twentieth century, systematic field exposure programs were implemented in the United Kingdom and the United States when it became obvious that commonly used metals, particularly steel, copper, zinc, and aluminum, suffered from corrosion when exposed in heavily polluted atmospheric environments. The environments were categorized into rural, marine, urban, and industrial, and it was recognized that the metals exhibited different corrosion behaviors in these environments. In the 1920s and 1930s Vernon performed his pioneering work that transformed the field from art to science. He investigated the effect of relative humidity in combination with SO_2 and discovered a rapid increase in atmospheric corrosion rates above a critical relative humidity.

In the decades to come, many important contributions were made by distinguished scientists, including U.R. Evans, J.L. Rosenfeld, and K. Barton, who, among others, could demonstrate the importance of electrochemical reactions in atmospheric corrosion. Further improvements were made by W. Feitknecht, who took into account the chemical properties of the solid products of the corrosion process. Electrochemical techniques thus became common tools for exploring the underlying mechanisms. The success was only partial, however, because of the obvious difficulties of reproducing the actual atmospheric exposure situation in an electrochemical cell in which the sample is completely immersed in an aqueous solution or covered by a relatively thick aqueous layer

In the 1960s and 1970s, atmospheric corrosion effects on electronic components and equipment were recognized. One of the first observations was made in the electronics of American aircrafts in the Vietnam War, which were not adequately protected from the tropical conditions of high humidity and high chloride concentration. It was soon recognized that even very small amounts of corrosion effects, detectable only by highly sensitive analytical techniques, could have detrimental effects on the reliability of electronics. This coincided with the advent of surface analytical techniques such as Auger electron spectroscopy and X-ray photoelectron spectroscopy, capable of providing information on the chemical composition of the outermost atomic layers of a corroded material. A new set of tools was thus available for the understanding of atmospheric corrosion mechanisms. They were complementary to the electrochemical techniques and able to provide more specific chemical information.

As a result of the increasing concern of acid deposition effects in general and the deterioration of the cultural heritage of various countries in particular, several national and international exposure programs were implemented in the 1980s and 1990s. The emphasis in some of these programs was not only on actual corrosion rates but also on a broader characterization of pollutant levels in the atmospheric environments. Efforts were made to correlate corrosion effects with levels of atmospheric constituents, mostly with limited success. Some exposure programs were also broadened to cover both metals and nonmetals and to indoor and outdoor exposures.

In the 1990s and the first decade of the new millennium, the focus was partially altered to consider also environmental consequences of atmospheric corrosion. Driven by new legislations primarily in Europe, the principal question from now on was not only what the environment does to the material but also what the material does to the environment as a result of corrosion. The last two decades have also seen a significant development in analytical tools based on, for example, vibrational spectroscopy, which can provide quantitative and qualitative information on the corrosion effects of a material during ongoing corrosion conditions.

1.4 CONTROLLED LABORATORY ENVIRONMENTS

Through exposures of materials in field environments, characterized by many atmospheric constituents having the potential to influence the corrosion behavior, Vernon and others soon felt the need to perform complementary exposures in laboratory environments, characterized by synthetic air with only a limited selection of atmospheric constituents. In designing such experiments a number of criteria have to be fulfilled, for example, How can a laboratory exposure be designed to simulate exposure in a given field environment? Do the same corrosion mechanisms occur in both types of environments? What ratio is obtained between the corrosion rates obtained in the laboratory and in the field?

Laboratory environments are usually characterized by constant relative humidity, constant temperature, and the addition of one or a few gaseous corrodents. For reproducibility reasons one usually tries to limit the number of gases to a maximum of four. Experience has shown that the levels of gases included in the laboratory environment should not be too high in comparison with the levels found in the field environments; otherwise the possibility exists of stimulating nonrealistic corrosion mechanisms. Earlier laboratory exposures frequently suffered from this problem. With the advent of new instrumental apparatuses, for example, measuring devices for continuous monitoring of levels of gases and permeation tubes for producing low and stable emission of gases, it is now possible to produce laboratory environments with almost the same gas concentrations as occurring in the field.

A further development has been the increased availability of experimental techniques that can be used to monitor under *in situ* conditions changes occurring on a metal surface in the laboratory environment, that is, during ongoing corrosion. As will be shown in later chapters, this greatly improves the possibilities of tracing the main processes responsible for atmospheric corrosion.

1.5 UNCONTROLLED FIELD ENVIRONMENTS

If laboratory environments represent the simplest form of atmospheric environment for corrosion studies, uncontrolled indoor environments definitively represent a higher level of complexity. An indoor environment is usually characterized by relatively constant humidity, temperature, and airflow conditions and also by a broad spectrum of gaseous and particulate constituents, mostly at moderate and relatively constant levels. The constituents may have been produced either outdoors or indoors. In the former case, the levels may be reduced during transport from the exterior to the interior environment because of absorption on walls or in air treatment or ventilation systems.

The experience gained so far from indoor studies is relatively limited. Nevertheless it appears that indoor corrosion effects normally can be explained by the presence of a large number of air constituents at low levels, rather than by a few dominant constituents.

Outdoor environments generally represent the most complex type of environment from an atmospheric corrosion point of view. They are characterized by diurnal variations in temperature and relative humidity, the presence of numerous gases and particles, strongly varying airflow rates, and seasonal variations in solar radiation, temperature, and precipitation, including rain, dew, fog, and snow.

Outdoor exposure programs in one form or another have been carried out during most of the twentieth century. One main result is that corrosion rates under outdoor exposures are strongly influenced by two dominant constituents, sulfur dioxide and chloride ions, in addition to the climatic factors humidity and temperature. Despite numerous attempts, the goal of predicting the corrosion effect of a given material in a given environment remains far from attainment. This is due to the difficulty of taking into account many complicating factors, including the extent of rain sheltering, airflow conditions, and solar radiation.

An important challenge is to predict future corrosion rates based on expected changes in air constituent levels. Whereas the emission of sulfur dioxide has decreased significantly in many urban areas of Europe and North America, the presence of other corrosion-stimulating pollutants, including nitrogen dioxide and ozone, still remains high.

In other parts of the world (e.g., parts of Asia, Africa, and Central and South America), the emission of many gases has increased to very high levels.

1.6 NEW APPROACHES TO ATMOSPHERIC CORROSION STUDIES

Over the last few decades, the new analytical techniques developed to study properties of solid surfaces, such as chemical composition, oxidation state, morphology, and electronic structure, have continued to increase and to improve in terms of resolution and sensitivity. Most techniques are based on photons, electrons, atoms, or ions as probing particles. The earlier surface probing techniques require high vacuum during their application. Their use is therefore restricted for *in situ* studies of a surface, that is, during ongoing corrosion. The more recent analytical techniques are both surface sensitive and able to provide information under *in situ* conditions. The most promising of these from an atmospheric corrosion point of view include atomic force microscopy, the quartz crystal microbalance, infrared reflection absorption spectroscopy, confocal Raman spectroscopy, and the Kelvin probe. It is anticipated that the number and variety of *in situ* techniques for probing surfaces will continue to increase.

In parallel with the increased availability of *in situ* information from corroding surfaces is the gradual availability of computer models for describing atmospheric corrosion. In the best circumstances these should include all of the most important physical, chemical, and other processes. At least two such models have been developed; they appear to describe the most important processes that occur during initial exposure of metals to laboratory environments.

1.7 AN OVERVIEW OF THIS BOOK

The intent in this book is to bring together the information from experimental and theoretical studies of atmospheric corrosion of primarily metallic materials in such a way that the current state of knowledge is presented to the reader in a pedagogically useful and technically accurate manner. We have not attempted to review all relevant work or to present a compendium of information. Rather, our intent is to guide the reader through the evidence leading to a consistent scientific picture of the atmospheric corrosion process.

To address this target, we begin by describing a framework for an understanding of atmospheric corrosion, followed by an overview of the atmospheric species responsible for the corrosion processes. Advanced stages of corrosion are described, and reaction sequences presented. Specific applications such as electronics and cultural artifacts are addressed, and the costs and dispersion of metals due to atmospheric corrosion are discussed. The book culminates with projections for the corrosion environments of the future. Individual materials and their corrosion susceptibilities are presented in a series of appendices.

In this complex field and in a changing world, much remains to be learned about the details of atmospheric corrosion processes. Much information is known, however, and its selective presentation constitutes the remainder of this book.

2

A CONCEPTUAL PICTURE OF ATMOSPHERIC CORROSION

2.1 INTRODUCTION

Atmospheric corrosion incorporates a wide spectrum of chemical, electrochemical, and physical processes in the interfacial domain from the gaseous phase to the liquid phase to the solid phase. What makes atmospheric corrosion so complex is the fact that important processes occur in all three phases and in the interfaces between them. In order to illustrate the numerous processes involved, this chapter provides a conceptual picture ranging from the initial stages of the corrosion process, occurring within far less than a second, to intermediate stages and to the final stages, which occur after many years or even decades of exposure. This generalized description is mainly based on the understanding of metals initially covered with an oxide or hydroxide layer of thickness a few nanometer (nm) ($1\ nm = 10^{-9}\ m$), although many of the processes operate on nonmetallic materials as well. To a large extent, this knowledge has emerged from the recent use of analytical techniques for detecting processes on a solid surface in contact with an atmospheric environment. Examples of such techniques and of the information extracted will be provided during the description of different stages involved in the atmospheric corrosion processes.

2.2 INITIAL STAGES OF ATMOSPHERIC CORROSION

2.2.1 Surface Hydroxylation

The first stage of interaction between the solid and the atmosphere is the instant reaction of water vapor with the solid. The water molecule may either bond in molecular form or in dissociated form. In the former case the bonding is through the oxygen

Atmospheric Corrosion, Second Edition. Christofer Leygraf, Inger Odnevall Wallinder, Johan Tidblad and Thomas Graedel.

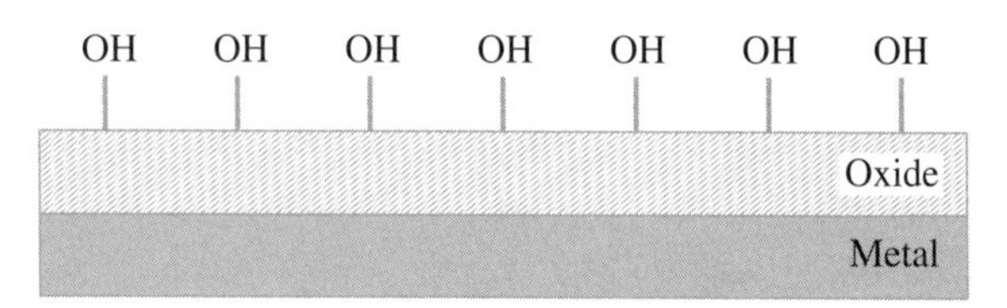

FIGURE 2.1 A schematic depiction of surface hydroxyl groups on a metal oxide surface.

atom to the metal or another positively charged surface constituent, a process that is associated with a net transfer of charge from the water molecule to the metal. The driving force for water dissociation is the formation of metal–oxygen or metal–hydroxyl bonds. Studies of well-characterized monocrystalline metal oxide surfaces have shown that the tendency for water dissociation increases with the number of lattice defects. Most polycrystalline materials used for engineering purposes are expected to adsorb water in dissociated form. Hence, surface hydroxyl groups are generated, and these act as sites for further water adsorption on most metal or metal oxide surfaces. A schematic illustration of surface hydroxyl groups is given in Figure 2.1.

2.2.2 Adsorption and Absorption of Water

The formation of the surface hydroxyl layer is a very fast process. It occurs within a small fraction of a second and results in a surface less conducive to rapid combination with water. Upon further exposure to the atmosphere, subsequent atmospheric water is adsorbed in the molecular form. The first layer of water has a high degree of ordering relative to the substrate because of its proximity to the solid surface. The second and third layers are more mobile with a higher degree of random orientation. Aqueous films thicker than three monolayers possess properties that are close to those of bulk water.

Studies performed on clean and well-defined metal surfaces covered by a thin oxide or hydroxide layer have shown that the amount of reversibly adsorbed water depends on the relative humidity, the time of exposure, and the nature of the substrate. Examples of the number of equivalent monolayers of water on different clean and oxidized metals at various relative humidities are shown in Figure 2.2. (The term "equivalent monolayer" refers to the amount of water present if it is uniformly distributed on the surface. The clustering of water into clusters or small droplets is common at early stages of adsorption, however, and the initially adsorbed water is generally present in cluster form.)

The variation in water adsorption characteristics among the metals can be caused by a number of factors, including the hydrophobic or hydrophilic properties of the solid surface and its density of defects. If the surface is exposed to an outdoor or indoor environment, rather than to a synthetic laboratory air, it will most likely contain adsorbed aerosol particles with more or less developed hygroscopic properties. The particles attract water and result in a higher number of adsorbed or absorbed

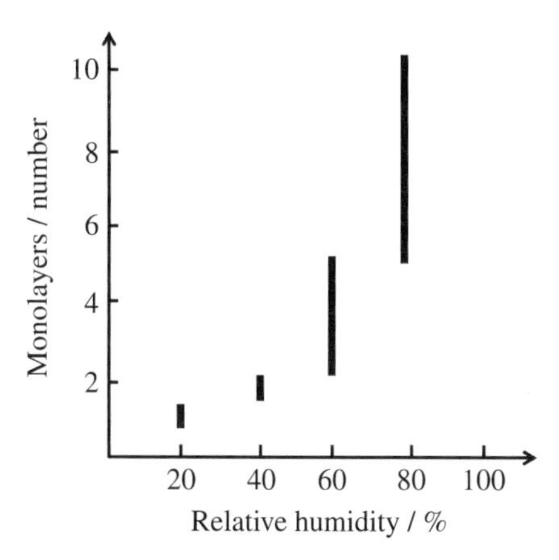

FIGURE 2.2 The number of adsorbed monolayers of water on different metals as a function of relative humidity.

water molecules at a given relative humidity than predicted by Figure 2.2. Other surface properties, such as porosity (of, e.g., stone surfaces) or roughness, also have a substantial influence on water adsorption characteristics.

In order to illustrate in more detail the interaction between humidity and a metal surface, Figure 2.3 shows how the total mass of a gold surface varies during an outdoor exposure in which the surface was exposed to the environment but protected from direct exposure to precipitation and wind. The mass response consists of a monotonous increase on which a periodic mass variation is superimposed. The periodic variation coincides with variation in relative humidity and is obviously caused by the adsorption or absorption of water. The linear mass increase, on the other hand, is due to the adsorption of aerosol particles. The results clearly demonstrate that the adsorbed aerosol particles have hygroscopic properties. The amount of reversibly absorbed water increases with the number of hygroscopic particles and reaches values in the range between 10^{-2} and $10^{-1}\,g\,m^{-2}$ (10^3 to $10^4\,ng\,cm^{-2}$). This range can be compared with the estimated amount of water on a metal surface covered by dew ($10^1\,g\,m^{-2}$) or wet from rain ($10^2\,g\,m^{-2}$), respectively.

Most metal surfaces are very heterogeneous due to the presence of lattice defects such as grain boundaries, steps, kinks, and terraces. Partly because of the substrate heterogeneity and partly because the bond strength between neighboring hydrogen bonded water molecules is similar to the bond strength between water molecules and the hydroxylated substrate surface, the water molecules tend to adsorb heterogeneously and form water clusters. As a result of this geometry in the heterogeneous substrate–water interaction, reaction products may frequently form as islands or as films that vary substantially in thickness over the surface.

Electrochemical reactions have long been recognized as playing a vital role in atmospheric corrosion. They occur at the interface between the solid substrate and the water layer. The heterogeneity of most surfaces results in the presence of some

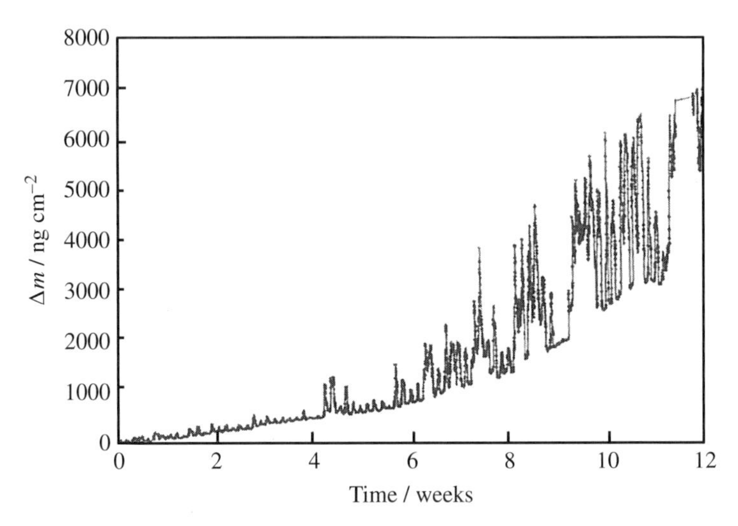

FIGURE 2.3 The variation of mass of a gold surface during outdoor exposure. The data is based on quartz crystal microbalance studies. The gold surface has been protected from direct exposure to precipitation and wind by means of a porous filter of polytetrafluoroethylene, having a pore size of 10 μm. (Reproduced with permission from The Electrochemical Society; M. Forslund and C. Leygraf, Humidity sorption due to deposited aerosol particles studied in situ outdoors on gold surfaces, *Journal of the Electrochemical Society, 144*, 105–113, 1997.)

surface sites conducive to predominantly anodic reactions (i.e., electron-producing reactions) and other surface sites conducive to predominantly cathodic reactions (i.e., electron-consuming reactions). The dominant anode and cathode reactions in atmospheric corrosion are normally written as

$$\mathrm{Me} \rightarrow \mathrm{Me}^{n+} + ne^{-} \quad \text{(anode reaction, metal dissolution)} \tag{2.1}$$

$$1/2\mathrm{O}_2 + \mathrm{H}^{+} + 2\mathrm{e}^{-} \rightarrow \mathrm{OH}^{-} \quad \text{(cathode reaction, oxygen reduction)} \tag{2.2}$$

The reaction product configuration displayed in Figure 2.4 provides direct experimental evidence of a clearly distinguishable area where the anode reaction has occurred and a surrounding area where the cathodic reaction has occurred. The example is from brass Cu20Zn (Cu with 20 wt% Zn) exposed to humidified air and with 120 ppbv (volume parts per billion) of acetic acid as a corrosion stimulator. The brass surface was heterogeneous before exposure start with some grains slightly more zinc rich than the surrounding matrix. Zinc is less noble than copper, and the more zinc-rich grains therefore act as anodes during the exposure and the surrounding matrix as cathode. In Figure 2.4 the inner circular part acted as anode where primarily zinc dissolved and then reacted with acetate ions from acetic acid to form zinc acetate. The outer ring-formed part acted as cathode where primarily dissolved copper ions reacted with hydroxyl ions from the cathode reaction to form Cu_2O. As shown in this figure, the sites for anodic and cathodic reactions are frequently separated from each other during atmospheric corrosion.

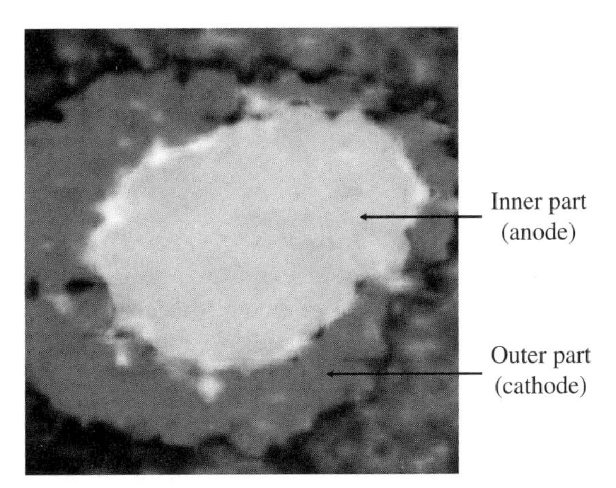

FIGURE 2.4 An image of an exposed brass surface as obtained with confocal Raman spectroscopy. The inner circular part has acted as anode during exposure to form zinc acetate. The outer ring-formed part has acted as cathode to form Cu_2O. The image size is $10 \times 10\,\mu m$. (Reproduced with permission from The Electrochemical Society; P. Qiu and C. Leygraf, Multi-analysis of initial atmospheric corrosion of brass induced by carboxylic acids, *Journal of the Electrochemical Society, 158*, C172–C177, 2011.)

Because of the ready access of atmospheric oxygen to the thin liquid layer on the surface, the anodic rather than the cathodic reaction tends to be the rate-limiting part of the process.

2.3 INTERMEDIATE STAGES OF ATMOSPHERIC CORROSION

2.3.1 Gas Deposition

The aqueous phase not only acts as a medium for electrochemical reactions but also as a solvent for atmospheric constituents, either gaseous or particulate, that deposit into the liquid layer. Important atmospheric constituents known to influence atmospheric corrosion rates include the gases nitrogen dioxide (NO_2), sulfur dioxide (SO_2), hydrogen chloride (HCl), hydrogen sulfide (H_2S), carbon dioxide (CO_2), ammonia (NH_3), nitric acid (HNO_3), oxygen (O_2), ozone (O_3), hydrogen peroxide (H_2O_2), and formaldehyde (HCHO), all in gaseous form, and sodium chloride (NaCl), ammonium sulfates ($(NH_4)_2SO_4$, NH_4HSO_4), ammonium chloride (NH_4Cl), and sodium sulfate (Na_2SO_4), all in particulate form. In some cases the constituents are incorporated into the liquid layer as components of rain, dew, fog, or snow. On the other hand, the deposition process involves impacting onto or diffusion to the surface.

When the aqueous layer and the atmosphere are in equilibrium, the chemical activity of any dissolved gaseous constituent in the aqueous layer is directly proportional to the partial pressure of the same constituent in the gas phase. In most atmospheric corrosion applications, however, the conditions are far from equilibrium.

One reason is the relatively slow deposition rate of most atmospheric constituents, controlled by factors that include aerodynamic properties, such as wind speed and type of wind flow, and surface specific properties, such as the thickness and chemical makeup of the liquid layer.

2.3.2 Change in Liquid Layer Chemistry

The deposition of constituents into the aqueous phase results in a number of chemical and electrochemical reactions. It may also eventually result in the emission of volatile reaction products from the aqueous phase into the atmosphere. Important chemical reactions that may occur include the transformation of SO_2 into sulfurous acid (H_2SO_3) or sulfuric acid (H_2SO_4), the transformation of NO_2 into nitric acid (HNO_3), and the dissociation of these acids.

Among the factors that make it unlikely for the liquid layer to be in chemical equilibrium with the atmosphere are the daily variations in atmospheric conditions. As seen clearly in Figure 2.3, the thickness of the liquid layer changes on a daily basis so, therefore, does its chemical composition. When the corroding surface undergoes these wetting and drying cycles, one can expect changes in ionic strength of many orders of magnitude. In this way, the pH of the aqueous layer may change from neutral or slightly acidic at high liquid layer thickness to highly acidic at low liquid layer thickness.

2.3.3 Proton- and Ligand-Induced Metal Dissolution

The ions formed in the liquid layer may each interact with the oxidized metal surface and thereby enhance the dissolution of the metal. Figure 2.5 provides a schematic illustration of the dissolution process as evidenced from both experimental studies and from computational modeling.

As described previously, the metal oxide surface terminates with surface hydroxyl groups next to the liquid layer under most initial atmospheric exposure

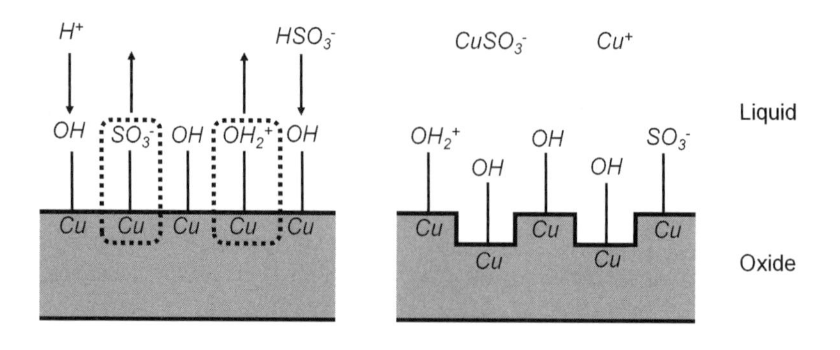

FIGURE 2.5 A schematic diagram of proton- and ligand-induced metal dissolution before (left) and after (right) surface corrosion reactions.

conditions. These groups have excellent ion exchange properties and may easily be replaced by ions from the liquid layer. Among the ions that commonly undergo exchange are protons (H^+) and ligands such as bisulfite (HSO_3^-) or bisulfate (HSO_4^-). As this replacement occurs, the bonds between the surface metal atoms and their immediate neighbors may be weakened, thereby promoting the dissolution of the metal. As discussed in more detail in later sections, this proton- and ligand-induced metal dissolution has been identified as one of the most important processes of atmospheric corrosion, at least in the initial stages. Although the dissolution process is not completely understood on a molecular level, it is believed that protons play the vital role. When released, they may easily move between different surface metal sites. A surface metal surrounded by two or more proton-bonded neighbors is more likely to be detached from the metal oxide lattice because of bond weakening through polarization of the bond than is a normally bonded atom. As the metal atom leaves the lattice, it enters into the liquid layer as a hydrated aquo-metal ion or aniono complex of the metal ion. The broken bond at the surface is immediately hydroxylated, and the protons involved in this process are released and can take part in the detachment of another surface metal atom. Simultaneously, as the dissolution of metal oxide proceeds, new metal oxide is formed. This oxide formation process is much faster than the dissolution process, at least under benign exposure conditions.

2.3.4 Ion Pairing

When the metal ion dissolves into the liquid layer, it can coordinate with counterions that are present. This ion pairing process depends on the nature of both the metal ion and the counterion. Although many combinations of ion pairing are possible, certain ones are more likely to occur than others. One way to rationalize ion pairing is by means of the Lewis acid–base concept. When an ion pair interacts, electrons from the counterion (the Lewis base) form a covalent bond to the metal ion (the Lewis acid). Lewis acids and bases possess different electron properties, and only those with similar properties are likely to coordinate with each other. Soft acids or bases have valence electrons that can more easily be removed or polarized, whereas hard acids or bases have valence electrons that are more tightly held and are not so easily polarized. Hard acids preferentially coordinate with hard bases and soft acids with soft bases.

Table 2.1 displays a selection of metal ions and counterions and their classifications into hard, intermediate, and soft Lewis acids and bases. The hardness of an ion generally increases with its oxidation state, for example, Fe^{3+} is a harder Lewis acid than Fe^{2+}. Table 2.1 suggests, for instance, that a soft acid such as Ag^+ will more likely coordinate with a reduced sulfur compound than with a sulfate ion. As will be shown later, this is in full agreement with the analysis of corroded silver surfaces, which frequently show silver sulfides as corrosion products but seldom any silver sulfate. Similarly, the table suggests that a hard acid, such as Al^{3+}, is more likely to coordinate with oxygen or sulfate ions than with reduced sulfur compounds, again in good agreement with observations of corrosion products.

TABLE 2.1 Classification of Selected Lewis Acids and Bases

Hard	Intermediate	Soft
Acids		
H^+, Na^+, Mn^{2+}, Al^{3+}	Fe^{2+}, Ni^{2+}, Cu^{2+}	Cu^+, Ag^+
Cr^{3+}, Fe^{3+}, Ti^{4+}	Zn^{2+}, Pb^{2+}	
Bases		
H_2O, OH^-, O^{2-}	SO_3^{2-}, NO_2^-	H_2S, HS^-
SO_4^{2-}, NO_3^-, CO_3^{2-}		

2.3.5 Photosensitivity

There is clear evidence for photosensitivity in at least some atmospheric corrosion processes. Experimental examples include the exposure of copper to H_2S gas in the presence of simulated sunlight; copper to O_3, humidity, and deposited NaCl; and silver to O_3, humidity, and deposited NaCl in ultraviolet (UV) radiation. The presence of UV radiation can both accelerate and retard the atmospheric corrosion process. On one hand, UV radiation induces photolysis of O_3. In the presence of humidity, this creates atomic oxygen (O) and the hydroxyl radical ($OH^\cdot$), both of which are much more reactive than O_3 toward a metal such as silver. But UV light also decomposes deposited NaCl, which slows down corrosion. The overall light-induced effect on atmospheric corrosion rate can be quite complicated as illustrated in Figure 2.6. The figure displays the amount of corroded copper exposed to 250 ppbv O_3 and deposited NaCl with and without UV radiation. The lower curve, obtained in the dark, shows how the corrosion rate increases with relative humidity because of the increase in the thickness of the aqueous phase. Above the point of deliquescence of NaCl around 75% relative humidity, the aqueous layer possesses a high NaCl concentration, which results in an accelerated amount of corroded copper. The upper curve shows how the amount of corroded copper accelerates upon UV radiation. This is attributed to two mechanisms: the photoinduced current caused by the creation of electron-hole pairs in the semiconducting cuprous oxide layer and the formation of atomic oxygen through photolysis of O_3. Another example of photosensitivity is the enhanced corrosion of iron in the presence of solar photons. In a wetted surface layer, iron ions form complexes with hydroxide ions. Upon absorption of photons with $\lambda < 400$ nm, the Fe(OH)-complex undergoes charge transfer to the metal ion, liberating the OH radical to inaugurate a series of aqueous chemical reactions including the sequential corrosion of iron. This particular photocorrosion process is unique to iron among the common industrial metals, since iron is the only one whose hydroxide complexes are sensitive to the solar photon wavelengths that penetrate the atmosphere and reach Earth's surface.

2.3.6 Nucleation of Corrosion Products

When the concentration of ion pairs in the liquid layer eventually reaches supersaturation, the ion pairs will precipitate into a solid phase. This precipitation process is complex, and the precipitated species may pass through the colloidal state before

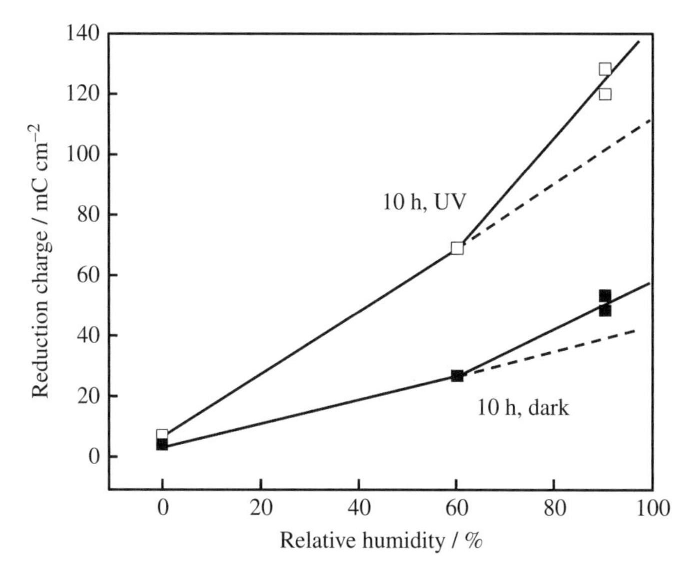

FIGURE 2.6 The amount of corroded copper (measured as the charge required during electrolytic cathodic reduction of the cuprous oxide formed) after 10 h exposure to 250 ppbv of O_3 and 4 μg cm^{-2} predeposited NaCl in air with various relative humidities. Upper: with UV radiation; lower: without UV radiation. The dashed lines are extrapolations of data from lower relative humidities to facilitate comparison with results at higher relative humidity. (Reproduced with permission from Maney Publishing; H. Lin and G.S. Frankel, Atmospheric corrosion of Cu by UV, ozone and NaCl, *Corrosion Engineering, Science and Technology, 48*, 461–468, 2013.)

they reach the solid state. The nucleation of precipitated species is facilitated by the heterogeneous nature of the substrate surface, in particular by solid state defects of various kinds, which can act as nucleation sites. The overall formation rate of the precipitate is therefore limited by the rate of growth rather than by the nucleation rate, from which it follows that many small precipitated nuclei rather than a few large ones are expected. This is in good agreement with experimental evidence, such as that of Figure 2.7, which shows the formation of Cu(I)-oxide at various sites on a copper substrate.

Figure 2.8 summarizes important concepts and processes described so far and their involvement during initial atmospheric corrosion. The example is based on brass (Cu20Zn) exposed to humidified air to which acetic acid (CH_3COOH) was added as corrosion stimulator. The initial atmospheric corrosion can be described by several steps. The first step (number 1 in the figure) is the dissociation of acetic acid into protons (H^+) and acetate ions (CH_3COO^-) upon entry into the aqueous film. Based on experimental findings and computational modeling, two reaction pathways have been identified, as already described in Section 2.3.3. One is the proton-induced dissolution of copper (steps 2 and 3) followed by the reaction of cuprous ions and hydroxyl ions from the cathodic reaction to form cuprite (Cu_2O, steps 4 and 5).

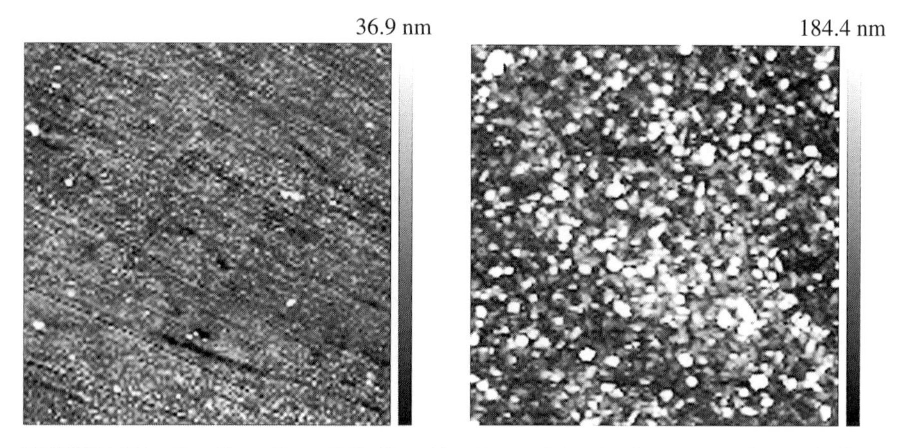

FIGURE 2.7 The formation of Cu(I)-oxide observed *in situ* by means of atomic force microscopy during exposure of a copper substrate to humidified air at 70% relative humidity. The left figure shows the diamond polished copper surface before exposure to humidified air, the right figure after 168 h of exposure. Each image is $10 \times 10\,\mu m$ in width. (Reproduced with permission from X. Zhang and X. Liu, KTH Royal Institute of Technology, Stockholm, Sweden.)

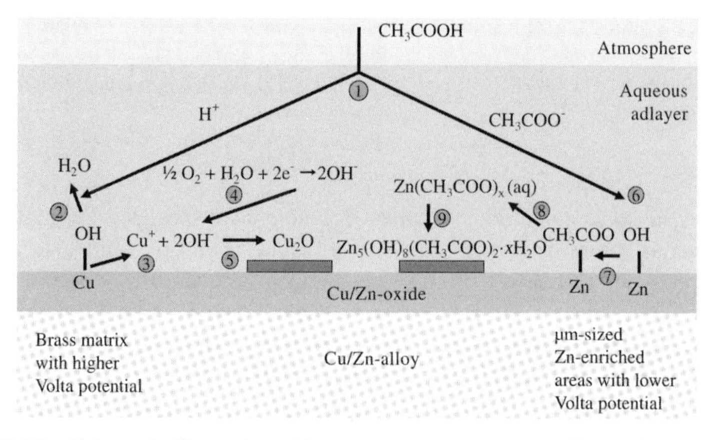

FIGURE 2.8 Schematic illustration of important concepts and reaction pathways identified during the initial atmospheric corrosion of brass (Cu20Zn) exposed to humidified air and addition of 120 ppbv of acetic acid. (Reproduced with permission from The Electrochemical Society; P. Qiu and C. Leygraf, Multi-analysis of initial atmospheric corrosion of brass induced by carboxylic acids, *Journal of the Electrochemical Society, 158*, C172–C177, 2011.)

The second step is the acetate ion-induced dissolution of zinc as $Zn(CH_3COO)_x$ (aq) (steps 6–8) followed by the precipitation of zinc hydroxyacetate, $Zn_5(OH)_8(CH_3COO)_2{\cdot}H_2O$ (step 9). The lateral distribution of these reaction products have already been discussed in Section 2.2.2 and displayed in Figure 2.4, and the first reaction pathway (steps 2–5) is associated with the cathodic reaction and the second pathway (steps 6–9) with the anodic reaction. A driving force for these reactions is the variation in electrochemical nobility along the brass surface from more Cu-rich grains, characterized by a slightly higher electrochemical nobility (Volta potential), to less Cu-rich grains, characterized by lower electrochemical nobility (Volta potential).

2.4 FINAL STAGES OF ATMOSPHERIC CORROSION

2.4.1 Coalescence of Corrosion Products

With prolonged exposure, the number of precipitated nuclei and their size increases until eventually they completely cover the metal surface, Figure 2.9. These precipitates at this stage are normally referred to as corrosion products, and they play a most important role for the behavior of any material in a given environment. The "brown rust" layer on steel, the "green patina" layer on copper, and the "white rust" layer on zinc are common designations for clearly visible and frequently observed corrosion products. Other corrosion products may not be visible to the eye. The most prominent examples are the passive films that form on stainless steel and are responsible for its excellent corrosion resistance in many environments

When the thin layer of corrosion products has grown to cover the whole surface, further growth requires that reactive species from the liquid layer must be transported inward through the layer, that metal ions are transported outward, or that both transport processes occur simultaneously. In general, protons and other singly charged ions such as chloride ions are transported more easily through the corrosion products than are doubly charged species such as sulfate ions. As long as the film of corrosion products is thin or porous, the transport of ions is not significantly hindered and the corrosion products have poor protective ability. Thicker and more

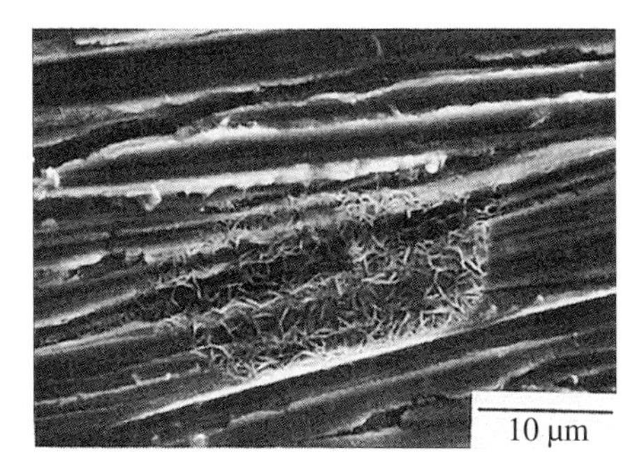

FIGURE 2.9 The growth and coalescence of corrosion products on a zinc substrate exposed to an urban atmosphere during 2 (left) and 30 (right) days, respectively. (Reproduced with permission from Elsevier; I. Odnevall and C. Leygraf, Formation of $Zn_4Cl_2(OH)_4SO_4 \cdot 5H_2O$ in an urban and an industrial atmosphere, *Corrosion Science, 36*, 1551–1668, 1994.)

dense films of corrosion products generally result in higher corrosion resistance because the transport of ions is increasingly restricted.

In addition to the transport of ions, the transport of electrons from the anodic to the cathodic reaction sites on the surface has to be considered. This transport is a necessary process so that the electrons produced in the anodic reaction can be consumed in the cathodic reaction. As long as the metal substrate is covered only by a thin oxide film, the transport of electrons through that film is generally not a rate-limiting step. However, when the corrosion products grow in thickness, especially if those products possess poor electron conduction properties, the electron transport may become rate limiting.

The eventual rate-limiting step caused by electron transport cannot be fully comprehended without considering the availability of oxygen for the cathodic reaction. Due to the ready availability of oxygen in thin liquid films, it is generally believed that the transport of oxygen from the atmosphere to the cathodic site is not a rate-limiting step. Hence, as long as the liquid layer and the film of corrosion products is thin enough, the rate-limiting step is most likely the anodic reaction, that is, the metal dissolution process. Exceptions occur if the liquid layer exceeds a thickness of several tenths of micrometers, in which case oxygen transport may become rate limiting, or if the film of corrosion products develops such that it hinders the electron or the oxygen transport.

2.4.2 Aging and Thickening of Corrosion Products

The long-term growth of the corrosion products is highly dependent on actual exposure conditions. In most outdoor exposures, characterized by daily variations in relative humidity and in liquid layer thickness, the sequential growth of the corrosion product layer follows a sequence of at least three steps. The first occurs during the increase of the liquid layer thickness when part of the corrosion product may dissolve, the second involves the coordination in the liquid layer of dissolved metal anions and counterions, and the third step occurs during the decrease of the liquid layer thickness when the newly coordinated ion pairs reprecipitate. The repeated cycles of dissolution–coordination–precipitation continuously cause the layer of corrosion products to age by changing its chemical composition, microstructure, crystallinity, thickness, and other properties. As will be discussed in Chapter 9, the corrosion products formed on zinc, copper, carbon steel, and aluminum are excellent illustrations of the gradual change in chemical composition that occurs during exposure in different environments. In a marine atmosphere, for instance, Cl^- ions from aerosols are the dominant atmospheric constituents although SO_2 gas and other aerosols are also important atmospheric corrosion stimulators. Within a few hours of marine exposure of zinc, islands of a weakly protective zinc hydroxychloride form, grow laterally, and coalesce. After weeks of exposure, a more protective zinc chlorohydroxysulfate has formed.

Another example of change in corrosion product characteristics during prolonged exposure is a gradual transformation from the amorphous state to more crystalline.

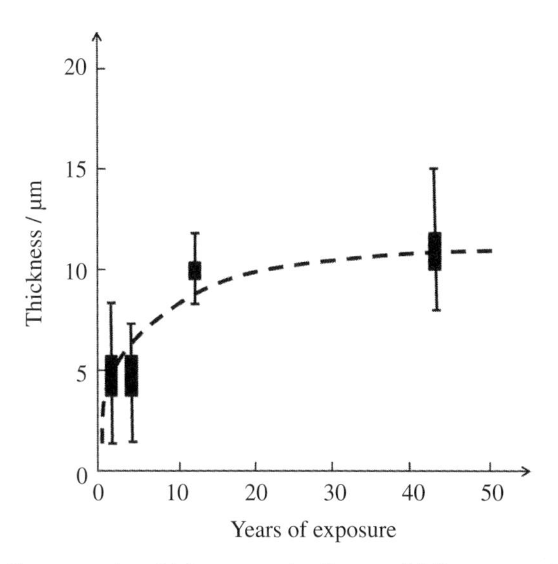

FIGURE 2.10 Copper patina thickness on the Statue of Liberty as a function of time. (Reproduced with permission from Elsevier; T.E. Graedel, Copper patinas formed in the atmosphere, *Corrosion Science, 27*, 741–769, 1987.)

This pattern is observed during the exposure of nickel by the formation of first an amorphous and later a crystalline nickel hydroxysulfate.

When the corrosion products eventually acquire characteristics that no longer change with time, the corroding material becomes characterized by a constant corrosion rate. This implies that the corrosion product finally reaches a constant thickness, with a certain amount of material corroding away during each unit of time and the same amount of material running off the corrosion products during that time. Figure 2.10 shows one of many pieces of evidence that the time needed to reach stationary atmospheric corrosion conditions may take several years or even decades; it summarizes the estimated copper patina thickness formed on the Statue of Liberty as a function of time. The results suggest that a constant patina thickness was reached after a few decades.

This gradual transformation is in agreement with the common observation that corrosion products formed after very long atmospheric exposure resemble or are identical to minerals formed as the result of natural processes in soil and rock. Many of the chemical, electrochemical, and physical processes applicable to the corrosion of materials and operating over long time periods are obviously also involved in the formation of natural minerals.

To summarize this chapter, Figure 2.11 illustrates the great difference in timescale involved in atmospheric corrosion processes. Although very approximate, the figure clearly demonstrates that these events occur over times that differ by many orders of magnitude.

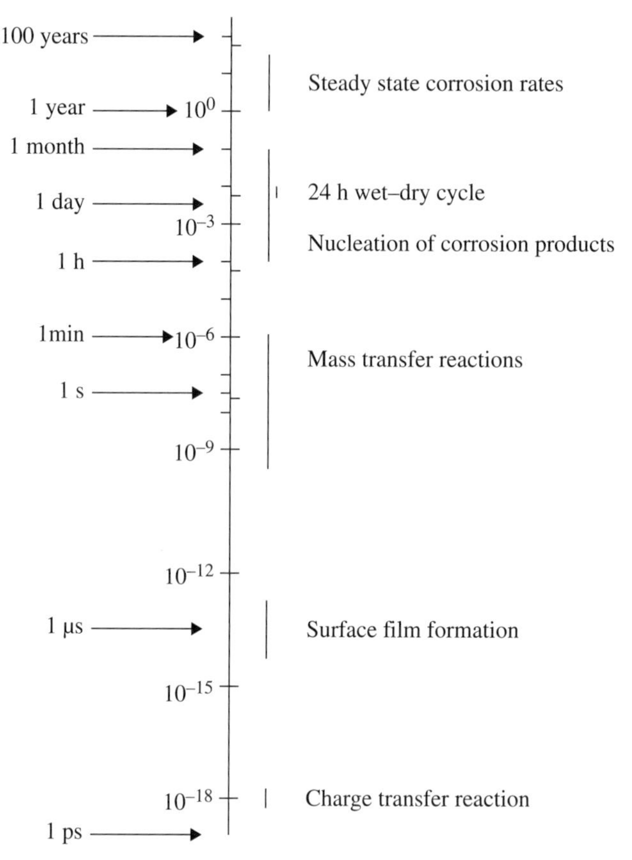

FIGURE 2.11 Orders of magnitude of times required for processes and events of relevance to atmospheric corrosion.

FURTHER READING

The following references provide more detailed reading on the concepts introduced in this chapter.

Surface Hydroxylation and Growth of Aqueous Layer on Metals

M. Salmeron, H. Bluhm, M. Tatarkhanov, G. Ketteler, T.K. Shimizu, A. Mugarza, X. Deng, T. Herranz, S. Yamamoto, and A. Nilsson, Water growth on metals and oxides: binding, dissociation and role of hydroxyl groups, *Faraday Discussions, 141,* 221–229, 2009.

E. Schindelholz and R.G. Kelly, Wetting phenomena and time of wetness in atmospheric corrosion: a review, *Corrosion Reviews, 30,* 135–170, 2012.

Proton- and Ligand-Induced Metal Dissolution

W. Stumm and G. Furrer, The dissolution of oxides and aluminum silicates; examples of surface-coordination controlled kinetics, in *Aquatic Surface Chemistry,* ed. W. Stumm, John Wiley & Sons, New York, p. 197, 1987.

Ion Pairing

R.G. Pearson, Hard and soft acids and bases, *Journal of American Chemical Society, 85*(22), 3533–3539, 1963.

Photosensitivity

H. Lin and G.S. Frankel, Atmospheric corrosion of Cu by UV, ozone and NaCl, *Corrosion Engineering, Science and Technology, 48,* 461–468, 2013.

Nucleation, Coalescence, and Aging of Corrosion Products

D. Persson and C. Leygraf, Initial interaction of sulfur dioxide with water covered metal surfaces: an in situ IRAS study, *Journal of Electrochemical Society, 142*, 1459–1468, 1995.

I. Odnevall and C. Leygraf, Reaction sequences in atmospheric corrosion of zinc, in *Atmospheric Corrosion,* ASTM STP 1239, eds. W.W. Kirk and H.H. Lawson, American Society for Testing and Materials, Philadelphia, pp. 215–229, 1995.

3

A MULTIREGIME PERSPECTIVE ON ATMOSPHERIC CORROSION CHEMISTRY

3.1 INTRODUCTION TO MOIST-LAYER CHEMISTRY

Earth and many processes on earth, including atmospheric corrosion, function as they do because of the presence and properties of water. Water, the universal solvent, is the medium in which most of nature's chemical reactions occur and in which many chemical products of interest and importance are produced. From both experimental and theoretical standpoints, the chemistry of bulk aqueous solutions is reasonably well understood. The framework involves statistical descriptions of the interactions of reactive constituents with water molecules in a spatially uniform, infinite matrix. This approach is suitable for analysis of a wide variety of processes taking place in an aqueous medium. It ignores, however, the circumstance that while surfaces are sometimes wet, they are perhaps more often moist, and the chemistry of these moist layers can differ from that of bulk water in a number of important respects. In particular, there is good evidence that, in the not uncommon scenario of wetting and drying of a surface, a common situation in atmospheric corrosion, the most interesting chemistry occurs during the short time when the aqueous layer is thin and the electrolyte concentrations are high.

Careful experiments have established that a moist surface is characteristic of metals exposed to the atmosphere. A perspective on the amount of water typically present on surfaces was provided by the data shown in Figure 2.2, where a number of different metals are seen to be covered by water equivalent to 2–10 monolayers for relative humidities exceeding 40% at normal room temperatures. In all cases the amount of water associated with the surface is a function of temperature, porosity, degree of oxidation, grain structure, and a variety of other surface-related properties. The simplest summary statement is that unless the local environment is at a

Atmospheric Corrosion, Second Edition. Christofer Leygraf, Inger Odnevall Wallinder, Johan Tidblad and Thomas Graedel.

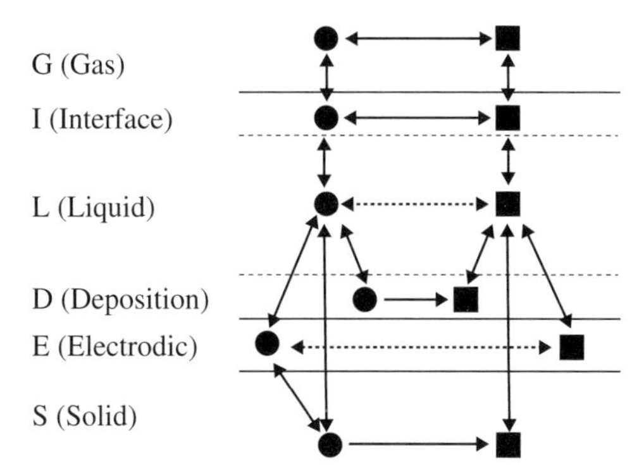

FIGURE 3.1 A schematic representation of the six regimes and the transitions and transformations that need to be considered in aqueous corrosion chemistry. In the figure, the circles indicate initial reactants and the squares indicate reaction products. Solid lines connecting reactants and/or products denote processes that can be readily described by formulations based on first principles. The dashed lines indicate that liquid-phase chemistry can be described by first principles at low ionic strength but must be parameterized at high ionic strength.

very low relative humidity, nearly all material surfaces are moist. Surfaces exposed to precipitation or to freshwater or seawater will, of course, be overlain with water amounts sufficiently thick to be considered as bulk water.

The water layer does not exist in chemical isolation. In fact, it is important to recognize that there are six distinct regimes that may need to be considered, as shown in Figure 3.1: gas (G), interface (I), liquid (L), deposition layer (D), electrodic regime (E), and solid (S), so models treating such systems may be designated as GILDES models. (Such models and their results will be described later in this book.) Within the regimes chemical reactions can occur to change the constituents. Transport of chemical species of interest between regimes can occur, as in condensation and dissolution, and must be assessed. If a regime is not well mixed, there may be concentration gradients within it, and transport within regimes is then of interest. The products of the chemical reactions are susceptible to transport and deposition or volatilization just as are the reactants.

For the six regimes, mathematical formulations can be specified to describe the transitions and transformations that occur. Some can be described from first principles, while in other cases parameterization from data is indicated. This conceptual framework draws on the knowledge of a number of different disciplines and incorporates the insightful work of many specialists in relevant scientific fields: gas layer—atmospheric chemistry; interface layer—mass transport engineering and interface science; liquid layer—freshwater, marine, and brine chemistry; deposition layer—colloid chemistry, surface science, and mineralogy; electrodic layer—electrochemistry; and solid layer—solid state chemistry.

3.2 THE GASEOUS REGIME

Gaseous regimes are often studied with atmospheric box models. In such a model, shown schematically in Figure 3.2, trace species are advected into and from the gaseous volume to be studied, entrainment and detrainment may cause reactive trace species to be exchanged out of the top of the volume, and volatilization and deposition processes dealing with the interface between the gaseous and liquid regimes change species concentrations as well. Potentially the largest changes are produced by chemical reactions occurring among the trace gaseous species.

Each species included in the gaseous regime must satisfy a continuity equation in which the rate of change of concentration is the sum of four contributing rates:

$$\frac{d[M_i]}{dt} = \Xi_i + Y_i - \Lambda_i + \Omega_i \tag{3.1}$$

where the terms are defined as follows:

Ξ_i = the sum of source and sink rates for species i, incorporating entrainment and detrainment of the species across the vertical boundaries of the regime. For the gas regime, this term is the sum of the entrainment and detrainment across the top of the regime boundary under consideration and the entrainment and detrainment across the gas–liquid interface at the bottom boundary of the regime.

Y_i = the rate of production of species i as a consequence of chemical reactions within the regime.

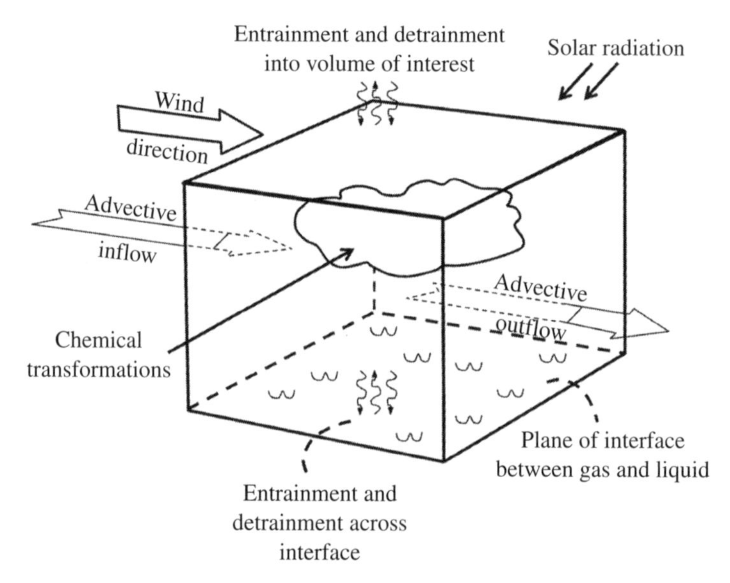

FIGURE 3.2 A schematic diagram of the basic elements in atmospheric corrosion gaseous regime.

Λ_i = the rate of loss of species i as a consequence of chemical reactions within the regime.

Ω_i = the rate of species concentration change as a result of horizontal advection into the regime volume from adjacent regions.

The complete formulation of the gas kinetics is thus a set of continuity equations of the form of Equation 3.1, derived from an appropriate set of elementary chemical reactions (i.e., reactions that cannot be broken down into two or more simpler reactions). For each reaction, the rate constant (and perhaps its temperature dependence) must be specified. In addition, source and sink terms and advection rates must be provided for all species of interest.

3.3 THE INTERFACE REGIME

Molecules in the gas phase that are to become involved in chemistry on a moist surface must, prior to that interaction, cross the interface that separates the gas and liquid phases. For a dilute solution of a gas i that dissolves but does not immediately react (O_2 is an example), the equilibrium distribution of the gas between the vapor and liquid phases is given by Henry's law, which states that the concentration of a gas in a liquid M_i is proportional to the partial pressure of the gas P_i:

$$[M_i] = H_i P_i \tag{3.2}$$

The constant of proportionality, H_i, is termed the Henry's law coefficient. The coefficient is temperature dependent, in the sense that gases are less soluble at higher temperature, a property generally expressed by writing

$$H(T) = C_1 \exp\left\{C_2\left(\frac{1}{T} - \frac{1}{298}\right)\right\} \tag{3.3}$$

where C_1 and C_2 are experimentally determined parameters referenced to a temperature of 298 K.

These expressions of the Henry's law equilibrium are not appropriate if any of the dissolved species is promptly transformed. An example is that of sulfur dioxide (SO_2), which (except in extremely acidic solutions) immediately ionizes to give a proton (H^+) and a bisulfite ion (HSO_3^-) or two protons and a sulfite (SO_3^{2-}) ion. In such a case, the effective Henry's law coefficient is given by

$$H_{S(IV)} = \frac{[S(IV)]}{p_{SO_2}} \tag{3.4}$$

which is equivalent to

$$H^*_{S(IV)} = \frac{\left([SO_2(aq)] + [HSO_3^-] + [SO_3^{2-}]\right)}{p_{SO_2}} \tag{3.5}$$

or

$$H_{S(IV)} = H_{SO_2}\left(1+\frac{K_1}{\left[H^+\right]}+\frac{K_1K_2}{\left[H^+\right]^2}\right) \tag{3.6}$$

where K_1 and K_2 are the first and second ionization constants of sulfurous acid. Thus, the effective Henry's law coefficient for SO_2 is pH dependent. If protonation and deprotonation in solution are treated explicitly, H is the appropriate factor to express transfer across the interface. If solution protonation and deprotonation are not so treated, use of H^* is appropriate.

Henry's law specifies relative concentrations where a water layer is exposed to a number of trace gases only if the attainment of equilibrium and the subsequent rate of chemical reactions of dissolved constituents are not limited by mass transport, either in the gas phase, the liquid phase, or across the gas–liquid interface. Otherwise, the Henry's law coefficient specifies the partitioning which the system approaches under rate-limiting constraints. The kinetic parameters can be addressed by the stagnant film model, in which the flux of gaseous species i from air to water is given by

$$\Phi_i = \frac{P_i}{\Pi_i} \tag{3.7}$$

where Π_i is the pathway resistance. The pathway resistance has the unit second per centimeter, which can also be obtained through a concentration–flux relationship. The inverse of this parameter, measured in centimeter per second, is denoted deposition velocity. This term represents three processes: transport from the ambient atmosphere to the surface boundary layer (r_{atm}), transport through the stagnant gaseous layer at the air–water interface ($r_{atm-interface}$), and transport through the stagnant liquid layer at the air–water interface ($r_{interface-water}$). Since the transport processes are sequential, the resistances are additive:

$$\Pi_i = r_{atm} + r_{atm-interface} + r_{interface-water} \tag{3.8}$$

Each of the processes involves diffusion, and each resistance can alternatively be expressed as the thickness of the layer through which transport occurs divided by the appropriate diffusion coefficient.

The numerical values of the first two resistances in Equation 3.8 are essentially independent of species and are given by measurement; typical values are $r_{atm} = 3.5\,\text{s cm}^{-1}$ and $r_{atm-interface} = 1.25\,\text{s cm}^{-1}$. The third resistance value can be derived from

$$r_{interface-water} = \frac{r_{surface}}{\left(H_i \cdot R \cdot T\right)} \tag{3.9}$$

where $r_{surface} = 170\,\text{s cm}^{-1}$ and R is the gas constant. This third resistance is often the limiting factor in air–water transport but the other factors may also be limiting, especially in indoor applications with stagnant conditions or conditions with very low air flow velocities.

3.4 THE LIQUID REGIME

Overview. The liquid regime receives reactive species from the gas phase and volatilizes species to it. The liquid phase resembles the gas phase in that describing its chemical processes involves the simultaneous solution of a set of nonlinear differential equations of the form of Equation 3.1. Transformations are influenced by the transport into the solution of species from the gas and solid regimes, and the conceptual picture of typical liquid regime models is shown in Figure 3.3.

Activity Coefficients. Much of the data and theoretical framework available for aqueous chemical kinetics studies refer to solutions of very high dilution. Many processes in thin water layers exposed to the atmosphere occur, however, at high ionic strength, μ, defined as

$$\mu = \frac{1}{2}\Sigma m_i \left[Z(i)\right]^2 \tag{3.10}$$

where m is the molality and Z the ionic charge. Measurable changes in solution parameters begin to be apparent at ionic strengths of order 10^{-3} M. A perspective for corroding surfaces exposed to various natural waters is provided by Figure 3.4, in which the typical ionic strengths of rivers, oceans, and fog droplets are displayed: all are above the ionic strength correction threshold. Rain is generally at or near the threshold. Also shown in Figure 3.4 is the ionic strength of an evaporating surface water film, as occurs in a corroding surface undergoing wetting and drying cycles. Ionic strength changes of several orders of magnitude can be expected under these circumstances, as can a transition from bulk water chemistry to moist surface chemistry.

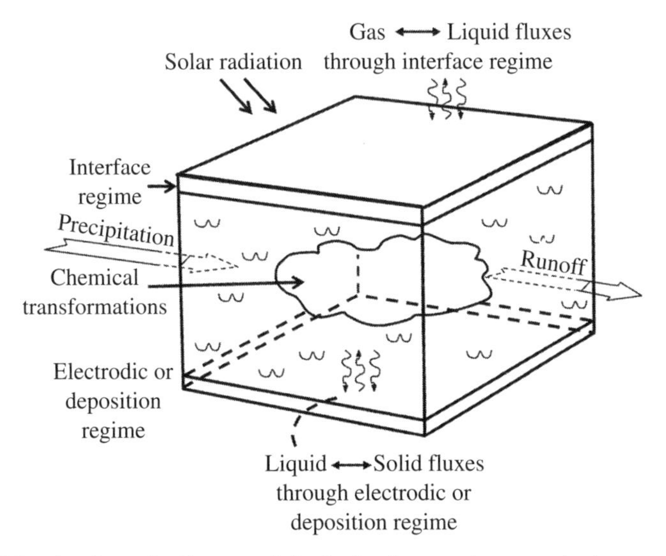

FIGURE 3.3 A schematic diagram of the basic elements in atmospheric corrosion liquid regime.

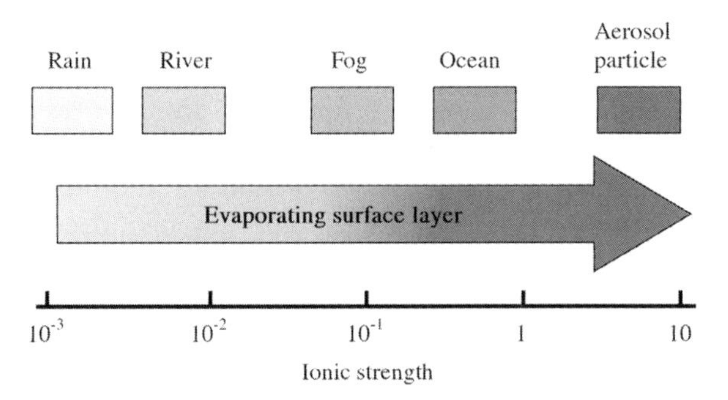

FIGURE 3.4 Ionic strengths of some natural waters and of an evaporating surface layer.

As a consequence of the high ionic strengths, the concentrations of the ions are replaced in the associated chemical formulas by their activities to indicate the degree to which their absolute concentrations are decreased by ion pairing and other solution processes. These activities, α_i, are given by

$$\alpha_i = m_i \gamma_i \tag{3.11}$$

where the activity coefficients γ_i are approximated by the Davies expression:

$$\log \gamma_i = -CZ_i^2 \left(\frac{\mu^{1/2}}{1+\mu^{1/2}} - 0.3\mu \right) \tag{3.12}$$

In Equation 3.12, C is a constant that depends on the dielectric properties of the solvent (an alternative approach is the explicit calculation of the ion-pairing solution constituents). At very high ionic strengths (μ>5–10), more detailed assessments of ionic forces are required to properly estimate activity coefficients. In any case, representations of concentrated solutions, such as those likely to be encountered in atmospheric corrosion, must employ activities and not concentrations if the results are to be realistic.

3.5 THE DEPOSITION REGIME

Overview. When a material covered with an aqueous solution undergoes corrosion, it generally becomes covered with one or more layers of deposited precipitates, often hydroxides or oxyhydroxides. The rate of crystal growth will control the overall crystallization rate. This growth rate may be either transport controlled or surface controlled; in the absence of large seed crystals, surface control is usually the dominant constraint. The rate R_G, with usual units of mole cm^{-2} s^{-1}, is given by

$$R_G = k_G \left(\Delta c - 1 \right)^n, \quad \Delta c > 1 \tag{3.13}$$

where k_G is the effective rate constant for crystal growth, sometimes defined as k_GS, the latter symbol referring to the effective surface area of the seed crystal. The magnitude of k_G must be determined empirically. Δc is the degree of supersaturation of the product species (dimensionless), given for a third-order process such as the ferrous hydroxide crystallization as

$$\Delta c = \frac{\left(\left[\mathrm{Fe}^{2+}\right]\cdot\left[\mathrm{OH}^{-}\right]^{2}\right)}{K_{\mathrm{sp}}} = \frac{(\text{Ion activity product})}{K_{\mathrm{sp}}} \tag{3.14}$$

where K_{sp} is the solubility product. n in Equation 3.13 is an order parameter dependent on crystal size and morphology; its value can range between 1 and 2 but is typically about 1.6.

Given the degree of supersaturation, Equation 3.13 is used to compute the rate of crystal growth. Where empirical values of k_G are not available (a frequent occurrence), k_G can be estimated by analogy with crystals of similar chemical structure and crystal habit.

Ion Transport in Deposition Layers. Once a layer of corrosion products has formed on a surface, generally trapping within or beneath it a portion of solution that may be called the inner electrolyte, any subsequent interaction between the surface and the overlying electrolyte solution requires that reactive species permeate inward through the deposition layer and that dissolution products be transported outward.

Deposition layers can be thought of as membranes, and all ions are not transported through membranes with the same efficiency. Normally, singly charged ions, especially if they have small dimensions, encounter less resistance than do multiply-charged ions. Thus, an anion-selective membrane will generally transport chloride and hydroxide ions more efficiently than nitrate ions and much more efficiently than sulfate ions. Similarly, a cation-selective membrane will transport protons more efficiently than metal ions. Some naturally formed corrosion films are strong inhibitors of ion transport and are termed passivating layers. Others are less than perfect membranes and transport ions such as sulfate relatively readily.

In order to assess the characteristics of ion selectivity of the deposition layer, the ionic flux across the membrane must be determined. The limiting rate for ion transport is given by

$$R_i(\lim) = \frac{2(\Delta a_i)^2 D_{\mathrm{i}}}{h_D \bar{X}} \tag{3.15}$$

where D_i is the diffusion constant for ion i in the membrane, Δa_i is the activity gradient of species i across the membrane, $\bar{X}$ is the concentration of fixed charge in the membrane, and h_D is the membrane thickness.

3.6 THE ELECTRODIC REGIME

The presence of a conducting or semiconducting surface rather than an electrically insulating surface changes several aspects of the perspective developed previously. The surfaces of such solids are charged, the magnitude and sign of the charge being functions of both solid and solution properties. Ions in the electrolyte are preferentially attracted by the surface charge and form a sheath layer with chemical and physical properties different from that of the bulk electrolyte. This surface layer influences chemistry and transport within the overall multiphase system.

When two dissimilar surfaces are present, one dominated by anodic reactions and the other by cathodic reactions, a corrosion cell is formed in which the electrolyte can function as a portion of an electrical circuit. The rate of electrons entering or leaving the liquid phase of a corrosion cell (in units of moles per unit area and unit time) is generally specified by a Butler–Volmer equation of the form

$$R = kc_{\mathrm{A}}^{+}\left[\exp\left(\frac{\alpha_a \eta F}{RT}\right) - \exp\left(\frac{-\alpha_c \eta F}{RT}\right)\right] \tag{3.16}$$

where the α terms are the charge transfer coefficients, F is the Faraday constant, k is the rate constant for charge transfer from surface to solution, c_{A}^{+} is the concentration of electron acceptor ions A^{+} on the solution side of the interface, and η is the overpotential.

Equation 3.16 is converted to the more common form used in electrochemistry by multiplying by the Faraday constant to give

$$i_c = kc_{\mathrm{A}}^{+}\left[\exp\left(\frac{\alpha_a \eta F}{RT}\right) - \exp\left(\frac{-\alpha_c \eta F}{RT}\right)\right] \tag{3.17}$$

where i is the electrode current density in units of coulombs per unit area and unit time. Formulating and solving this equation, where an applied potential is present, is the central topic in electrochemistry.

It is also possible for electrochemistry to be incorporated by including sources and sinks of charged species in the chemical equations, rather than, as is traditional, by stoichiometric balancing of oxidizing agents, reducing agents and electrons. For example, in iron corrosion reactions in oxygenated water, one balances charge in the corrosion cell with the following anodic and cathodic reactions:

$$2\left[\mathrm{Fe} \rightarrow \mathrm{Fe}^{2+} + 2(\mathrm{e}^{-})_{\mathrm{aq}}\right] \tag{3.18}$$

$$4(\mathrm{e}^{-})_{\mathrm{aq}} + 4\mathrm{OH}^{-} \rightarrow 2\mathrm{H_2O} + \mathrm{O_2} \tag{3.19}$$

In any but the simplest situations, however, competing reactions and kinetic constraints require that more detailed assessments of the chemistry be invoked, such as alternative sources and sinks of the ferrous and hydroxide ions.

Electrons entering an electrolyte solution do so as solvated ions. In pure water, solvated electrons have a lifetime of about a quarter of a second, largely due to the reaction

$$(e^-)_{aq} + H^+ + O_2 \rightarrow HO_2 \tag{3.20}$$

In more chemically diverse solutions, the solvated electrons may react as well with such substances as hydrogen peroxide, nitrate ion, and ammonium ion to produce a variety of chemically active free radicals. Such complexities require a full kinetic treatment of the electrochemical cell if the cell process is to be understood in detail.

3.7 THE SOLID REGIME

The solid regime enters into the chemical interactions of the system as a consequence of the attack of the liquid-phase constituents upon it and of its subsequent dissolution. The dissolution rate is generally acid dependent:

$$k_{\text{diss}} = C\left[a_{H^+}\right]^n \tag{3.21}$$

n for most minerals being between 0 and 0.5 and approaching unity only for hydroxide minerals and for very high proton concentrations.

In the initial step of proton-assisted surface dissolution, M—O bonds are polarized by the rapid attachment of protons to surface OH groups. The metal ion then slowly detaches itself from the surface at energy-favorable sites such as kinks or steps. Ligands in the solution may influence the dissolution rates, since $ML^{(n-1)+}$ complexes are more easily detached than are the uncomplexed metal ions. While at least minor enhancement in the dissolution rate is seen with many different ligands, the most effective are the bidentate organics such as oxalate and multiply-charged inorganics such as sulfate, which bind efficiently to the metal in the lattice. As ligand concentrations increase, the rate of dissolution increases up to the point where ligand effects on the metal lattice atoms are overwhelmed by supersaturation effects in the solution.

A complication with proton- and ligand-assisted dissolution is that the transfer of ions from the solution to the surface induces a surface charge, and this charge in turn influences the equilibrium constants for the transfer reactions. The effective equilibrium constant can be expressed as

$$K_{\text{eff}} = K \cdot \exp\left(\frac{-\Delta ZF\Psi_0}{RT}\right) \tag{3.22}$$

where ΔZ is the stoichiometric change in surface charge due to the reaction under consideration, F is the Faraday constant, Ψ_0 is the surface potential, R is the gas constant, and T is the absolute temperature. Sample calculations indicate that the exponential correction term for the equilibrium constants can be of order 0.4–0.7, large enough to be significant but probably within the uncertainties in other aspects of the corrosion kinetics.

In selecting dissolution rate values, it is preferable to utilize laboratory data on the dissolution of particular minerals in solutions of specified pH and constituent concentrations. Such data are often unavailable, but it is generally satisfactory to estimate dissolution rates by the use of the following relationship:

$$R_{\mathrm{diss}} = k_{\mathrm{diss}} x_{\mathrm{a}} P_j S \tag{3.23}$$

where R_{diss} is the proton- or ligand-promoted dissolution rate (mole cm^{-2} s^{-1}), k_{diss} is the rate constant (s^{-1}), x_{a} is the mole fraction of dissolution active sites (dimensionless), P_j represents the probability of finding a specific site in the coordinative arrangement of the precursor complex (dimensionless), and S is the surface concentration of sites (mole m^{-2}). As has been pointed out, relevant data on thermodynamic and, more importantly, kinetic parameters for fundamental processes on the metal and metal-oxide surfaces are often missing. However, recent developments in the field of atomistic modeling show promising results for systematic determination of these parameters.

3.8 THE MULTIREGIME PERSPECTIVE

This overview of the six chemical regimes involved in atmospheric corrosion begins to suggest why the field has waited so long for mechanistic and quantitative understanding—it is inherently more multidisciplinary than almost any other topic in science. Moreover, the relevant disciplines tend not to interact extensively, so topics that combine atmospheric chemistry and surface science, for example, tend to go unaddressed. As this chapter demonstrates, however, a reasonably comprehensive theoretical framework for the process involved in atmospheric corrosion has now been constructed. In subsequent chapters we will fill in this framework with observations in the atmosphere, at field sites, and on corroding surfaces outdoors and in the laboratory. The result will be a coherent physical and chemical picture of atmospheric corrosion phenomena.

FURTHER READING

Multiphase Modeling

B.R. Brady and L.R. Martin, Use of SURFACE CHEMKIN to model multiphase atmospheric chemistry: application to nitrogen tetroxide spills, *Atmospheric Environment, 29,* 715–726, 1995.

T.E. Graedel, GILDES model studies of aqueous chemistry. I. Formulation and potential applications of the multi-regime model, *Corrosion Science, 38,* 2153–2180, 1996.

The Solid Water Interface

W. Stumm, *Chemistry of the Solid Water Interface*, John Wiley & Sons, Inc., New York, 1992.

C.D. Taylor, Atomistic modeling of corrosion events at the interface between a metal and its environment, *International Journal of Corrosion,* 2012, 204640, 2012.

Transport Processes Through Solid Phases

J.H. Payer, G. Ball, B.I. Rickett, and H.S. Kim, Role of transport properties in corrosion product growth, *Materials Science and Engineering, A198,* 91–102, 1995.

Engineering Principles

Y.N. Mikhailovsky, Theoretical and engineering principles of atmospheric corrosion of metals, in *Atmospheric Corrosion,* ed. W.H. Ailor, John Wiley & Sons, New York, pp. 85–105, 1982.

(See also end of Appendix B for additional further reading about computer models.)

4

ATMOSPHERIC GASES AND THEIR INVOLVEMENT IN CORROSION

4.1 CHEMICAL SPECIES OF INTEREST

Atmospheric chemistry has been a distinct field of study and specialization only since about 1970, but its central role in such important topics as global climate change, urban air quality, and corrosion has contributed to its rapid development. In this effort, several thousand distinct chemical species have been identified as atmospheric gases or as constituents of atmospheric particles. Photochemistry plays an important role in many transformations, and the rates of key reactions govern the significance of a myriad of atmospheric processes.

Fortunately for the student of corrosion, not all of this complexity is of interest in connection with the atmospheric degradation of metallic materials. Over the course of decades of experimentation, a relatively small number of constituents have been identified as corrosive species, that is, as constituents capable of corroding or otherwise degrading metallic materials exposed to them. A particular corrosive species does not necessarily degrade all metals, however, nor is a particular material necessarily susceptible to all corrosive species. Table 4.1 presents a summary of corrosion species–metal susceptibilities. Eight corrosion species or pairs of corrosive species are identified. In the case of the pairs, either the gaseous form or the ionic form or both may be active as corrosive species (the data is not always definitive). The table is indicative rather than quantitative; that is, identical sensitivity designations do not imply identical corrosion rates.

Concentration of gases will usually be presented in parts per million by volume (ppmv) or parts per billion by volume (ppbv) but occasionally also in mass per volume ($\mu g\, m^{-3}$). Table 4.2 lists conversion factors of the gaseous corrosion stimulators discussed in this chapter.

Atmospheric Corrosion, Second Edition. Christofer Leygraf, Inger Odnevall Wallinder, Johan Tidblad and Thomas Graedel.

TABLE 4.1 Material Sensitivities to Atmospheric Corrosive Species

Corrosive Species	Ag	Al	Cu	Fe	Ni	Pb	Sn	Zn
CO_2/CO_3^{2-}	L			M	L	M		M
NH_3/NH_4^+	M	L	M	L	L	L	L	L
NO_2/NO_3^-	N	L	M	M	M	M	L	M
H_2S	H	L	H	L	L	L	L	L
SO_2/SO_4^{2-}	L	M	H	H	H	M	L	H
HCl/Cl^-	M	H	M	H	M	M	M	M
$RCOOH/COOH^-$	L	L	M	M	M	H	L	M
O_3	M	N	M	M	M	M	L	M

An unfilled location indicates that no experimental assessments are known to have been performed.
H, high sensitivity; L, low sensitivity; M, moderate sensitivity; N, no sensitivity.

TABLE 4.2 Conversion Factor (1 ppbv to × μg m^{-3}) for Gaseous Species in Table 4.1 (25°C and 1 atm Gas Pressure)

Gas	μg m^{-3}
CO_2	1.80
NH_3	0.70
NO_2	1.88
H_2S	1.39
SO_2	2.62
HCl	1.49
HCOOH	1.88
O_3	1.96

4.2 ATMOSPHERIC CORROSIVE GASES

4.2.1 Carbon Dioxide

Carbon dioxide (CO_2) is a natural constituent of the atmosphere. It is absorbed by vegetation as part of the photosynthesis process in the spring of each year and given back in the fall as the vegetation decays. A cyclic pattern occurs with the oceans, which absorb CO_2 when cool and release it when warm. The natural cycle, as it is understood, is in balance.

CO_2 is also emitted by anthropogenic activities. This flux is only about 3% of the natural flux but is not completely balanced by loss processes. As a result, the average CO_2 concentration in the atmosphere increases each year by about half a percent. Most of this CO_2 results from the combustion of fossil fuels, although several percent of the emissions are attributable to the manufacture of cement. The spatial distribution of the anthropogenic emissions largely reflects the locations of the more highly industrialized nations, North America, Europe, and China being large emitters.

CO_2 is unreactive in the atmosphere, so it has a long atmospheric lifetime and tends to be well mixed by atmospheric motions. The average concentration as of early 2015 was about 400 ppmv; near urban areas where sources are large, concentrations within the range of 300–600 ppmv can be observed.

Because CO_2 is weakly soluble in water, any body of water, water droplet, or a surface water layer will absorb it. A portion of the dissolved gas dissociates to produce the moderately corrosive carbonate ion (CO_3^{2-}):

$$CO_2(g) \rightarrow CO_2(aq) \xrightarrow{H_2O} 2H^+ + CO_3^{2-} \quad (4.1)$$

4.2.2 Ozone

Most of the gaseous transformations that occur in the atmosphere are initiated by either the hydroxyl radical (OH·) or the ozone molecule (O_3). Neither is emitted directly from any sources, natural or anthropogenic, but rather they are products of atmospheric photochemistry. The dominant source reaction is the photolysis of nitrogen dioxide (NO_2), followed by combination with molecular oxygen:

$$NO_2 \xrightarrow{\text{visible and ultraviolet light}} NO + O \quad (4.2)$$

$$O + O_2 \rightarrow O_3 \quad (4.3)$$

O_3 is itself sensitive to ultraviolet radiation, generating an energetic oxygen atom that combines with a water molecule to form the hydroxyl radical

$$O_3 \xrightarrow{\text{ultraviolet light}} O_2 + O \quad (4.4)$$

$$O + H_2O \rightarrow 2OH\cdot \quad (4.5)$$

The reactions in which O_3 and OH· participate, and the chemical products that result, comprise the phenomenon known as photochemical smog (the word is shorthand for "smoke and fog").

The hydroxyl radical reacts promptly with a very wide variety of atmospheric gases. As a result, its concentrations are low. O_3 is significantly less reactive and is rather poorly soluble in water; its atmospheric concentrations are thus much higher. In regions where major episodes of photochemical smog develop, ozone concentrations can reach several hundred ppbv.

A small amount of the ambient O_3 will dissolve in any water droplets or surface films that are present. Ozone is known to be chemically active as an oxidizer and will take part in a variety of corrosion processes.

4.2.3 Ammonia

Ammonia (NH_3) is the only common atmospheric gas that is a base; it thus serves as a counterweight to the several that are acidic. Natural ecosystems and the sea surface emit some NH_3, but the global NH_3 cycle is dominated by anthropogenic sources. Agriculture is by far the major industrial sector for NH_3 emissions, with animal excrement, synthetic fertilizer, and cropland emissions all being important, as is the burning of biomass to clear farmland. Domestic animals are the largest single NH_3 source, and geographical regions with high animal populations (India, Europe, etc.) have the largest NH_3 emission rates worldwide.

NH_3 is highly soluble, with the result that it is lost to water-covered aerosol particles and cloud droplets at high rates, a process termed wet deposition. The end product after combination of NH_3 with oxidized sulfur dioxide is generally ammonium sulfate $(NH_4)_2SO_4$. A substantial amount of NH_3 is lost to the atmosphere as well by deposition to the planet's surface, a process termed dry deposition in which the gaseous molecule encounters the surface and is absorbed.

The high solubility of NH_3 produces substantial variations in its atmospheric concentrations, which in urban areas range from about 1 to more than 50 ppbv.

4.2.4 Nitrogen Dioxide

The primary source of nitrogen dioxide (NO_2) is high-temperature combustion, in which nitric oxide (NO) is formed and then rapidly oxidized by ambient O_3:

$$N_2 + O_2 \xrightarrow{\Delta} 2NO \xrightarrow{2O_3} 2NO_2 \tag{4.6}$$

Δ indicates that the reaction needs extra energy. Fossil fuel combustion is the largest NO source, but biomass burning is significant as well.

Because the principal anthropogenic source of NO_2 is the same as for CO_2, the spatial emission distribution is similar as well. North America, Europe, and Southeast Asia dominate the emission map. In these urban areas, NO_2 concentrations can be as high as 300 ppbv during periods of intense photochemical smog.

Nitrogen dioxide is quite insoluble in water, so it is not lost to water droplets or surfaces to any significant degree. The only operable sink is gas-phase reaction with the hydroxyl radical to produce nitric acid (HNO_3):

$$NO_2 + OH\cdot \xrightarrow{M} HNO_3(g) \tag{4.7}$$

where M is any gaseous third body that carries away excess energy. HNO_3 is extremely soluble in water. It is therefore lost to the atmosphere in the same manner as is NH_3, by incorporation into aerosol particles and cloud droplets and by wet and dry deposition.

4.2.5 Hydrogen Sulfide

Hydrogen sulfide (H_2S) is a product of the anaerobic degradation of organic sulfur compounds. The sources are largely natural or are anthropogenic processes similar to natural ones: decay of vegetation in wetlands and soils, emission of excess sulfur from vegetation, emission from sewage treatment plants, and the like. H_2S emissions that are small on a global basis but large locally can also be generated from kraft pulp mills and some oil fields.

H_2S concentrations in urban areas are generally quite low. Most atmospheric corrosion involving H_2S occurs in the vicinity of pulp mills or in oil fields pumping "sour crude" (i.e., oil, containing significant amounts of H_2S) or near poorly operated oil refineries.

H_2S is slightly soluble in water, and the likely route to corrosion is dissolution into aqueous surface films followed by dissociation:

$$H_2S(g) \rightarrow H_2S(aq) \rightarrow H^+ + HS^- \tag{4.8}$$

where the HS^- ion is the active corrosion agent.

4.2.6 Sulfur Dioxide

Sulfur dioxide (SO_2) is perhaps the most important of the atmospheric corrosive gases. It is formed during the combustion of all fossil fuels containing sulfur and is also emitted during metal smelting processes. Many metals occur as sulfides in their geological deposits, and the smelting process involves heating the ore at high temperatures to liberate the sulfur:

$$MS \xrightarrow{O_2} SO_2 \uparrow + M \tag{4.9}$$

As with CO_2 and NO_2, the emission of SO_2 gas tends to reflect the level of industrial development of the different continents and regions. A difference, however, is that it is much easier to capture sulfur emissions. As a result, regions that have implemented extensive air pollution control technologies have been able to control SO_2 emissions to a substantial degree. China and Eastern Europe tend to have the highest emission fluxes.

SO_2 urban concentrations cover a very wide range and can be higher than any other atmospheric corrosive gas, with the exception of CO_2. The difference in concentration within a single city can be great, however, and that between cities can be very great as well. Such a pattern is generally true for all the corrosive gases, CO_2 excepted.

SO_2 is moderately soluble in water, so that a significant fraction is absorbed into aerosol particles. There it is oxidized into the sulfate ion (SO_4^{2-}):

$$SO_2(g) \rightarrow SO_2(aq) \rightarrow \text{several steps} \rightarrow SO_4^{2-} \tag{4.10}$$

Essentially identical processes occur when SO_2 undergoes deposition to surfaces. The molecule is reasonably reactive to the hydroxyl radical, so there is also a gas-phase process ending in sulfuric acid (H_2SO_4):

$$SO_2(g) + OH\cdot \rightarrow HSO_3\cdot \rightarrow \text{several steps} \rightarrow H_2SO_4 \tag{4.11}$$

The sulfuric acid rapidly dissolves in any available water and dissociates, the resulting corrodent again being the sulfate ion.

4.2.7 Hydrogen Chloride

Most gaseous hydrogen chloride (HCl) enters the atmosphere through the dechlorination of airborne sea-salt particles. These particles are very rich in chlorine, which they liberate when a strong acid (nitric or sulfuric) is deposited on them:

$$HCl(aq) + HNO_3(g) \rightarrow HCl(g) + HNO_3(aq) \tag{4.12}$$

$$HCl(aq) + H_2SO_4(g) \rightarrow HCl(g) + H_2SO_4(aq) \tag{4.13}$$

Because the strong acid gases tend to be emitted near industrial and energy-consuming regions, the highest HCl fluxes from dechlorination are seen at the downwind edges of highly industrialized continents. The three areas with the highest HCl emission fluxes are the North Sea between England and Norway, Northeastern Europe, and the Southwest coast of Australia.

The atmospheric concentrations of HCl are moderately high near the source regions mentioned previously and quite low elsewhere.

HCl has an aqueous solubility similar to that of NH_3, so it dissolves promptly in atmospheric water droplets or deposits onto moist surfaces. In either case, it dissociates readily to form the chloride ion (Cl^-):

$$HCl(g) \rightarrow HCl(aq) \rightarrow H^+ + Cl^- \tag{4.14}$$

Here, both $H^+ + Cl^-$ ions are active corrosive species.

4.2.8 Organic Acids

Formic acid (HCOOH) and acetic acid (CH_3COOH), the simplest and most common of the organic acids, are known to be corrosive to several of the industrial metals. Many technological processes use and emit these acids to some degree, but the most prolific sources appear to be the incomplete combustion of wood or other vegetation and the reactions of photochemical smog. Freshly cut wood outgasses organic acids, sometimes at a fairly high rate, and can be an important indoor source in museums or other indoor locations. No global or regional emission data is available for the organic acids, and atmospheric concentrations are infrequently measured, but urban areas apparently observe RCOOH ($HCOOH + CH_3COOH +$ small amounts of other organic acids) concentrations in the 1–10 ppbv range.

The organic acids are highly soluble in water, where they are partly dissociated, e.g., formic acid:

$$HCOOH(g) \rightarrow HCOOH(aq) \rightarrow H^+ + COOH^- \tag{4.15}$$

4.2.9 Corrosive Species in Aerosol Particles and Precipitation

Precipitation (rain, fog, snow, cloud, dew) is of vital importance in atmospheric corrosion. By the simple process of surface flushing, it can retard corrosion instigated by deposited gases or particles; this is especially significant for surfaces that are well drained. Conversely, it can bring dissolved reactive species onto a surface to initiate corrosion, a process especially significant for surfaces shaped so that atmospheric water is retained.

In many climates, rain is of most concern from a corrosion perspective. Raindrops vary in diameter from about 0.1 to 2 mm, or even larger. Most, however, have diameters within the range of 0.2–0.5 mm. The chemical content tends to vary with size. The smallest drops generally have the highest concentrations of the ions involved in corrosion, but most ions of concern are present in most raindrops as a result of the nucleation of the drops around small particles made up of corrodent substances such as $(NH_4)_2SO_4$, as well as the dissolution of gaseous corrosive species into the drop, once formed.

Atmospheric corrosion cannot occur until the corrosive species comes into contact with the surface of interest and interacts chemically with it. The most common such process is dry deposition, in which surface contact occurs primarily by turbulent diffusion. The relationship between the concentration of a species i and the vertical flux Φ_i to a surface is often expressed by

$$\Phi_i = v_{\mathrm{d}} C_i \tag{4.16}$$

where C_i is the concentration of the constituent at some reference height and v_{d} is a parameter with units of length divided by time called the deposition velocity (see further in Section 3.3).

The efficiency of deposition to corroding surfaces is a strong function of aqueous solubility, since the surfaces are generally water covered. The equilibration of a gaseous species between air and water phases is given by its Henry's constant (H), which was defined in Section 3.3. Deposition velocities are rather strong functions of local atmospheric conditions as well as the particular surface involved; the relative v_{d} values are in approximately the same ranking as the H values.

4.3 HISTORIC TRENDS IN ATMOSPHERIC CORROSIVE GAS CONCENTRATIONS

Concentration trends are, of course, related to the existence or lack thereof of trends in activity of the respective sources of the gases. In most cases, the trends in concentration that have been determined reflect changes in the magnitude of activity of their anthropogenic sources. The histories of emissions from 1990 to 2010 are shown in Figure 4.1. For each of the gases discussed earlier in this chapter, specific information on historic source activity and related concentration trends is given in the following.

4.3.1 Carbon Dioxide

The trend in CO_2 is very well quantified, for two reasons: CO_2 is a very important greenhouse gas, and its life in the atmosphere is long. Because of the latter, it has the opportunity to become well mixed, and measurements made at sites throughout the planet can be related to global source activity.

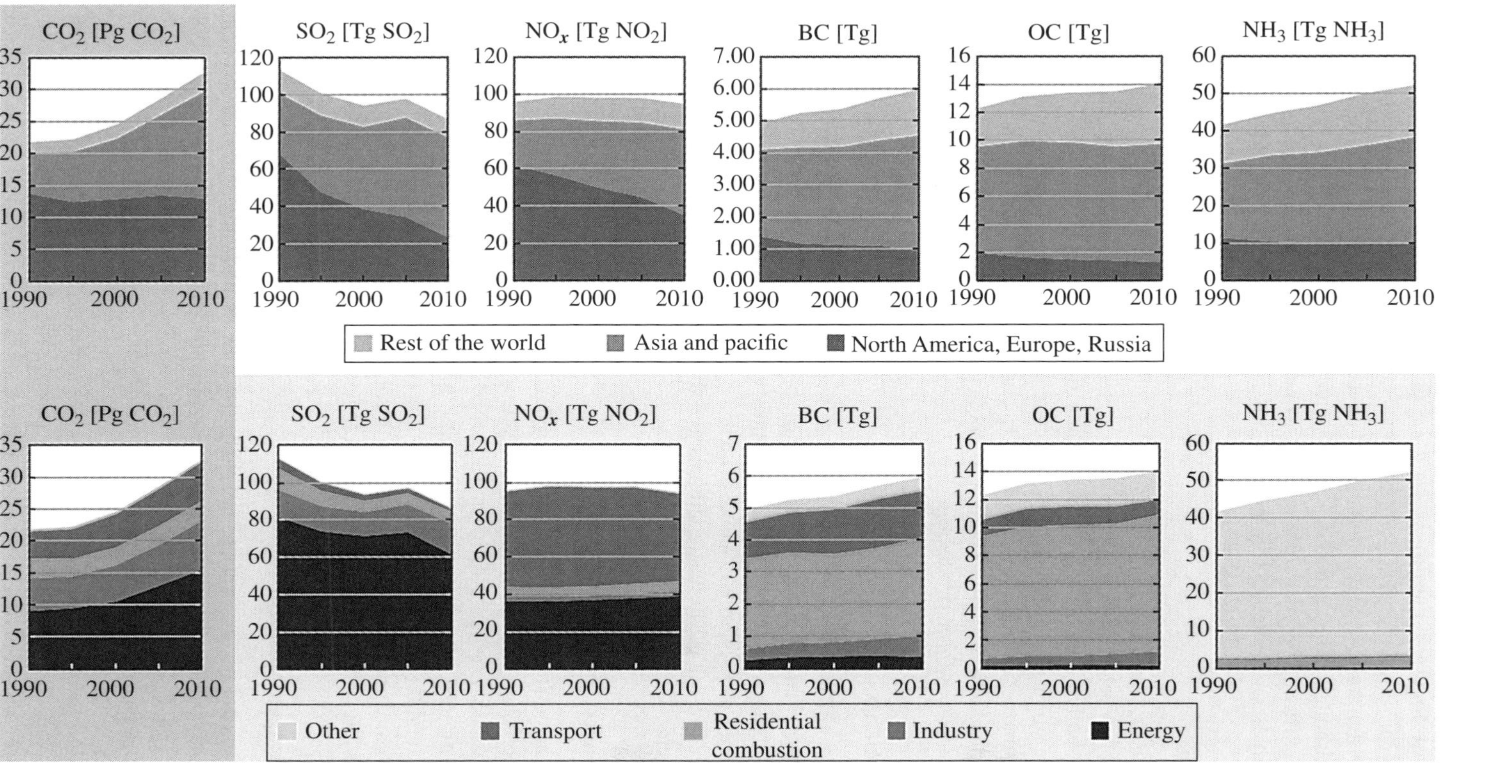

FIGURE 4.1 The evolution of anthropogenic emissions of CO_2 and key air pollutants, 1990–2010, by world regions (top) and source sectors (bottom), excluding international shipping and aviation. BC, black carbon; OC, organic carbon; Pg, petagram; Tg, teragram. (Reproduced with permission from Annual Reviews; M. Amann, Z. Klimont, and F. Wagner, Regional and global emissions of air pollutants: recent trends and future scenarios, *Annual Review of Environment and Resources, 38*, 31–55, 2013.) (*See insert for color representation of the figure.*)

In the past 40 years, the global average CO_2 concentration has increased about 0.5% each year. From the standpoint of corrosion rates, CO_2 trends are probably not particularly important. One reason is that the rate of change, while readily determinable, is not dramatic. The second is that the influence of CO_2 on corrosion appears not to be very significant, because carbonic acid is one of the weaker acids and its influence will generally be overwhelmed by more aggressive atmospheric corrosive species.

4.3.2 Ozone

Atmospheric O_3 concentrations have been monitored in many locations for decades. They are clearly related to emissions of nitrogen oxides and volatile organic carbon compounds, both products of fossil fuel combustion. Where energy use has increased, especially motor vehicle-related use, and where abundant sunshine is available to drive photochemical reactions, average O_3 concentrations have climbed rapidly. In some tropical and midlatitude cities, average ozone levels have doubled or tripled over the past two or three decades.

4.3.3 Ammonia

NH_3 is a very difficult compound to measure with accuracy. In addition, it is highly soluble in water and thus has a short atmospheric lifetime. As a consequence, no reliable trend data are available. However, most of NH_3's sources are related to agricultural activity, so local trends will depend more on the evolution of local agriculture, especially the raising of domestic animals and the use of synthetic fertilizer, than on any other factor.

4.3.4 Nitrogen Dioxide

NO_2 is produced largely by fossil fuel combustion, especially near the populated areas in which corrosion is likely to be of interest. As a result, historic records of NO_2 concentrations in Europe, North America, and Southeast Asia show patterns that reflect energy use, especially the quantity of fuel burned by motor vehicles. Smokestack and tailpipe emission controls for NO_2 are less effective than for some other pollutants, and the best efforts at control have produced NO_2 concentration patterns that are no better than roughly stable with time.

4.3.5 Hydrogen Sulfide

H_2S measurements are uncommon, and concentrations are low enough to make reliable measurements difficult. As a result, there is no reliable data on H_2S temporal trends.

4.3.6 Sulfur Dioxide

Trends in SO_2 concentrations are an important concern for atmospheric corrosion, because SO_2 and its related acidic compounds are the dominant corrosive species. Although SO_2 emissions from metal smelters can have severe local impacts, most

SO_2 is related to the combustion of sulfur-containing fossil fuels. Two types of trends have been noticed. One is an increase in atmospheric SO_2 in countries where rapid industrialization is combined with a lack of emission controls. The second is a decrease in countries that have implemented emission controls, especially where that action has been combined with the use only of coal and oil that has low sulfur content.

4.3.7 Hydrogen Chloride

Gaseous HCl concentrations are determined only with great difficulty, and no temporal trends have been measured. Since most gaseous HCl is generated by dechlorination of sea-salt aerosol as a result of HNO_3 or H_2SO_4 incorporation, it is anticipated that rapidly industrializing regions in coastal locations can anticipate increasing HCl concentrations, an anticipation thus far largely unconfirmed.

4.3.8 Organic Acids

The data on atmospheric organic acids are sparse, much too sparse to determine any temporal trends.

4.3.9 Summary of Historic Trends

The two decades between 1990 and 2010 have largely been encouraging from an atmospheric corrosion perspective. The emission histories for a number of atmospheric species are shown in Figure 4.1. On a global basis, emissions of the corrosive gases SO_2 and NO_x ($NO+NO_2$) declined significantly, primarily because of major reductions in Europe and North America. Black carbon and organic carbon emissions increased, however, residential-related emissions worldwide being the major factor. Ammonia, widely used in fertilizer, saw emission increases in Asia. Finally, emissions in Asia resulted in a major CO_2 increase.

4.4 PREDICTED FUTURE EMISSIONS OF CORROSIVE SPECIES

Projections of the future emissions of corrosive atmospheric species suggest a decline at the global level, although regional and local levels will surely differ from that general conclusion. Figure 4.2 illustrates the predictions. SO_2 emissions are anticipated to continue their gradual decrease, and emission controls will probably decrease black carbon and organic carbon emissions as well. NO_x emissions are expected to decrease in Europe and North America but stabilize globally because of increased Asian emissions. Enhanced agricultural use worldwide will sharply increase NH_3 emissions, and energy use (especially in Asia) will continue to increase CO_2 emissions. There are no emission projections for chlorinated species, but those emissions are largely the result of marine atmospheric influences rather than industrial activities.

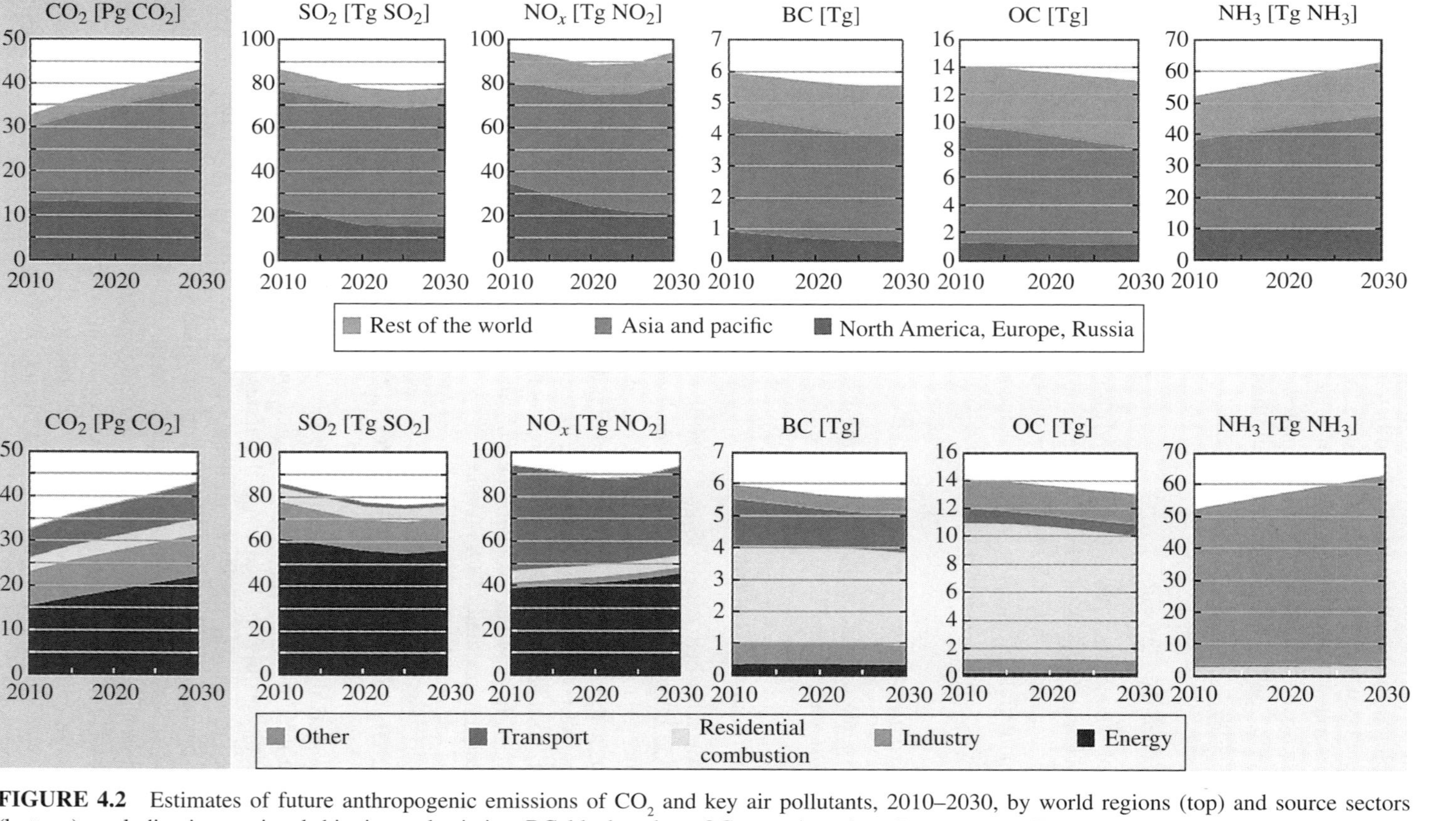

FIGURE 4.2 Estimates of future anthropogenic emissions of CO_2 and key air pollutants, 2010–2030, by world regions (top) and source sectors (bottom), excluding international shipping and aviation. BC, black carbon; OC, organic carbon; Pg, petagram; Tg, teragram. (Reproduced with permission from Annual Reviews; M. Amann, Z. Klimont, and F. Wagner, Regional and global emissions of air pollutants: recent trends and future scenarios, *Annual Review of Environment and Resources, 38*, 31–55, 2013.) (*See insert for color representation of the figure.*)

From an atmospheric corrosion standpoint, these emission reductions or stabilizations over time should be largely beneficial. SO_2 is generally the most aggressive corrosive species, so control of its emissions will have significant benefits, and reductions in atmospheric particle concentrations should further constrain atmospheric corrosion. The increase in NH_3 and the stabilization of NO_x are not particularly worrisome from a corrosion standpoint, as these species are not vigorous corrosive species. The corrosive effects of CO_2, even at elevated concentrations, should not be very significant.

FURTHER READING

Materials Degradation

T.E. Graedel and R. McGill, Degradation of materials in the atmosphere, *Environmental Science and Technology, 20*, 1093–1100, 1986.

Trends in Emissions of Corrosive Species

M. Amann, Z. Klimont, and F. Wagner, Regional and global emissions of air pollutants: recent trends and future scenarios, *Annual Review of Environment and Resources, 38*, 31–55, 2013.

J. Kurokawa, T. Ohara, T. Morikawa, S. Hanayama, G. Janssens-Maenhout, T. Fukui, K. Kawashima, and H. Akimoto, Emissions of air pollutants and greenhouse gases over Asian regions during 2000–2008, *Atmospheric Chemistry and Physics, 13*, 11019–11058, 2013.

J.E. Sickles II and D.S. Shadwick, Air quality and atmospheric deposition in the eastern US: 20 years of change, *Atmospheric Chemistry and Physics, 15*, 173–197, 2015.

Gas-Surface Reactions

J. Liu, X. Zhang, E.T. Parker, P.R. Veres, J.M. Roberts, J.A. de Gouw, P.L. Hayes, J.L. Jimenez, J.G. Murphy, R.A. Ellis, L.G. Huey, and R.J. Weber, On the gas-particle partitioning of soluble organic aerosol in two urban atmospheres with contrasting emissions: 2. Gas and particle phase formic acid, *Journal of Geophysical Research, 117*, 2012, doi: 10.1029/2012JD017912

G. Rubasinghege and V.H. Grassian, Surface catalyzed chlorine and nitrogen activation, *Journal of Physical Chemistry A, 116*, 5180–5192, 2012.

Atmospheric Organic Acids

P. Khare, N. Kumar, K.M. Kumari, and S.S. Srivastava, Atmospheric formic and acetic acids: an overview, *Reviews of Geophysics, 37*, 227–248, 1999.

Organic Acid Corrosion

A. Niklasson, J.-E. Svensson, S. Langer, K. Arrhenius, L. Rosell, C.J. Bergsten, and L.-G. Johansson, Air pollutant concentrations and atmospheric corrosion of organ pipes in European church environments, *Studies in Conservation, 53*, 24–40, 2008.

5

ATMOSPHERIC PARTICLES AND THEIR INVOLVEMENT IN CORROSION

5.1 INTRODUCTION

The term aerosol refers to a suspension of small liquid and/or solid particles in a gaseous medium. The particles are generated in one of two ways. The first process, which produces very small particles, is the condensation of low-volatility gases, either homogeneously or on the surface of atmospheric water droplets (the nuclei mode and Aitken mode for the small particle group with diameter <0.1 μm), followed by the coalescence of a number of these nuclei to create larger (but still small) particles (the accumulation mode for the small particle group with diameter 0.1–2.5 μm). The second process, which generates the coarse mode group comprised of larger particles, includes a variety of mechanical processes such as the windblowing of dust or industrial machining. Figure 5.1 depicts the source types and the resulting size distributions. There is no feasible way to convert the smaller particles into larger particles or vice versa. As a result, the physics and chemistry of the two modes can be treated quite independently.

The number of particles within the two modes is generally very different as well. Figure 5.2 shows typical size spectra for urban and rural aerosols and for individual particles incorporated into rainwater. Accumulation mode particles outnumber those in the coarse mode by a factor of a thousand or more and have around 10 times the surface area. Particle volume, however, is proportional to the cube of the radius, so the ensemble volume of the coarse mode particles approaches that of the accumulation mode.

As with gases, aerosol particles are lost to the atmosphere when they come into contact with a surface and are retained by that surface (dry deposition). Unlike gases, particle deposition has the additional aspect of size dependence, because very small particles that possess high diffusion rates behave quite differently from larger

Atmospheric Corrosion, Second Edition. Christofer Leygraf, Inger Odnevall Wallinder, Johan Tidblad and Thomas Graedel.

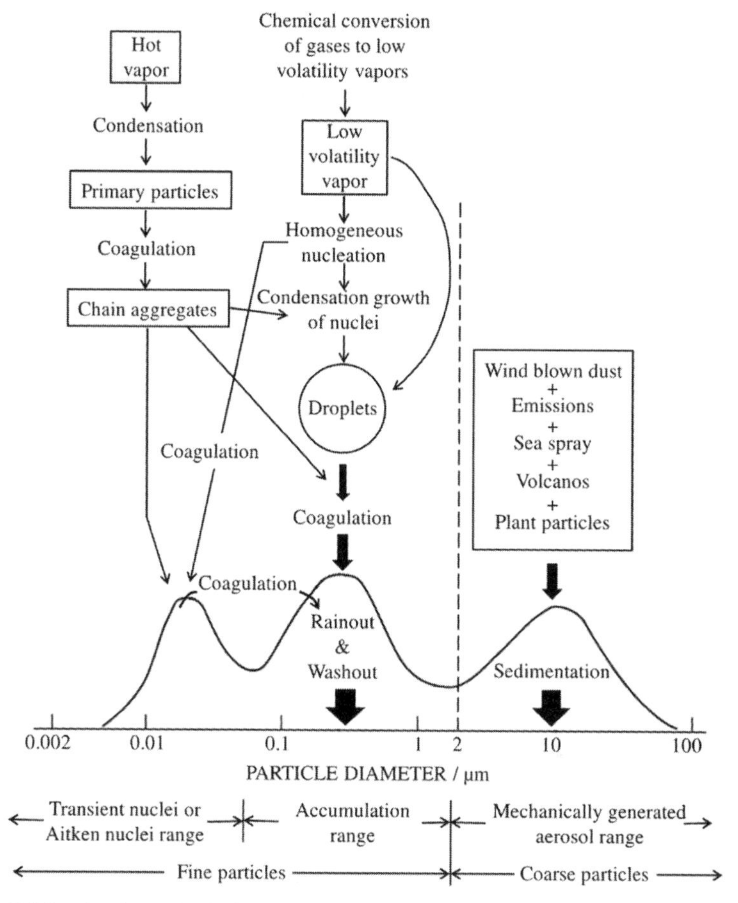

FIGURE 5.1 A schematic depiction of the surface area distribution of a typical atmospheric aerosol. The sources, interactions, and removal mechanisms for the two modes, accumulation and coarse modes, are shown. (Reproduced with permission from John Wiley & Sons; J. Seinfeld and S. Pandis, *Atmospheric Chemistry and Physics*: From Air Pollution to Climate Change, John Wiley & Sons, Inc., Hoboken, 2006.)

particles. The momentum of the former constrains them to follow the motions of the atmospheric parcels in which they are embedded. The larger particles have enough momentum of their own to be relatively unaffected by air movements.

Deposition velocities can vary by as much as three orders of magnitude for particles of different diameters. Figure 5.3 illustrates this behavior. Particles larger than about 1 μm have too much momentum to follow the deflected air and forcefully impact on surfaces toward which they are heading. Much smaller particles are light enough to behave rather like gases, diffusing toward outdoor surfaces at a rather high speed. Air motions indoors, however, are much less vigorous, and indoor deposition velocities are 10–100 times smaller than the outdoor values shown in Figure 5.3.

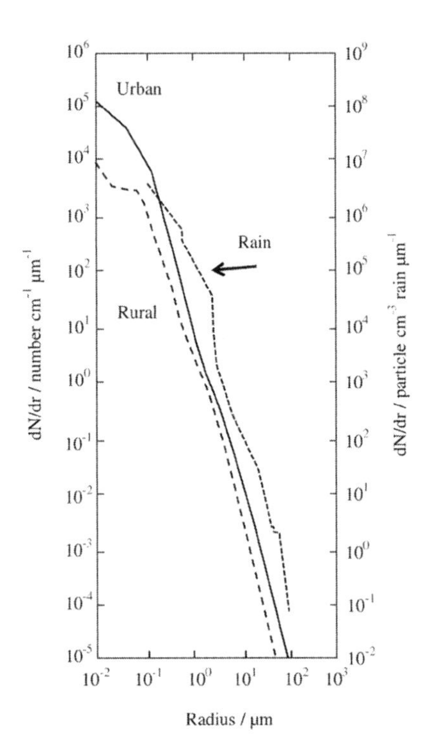

FIGURE 5.2 Representative particle size spectra of average rural and urban aerosols and of insoluble particles in rainwater. (Reproduced with permission from Springer; L. Schütz and M. Krämer, Rainwater composition over a rural area with special emphasis on the size distribution of insoluble particulate matter, *Journal of Atmospheric Chemistry*, *5*, 173–184, 1987.)

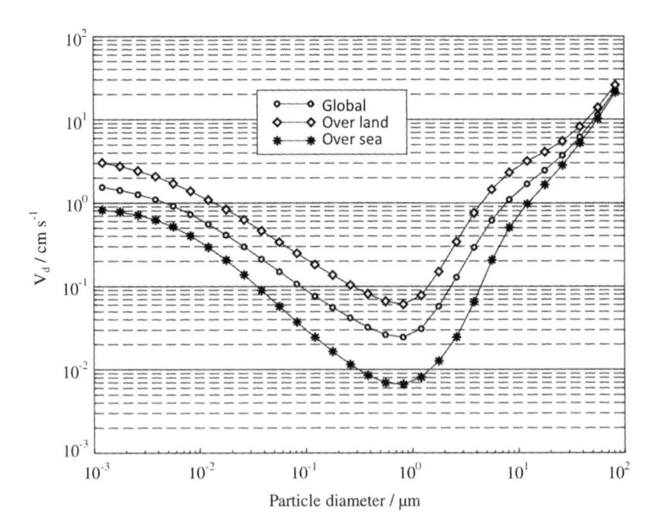

FIGURE 5.3 Global annual mean deposition velocities versus particle size and surface type. The model data have been parameterized assuming a particle density of 1000 kg m^{-3}. A minimum in deposition velocity is seen in the diameter range 0.5–1.0 µm. For all particle sizes the deposition velocity is higher over land than over sea. (Reproduced with permission from Elsevier; E.-Y. Nho-Kim, M. Michou, and V.-H. Peuch, Parameterization of size-dependent particle dry deposition velocities for global modeling, *Atmospheric Environment, 38*, 1933–1942, 2004.)

In both indoor and outdoor settings, the deposition velocity minimum occurs where particles have diameters of a few tenths of a micrometer, so both the impaction and diffusion processes are inefficient and particle lifetimes are long. This situation holds with particle filtration in clean rooms, which explains why it is so difficult to achieve low indoor concentrations of submicrometer particles.

The difference in particle momentum shows up not only in deposition rates but also in deposition to differently oriented surfaces. Coarse particles readily deposit on top-facing surfaces but have little access to bottom-facing surfaces. For fine particles, which tend to move more like gases, surface orientation is much less a factor in deposition.

In wet deposition, interactions occur between airborne particles and atmospheric water droplets, and these interactions can be described in reasonable detail. Each individual particle is exposed repeatedly to raindrops, each time with some probability of inertial capture. By the same token, each raindrop has many opportunities to scavenge (i.e., to incorporate) particles. Individual raindrops reaching the ground may contain as many as 10,000 small particles that have been scavenged during the fall from the base of the clouds.

An important consideration for atmospheric corrosion is that the rapid sedimentation of coarse mode particles, which possess an ensemble volume within a factor of perhaps 10 or less of the accumulation mode, means that one should anticipate the presence of both modes on exposed surfaces. Typically, deposited accumulation mode particles will outnumber those of the coarse mode, but the volumes of the two classes of deposited particles are likely to be similar. Precipitation falling onto these surfaces may add or subtract from the surface particles depending on their tendency to be washed off and on the presence of new particles incorporated into the raindrops.

There are few long-term studies of accumulation mode concentration and none for coarse mode concentration in the atmosphere. For the former, observations in remote areas indicate an increase in concentration of a few percent over the past decade.

5.2 CHEMICAL SPECIES OF INTEREST

Table 4.1 in the preceding chapter identified a number of ionic species of interest to corrosion studies: the carbonate ion (CO_3^{2-}); the chloride ion (Cl^-); the ammonium ion (NH_4^+); the nitrate ion (NO_3^-); the sulfate ion (SO_4^{2-}); two organic ions, formate ($COOH^-$) and acetate (CH_3COO^-); and the hydrogen ion (H^+). Because aerosol particles are chemical mixtures, all these species are potentially of concern. However, aerosol particle sources generally produce chemically distinct particles, and locations near sources of corrosion-enhancing particles (see next section) may need to take special precautions against corrosion. The particles gradually evolve chemically as they are agglomerated with other particles, as gases are absorbed onto them, and as chemical reactions take place. Thus, chemically mixed particles are common away from aerosol sources.

TABLE 5.1 Global Natural Emissions of Aerosols (A) and Aerosol Precursors (P) in the Year 2000 (Tg year^{-1})

	Minimum	Maximum
Sea spray (A)	1400	6800
Mineral dust (A)	1000	4000
Terrestrial primary biological aerosols (A)	50	1000
Dimethylsulfide (P)	10	40
Biogenic volatile organic compounds (P)	20	380
Monoterpenes (P)	30	120
Isoprene (P)	410	600

Minimum and maximum values from a range of inventories.
The estimates are from O. Boucher, D. Randall, P. Artaxo, C. Bretherton, G. Feingold, P. Forster, V.-M. Kerminen, Y. Kondo, H. Liao, U. Lohmann, P. Rasch, S.K. Satheesh, S. Sherwood, B. Stevens, and X.Y. Zhang, Table 7.1 in Clouds and aerosols, in *Climate Change 2013: The Physical Science Basis*. Contribution of Working Group I to the Fifth Assessment Report of the Intergovernmental Panel on Climate Change, eds. T.F. Stocker, D. Qin, G.-K. Plattner, M. Tignor, S.K. Allen, J. Boschung, A. Nauels, Y. Xia, V. Bex, and P.M. Midgley, Cambridge University Press, Cambridge/New York, 2013.

5.3 SOURCES OF ATMOSPHERIC AEROSOL PARTICLES

Particles are directly emitted from sources that are both natural and anthropogenic. In addition, gases produced from these sources can undergo transformation into particles, in which case the sources are considered indirect so far as particles are concerned. All of these sources, direct and indirect, natural and anthropogenic, must be considered. In general, most coarse mode particles are generated from direct sources and most accumulation mode particles from indirect sources.

Natural source fluxes for atmospheric particles are given in Table 5.1. Two sources dominate: sea spray and mineral dust, the former involving breaking ocean waves and the latter windblown surface soil. Biological debris also adds to the overall composition of atmospheric particles, and several studies have shown that marine aerosols contain a significant amount of organic matter. The compositions of mineral dust in particular are influenced as well by emissions of gases from natural sources, the most important of which are dimethylsulfide, emitted by the oceans; biogenic volatile organic compounds, emitted mainly by the terrestrial biosphere; and monoterpenes and isoprene, emitted mainly from trees and other plants.

Anthropogenic source fluxes are given in Table 5.2. Directly emitted (primary) particles originate from industrial processes, vehicle traffic, and construction activity. These particles will have local impacts, but on a global scale the natural sources of directly emitted particles are far more important. A different situation exists for precursors, such as volatile organic compounds, SO_2, and NH_3. They result in the subsequent formation of particles, where anthropogenic sources significantly exceed natural sources. For these secondary particles, whose low gravitational setting velocities give them atmospheric lifetimes of several days, anthropogenic sources can be expected to control concentrations over most land areas (except perhaps those barren of people, industrial activity, and hence surfaces able to corrode).

TABLE 5.2 Global Anthropogenic Emissions of Aerosols (A) and Aerosol Precursors (P) in the Year 2000 (Tg year^{-1} or Tg S year^{-1} for SO_2)

	Minimum	Maximum
Biomass burning aerosols (A)	29.0	85.3
Soot (A)	3.6	6.0
Marine primary organic aerosols (A)	6.3	15.3
Nonmethane volatile organic compounds (P)	98.2	157.9
SO_2 (P)	43.3	77.9
NH_3 (P)	34.5	49.6

Minimum and maximum values from a range of inventories.
The estimates are from O. Boucher, D. Randall, P. Artaxo, C. Bretherton, G. Feingold, P. Forster, V.-M. Kerminen, Y. Kondo, H. Liao, U. Lohmann, P. Rasch, S.K. Satheesh, S. Sherwood, B. Stevens, and X.Y. Zhang, Table 7.1 in *Clouds and aerosols, in Climate Change 2013*: The Physical Science Basis. Contribution of Working Group I to the Fifth Assessment Report of the Intergovernmental Panel on Climate Change, eds. T.F. Stocker, D. Qin, G.-K. Plattner, M. Tignor, S.K. Allen, J. Boschung, A. Nauels, Y. Xia, V. Bex, and P.M. Midgley, Cambridge University Press, Cambridge/New York, 2013.

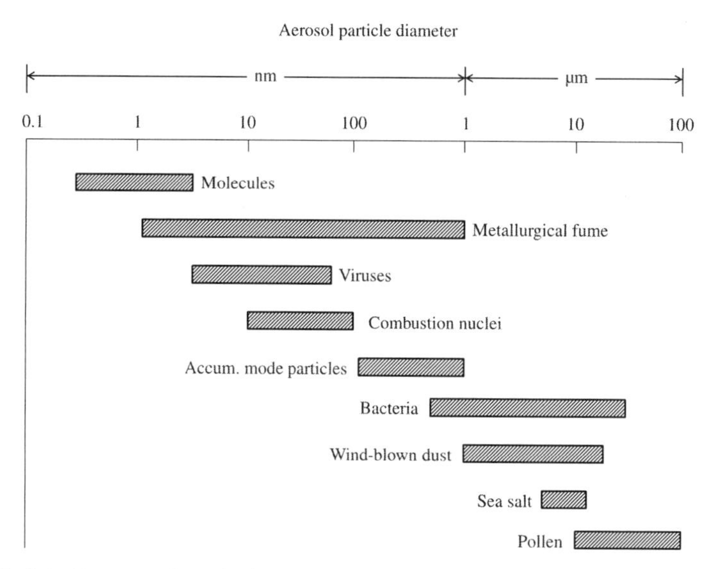

FIGURE 5.4 Ranges of equivalent diameter for some types of aerosol particles. For perspective, the diameters of molecules are also shown.

Typical size ranges for particles from a number of anthropogenic and natural sources are shown in Figure 5.4.

5.4 AEROSOL PARTICLE PHYSICS AND CHEMISTRY

The physical and chemical complexity of atmospheric aerosol particles manifests itself in a variety of ways. One is the ability of most of the particles to sorb water, transforming themselves in the process from agglomerated solid particles to highly

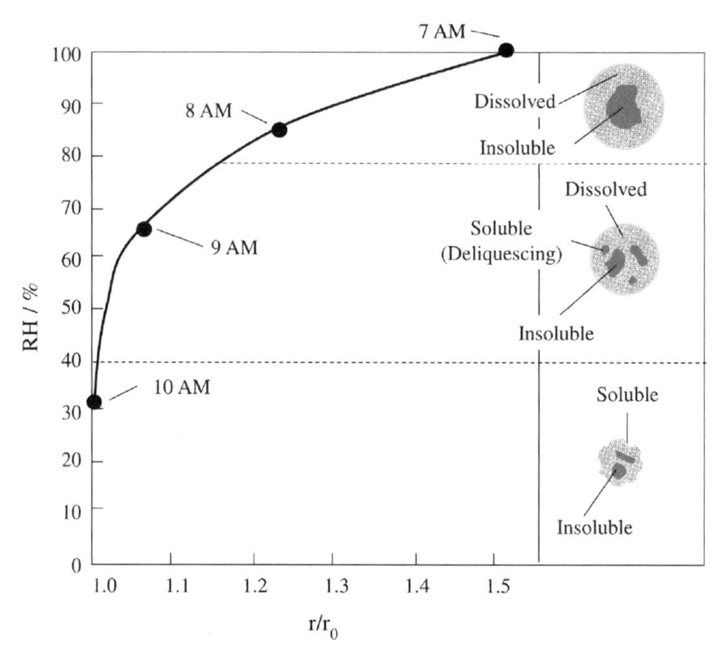

FIGURE 5.5 A schematic description of the deliquescence of a mixed aerosol particle. The general pattern shown here is common, but the exact shape of the response curve depends on the particle's chemical content. The times indicated suggest the evolution of the particle from a humid early morning to a dryer midday. (Reproduced with the permission from Springer Verlag; P. Winckler, Relations between aerosol acidity and ion balance, in *Chemistry of Multiphase Atmospheric Systems*, ed. W. Jaeschke, Springer Verlag, Berlin, pp. 269–298, 1986.)

concentrated aqueous solution droplets. Figure 5.5 suggests the process. A particle, initially containing a mixture of soluble and insoluble compounds, is exposed to air with increasing relative humidity. The soluble compounds deliquesce, acquiring water from the water vapor in the surrounding air. As the humidity continues to increase, the soluble species eventually become fully dissolved, having incorporated substantial amounts of added water. At relative humidities above 90%, a typical atmospheric particle may have a radius some 1.5 times its dry radius, and since volume is proportional to the cube of the radius, the volume of the particle will have more than tripled.

The interactions of atmospheric gases with aerosol particles generate highly complex entities, as pictured in Figure 5.6. Several stages are typically involved: emission of corrosive gases, transformation in the gas phase by atmospheric reactions, deposition onto the wetted aerosol surface, and (potentially) transformations in the aqueous phase. Because sources of corrosive gases are spatially diverse and source specific, the resulting particles can be expected to be chemically diverse as well.

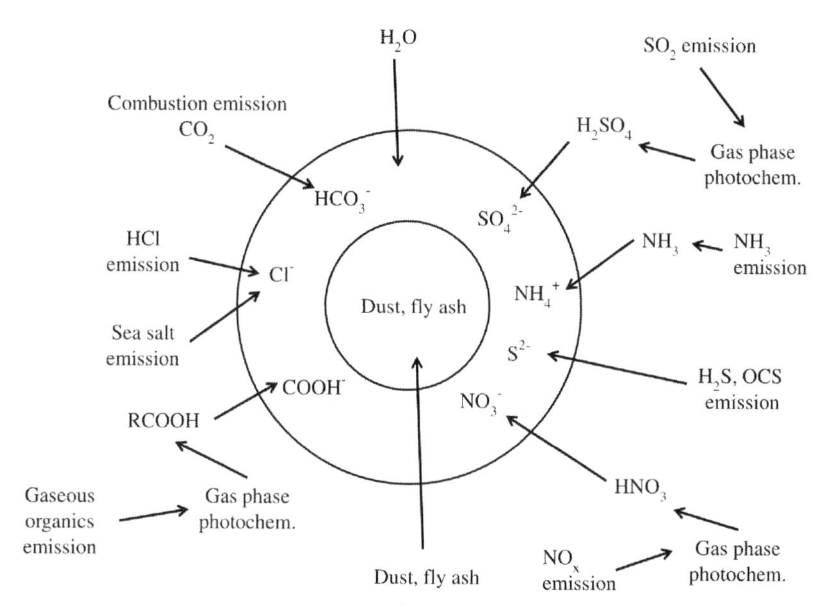

FIGURE 5.6 An atmospheric aerosol particle or corrodible surface, its aqueous envelope, and its interactions with gaseous atmospheric corrosive species. ("R" in RCOOH indicates either H or CH_3).

Determination of the chemical contents of particles was traditionally performed by collecting large numbers of particles and conducting chemical analyses of the ensemble. In recent decades a number of analytical techniques have become available for more detailed analysis, such as laser fragmentation of individual particles followed by mass spectrometric analysis. Such a technique can provide information not only on the chemical composition of individual particles but can also indicate the source from which the particle originates. Other analytical tools for analyzing aerosols include focused ion beam–scanning electron microscopy with energy dispersive spectroscopy, transmission electron microscopy, Figure 5.7, X-ray diffraction, gas chromatography–mass spectrometry, ion chromatography, aerosol mass spectrometry, and spectroscopy techniques based on synchrotron radiation.

It is of interest to ask whether the aerosol particle's constituents are well mixed or not. The evidence is not unequivocal, but several lines of reasoning suggest a common picture like that of Figure 5.8, in which the organic molecules form a thin film at the outer surface of the wetted particle, the hydrophilic ends of the molecules toward inside and the hydrophobic ends toward the outside. The film may or may not be complete, depending on the availability of organic constituents and the chemistry involving the inorganic constituents of the particle. In regions having high atmospheric concentrations of long-chain (greater than about C_8) molecules, such as densely polluted urban areas or heavily forested areas with high emission fluxes of

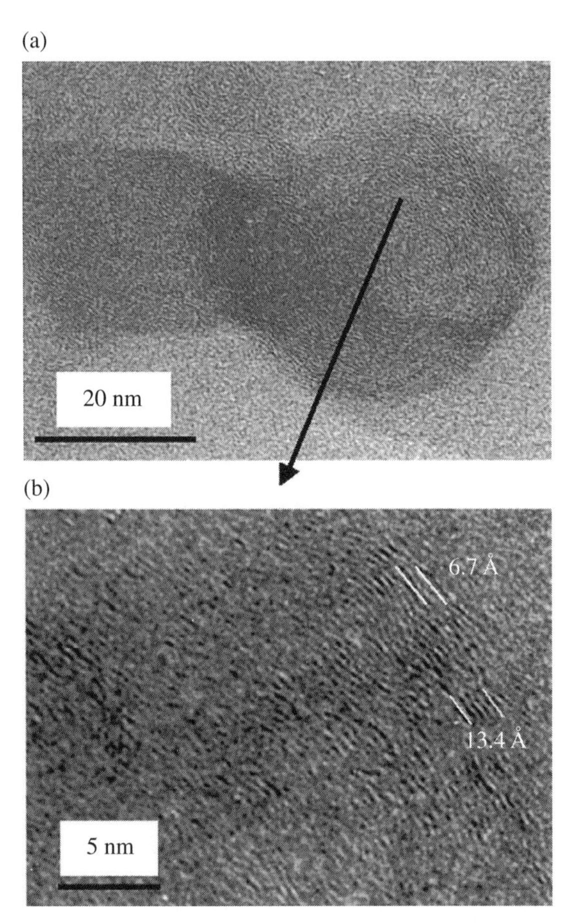

FIGURE 5.7 Transmission electron microscopy image of diesel soot particles. They consist of nanocrystalline graphite with an onion-shell structure and a domain size between 2 and 3 nm. (a) Agglomerate of particles. (b) Enlarged view showing the onion-shell structure. (Reproduced with permission from Elsevier; M. Wentzel, H. Gorzawski, K.-H. Naumann, H. Saatho, and S. Weinbruch, Transmission electron microscopical and aerosol dynamical characterization of soot aerosols, *Aerosol Science, 34*, 1347–1370, 2003.)

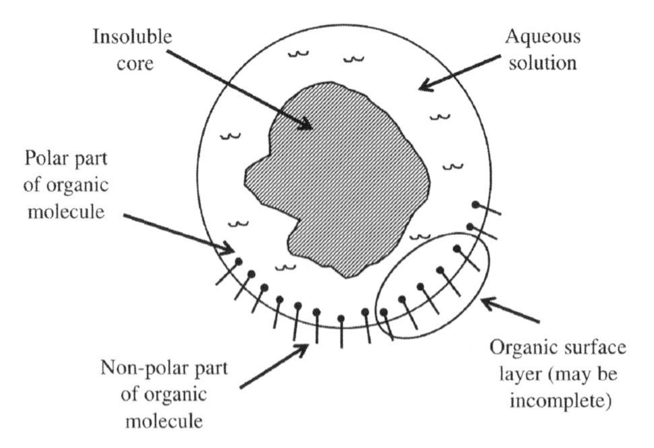

FIGURE 5.8 A schematic picture of an aerosol particle covered by an organic surface film.

terpenes or isoprene, relatively complete films may be likely. Such films have the potential to inhibit the transport of water and other molecules into and out from the particle, essentially isolating its chemistry from its surroundings.

Related to these mixtures of aerosol compositions is the observation that urban outdoor surfaces commonly are covered by an urban surface film in the thickness range from tens to hundreds of nanometers, so-called urban grime, whose chemical composition reflects the aerosol composition found in urban environments. More detailed studies have shown that surfaces on different outdoor constructions rapidly become coated with a mixture of constituents. Most of the identified mass of these films consists of inorganic compounds, such as sulfates, nitrates, and various metals, whereas a minor part consists of organic constituents originating from various natural and anthropogenic sources.

5.5 IMPLICATIONS OF AEROSOL PARTICLES FOR ATMOSPHERIC CORROSION

The principal goal of this chapter, of course, is to consider not only the sources, concentrations, and properties of atmospheric particles but also the potential impacts of the particles on atmospheric corrosion. In this connection, the relationships among particle number, area, and volume are of interest. Typical distributions appear in Figure 5.9. As mentioned before, although the number of coarse mode particles is very small compared to that of smaller particles, the coarse mode volume is nearly as large as that of the accumulation mode. This has implications for particles deposited on a corroding surface.

Assuming typical urban concentrations of 10^4 particles cm^{-3} in accumulation and 10^1 particles cm^{-3} in coarse mode and using appropriate deposition velocities from Figure 5.3, one can estimate the particle flux to an unsheltered surface. Depending on the particle size of interest, the surface is completely covered in a few hours if all particles are retained (an unlikely occurrence). In any case, coverage within a few days seems likely.

As seen earlier in this chapter, the particles will sorb water. The water amount that can potentially be sorbed by particles that accumulated within a few days, if distributed as a thin film, may comprise some 1000 monolayers, easily enough to support bulk aqueous-phase chemistry. As the ambient humidity fluctuates, so will the amount of water on the surface. When humidity is low, condensable species will precipitate onto the surface, as described in Chapter 3 and experimentally verified as described in Chapter 8. When humidity increases, soluble species dissolve. The deliquescent particles thus provide a means by which a corrosive chemical environment can be augmented and through which wet–dry cycles promote chemical transformation and the deposition of corrosion products. Although quite approximate, such estimates demonstrate that aerosol particles almost certainly have the potential to play a major role in atmospheric corrosion.

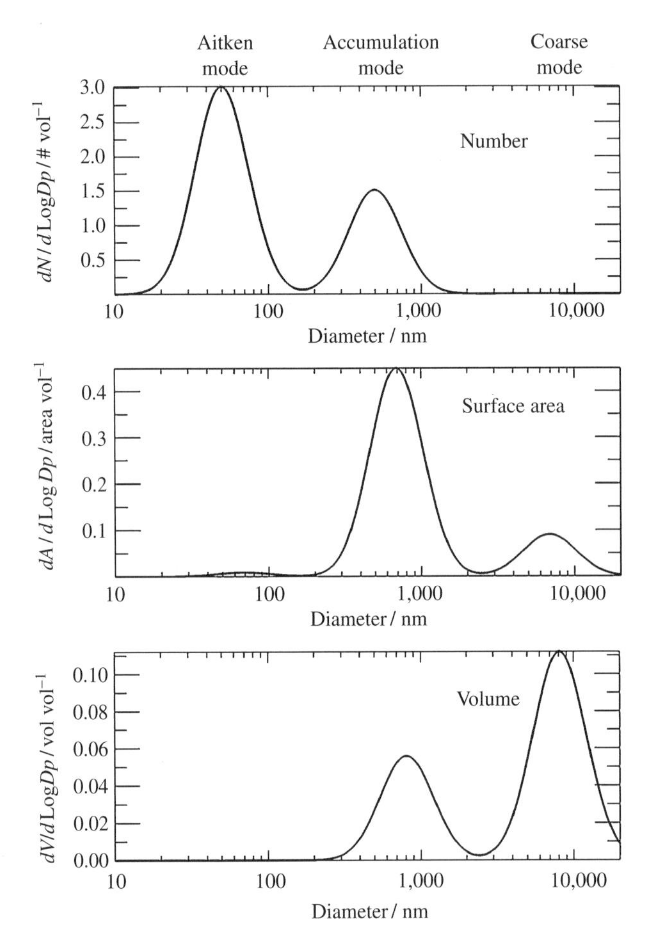

FIGURE 5.9 Normalized frequency plots of the number, surface, and volume distributions of aerosols. From the standpoint of surface chemistry and optical effects, area distribution is of greatest interest. From the standpoint of chemical mass balances and atmospheric corrosion, physiological volume (mass) distribution is of interest. (Source: *Wikipedia.*)

Recent studies have shown clear evidence of the major impact of both marine and industrial aerosols on the atmospheric corrosion of metals. A detailed analysis reveals that marine aerosols acidified through absorption of gases (SO_2, NO_2, HCl, and HNO_3) and fine acid aerosols represent a particularly aggressive type of aerosols that can cause atmospheric corrosion effects in large areas because of their tendency to travel long distances before interaction with metallic surfaces.

To illustrate the complexity in understanding the relationship between aerosol chemistry and atmospheric corrosion effects, Figure 5.10 displays the summary of results from the analysis of particulate matter collected inside a road tunnel environment.

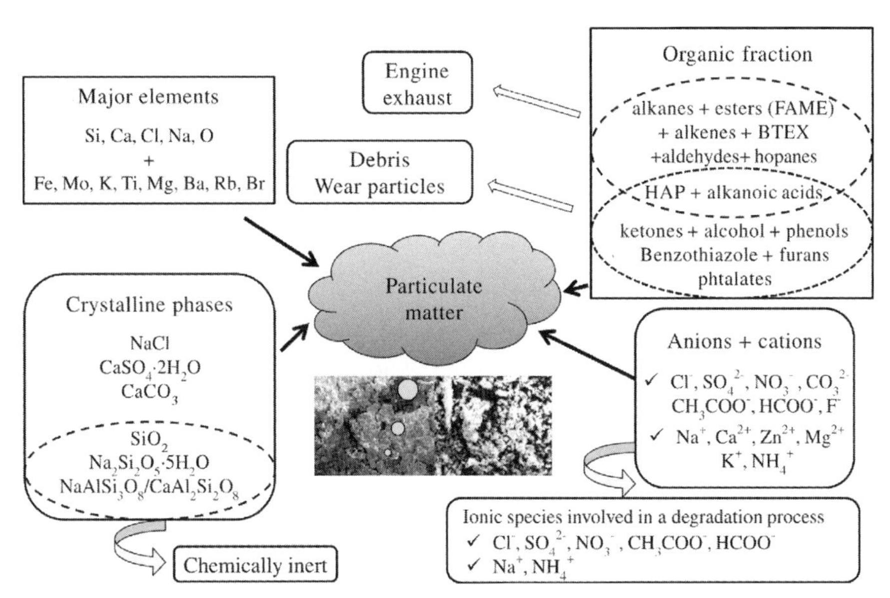

FIGURE 5.10 Identified species in particulate matter, collected in the Grand Mare road tunnel in Rouen, France, by means of a multianalytical approach. (Reproduced with permission from Taylor & Francis; I. Ameur-Boudabbous, J. Kasparek, A. Barbier, and B. Hannoyer, Transverse approach between tunnel environment and corrosion: particulate matter in the Grand Mare tunnel, *Journal of the Air & Waste Management Association, 64*, 198–218, 2014.)

FURTHER READING

Fundamentals of Aerosol Particles

J. Heintzenberg, Fine particles in the global troposphere: a review, *Tellus, 41B*, 149–160, 1989.

H. Horvath, M. Kasahan, and P. Pesava, The size distribution and composition of the atmospheric aerosol at a rural and nearly urban location, *Journal of Aerosol Science, 27*, 417–435, 1996.

W.W. Nazaroff and G.R. Cass, Mass-transport aspects of pollutant removal at indoor surfaces, *Environment International, 15*, 567–584, 1989.

A. Schenkel and K. Schaber, Growth of salt and acid aerosol particles in humid air, *Journal of Aerosol Science, 26*, 1029–1039, 1995.

Sources of Aerosol Particles

M.O. Andreae, D.A. Hegg, and U. Baltensperger, Sources and nature of atmospheric aerosols, in *Aerosol Pollution Impact on Precipitation: A Scientific Review*, eds. Z. Levin and W.R. Cotton, Springer Science+Business Media B.V., Dordrecht/London, pp. 45–89, 2009.

O. Boucher, D. Randall, P. Artaxo, C. Bretherton, G. Feingold, P. Forster, V.-M. Kerminen, Y. Kondo, H. Liao, U. Lohmann, P. Rasch, S.K. Satheesh, S. Sherwood, B. Stevens, and X.Y. Zhang, Clouds and aerosols, in *Climate Change 2013: The Physical Science Basis. Contribution of Working Group I to the Fifth Assessment Report of the Intergovernmental Panel on Climate Change*, eds. T.F. Stocker, D. Qin, G.-K. Plattner, M. Tignor, S.K. Allen, J. Boschung, A. Nauels, Y. Xia, V. Bex, and P.M. Midgley, Cambridge University Press, Cambridge/New York, pp. 571–658, 2013.

Aerosol Particle Physics and Chemistry

D.J. Donaldson, T.F. Kahan, N.O.A. Kwamena, S.R. Handley, and C. Barbier, Atmospheric chemistry of urban surface films, *American Chemical Society Symposium Series, 1005*, 79–89, 2009.

L.A. Gundel and H. Destaillats, Aerosol chemistry and physics: an indoor perspective, in *Aerosols Handbook: Measurement, Dosimetry, and Health Effects*, 2nd edition, eds. L.S. Ruzer and N.H. Harley, CRC Press, Boca Raton, pp. 129–178, 2013.

K.A. Prather, C.D. Hatch, and V.H. Grassian, Analysis of atmospheric aerosols, *Annual Review of Analytical Chemistry, 1*, 485–514, 2008.

Implications of Aerosol Particles for Atmospheric Corrosion

I.S. Cole, N.S. Azmat, A. Kanta, and M. Venkatraman, What really controls the atmospheric corrosion of zinc? Effect of marine aerosols on atmospheric corrosion of zinc, *International Materials Review, 54*, 117–133, 2009.

6

CORROSION IN LABORATORY EXPOSURES

6.1 THE NEED FOR WELL-DEFINED LABORATORY EXPERIMENTS

As is evident from previous chapters, there are many parameters that can influence the atmospheric corrosion of a given material. In order to systematically study these influences, it is necessary to perform studies in which only a few selected parameters are varied. One advantage of laboratory environments is that individual parameters that are believed to be important can be studied under controlled conditions, and their concentrations or magnitudes varied over a wide range. Another advantage is that comparisons can be made of the corrosion resistance of different materials under well-defined conditions. Over the years a very large number of laboratory experiments have been performed, some of which have resulted in new insights into actual corrosion mechanisms. This chapter discusses general design considerations of laboratory experiments, depicts important examples performed for either scientific or industrial purposes, explores the capability of laboratory experiments to mimic exposure in service conditions, and describes examples of laboratory exposures, combined with computational model studies, that have provided new mechanistic insight into atmospheric corrosion processes.

6.2 CONSIDERATIONS FOR SPECIFIC METALS

Because a laboratory exposure only includes a small number of corrosive substances, sometimes referred to herein as corrodents, the unique susceptibility of each material to corrode in the presence of a given corrosive substances can be examined. The specific corrosion behavior of different materials is described in some detail in the corresponding appendices. Marked variations in corrodent-material susceptibilities

Atmospheric Corrosion, Second Edition. Christofer Leygraf, Inger Odnevall Wallinder, Johan Tidblad and Thomas Graedel.

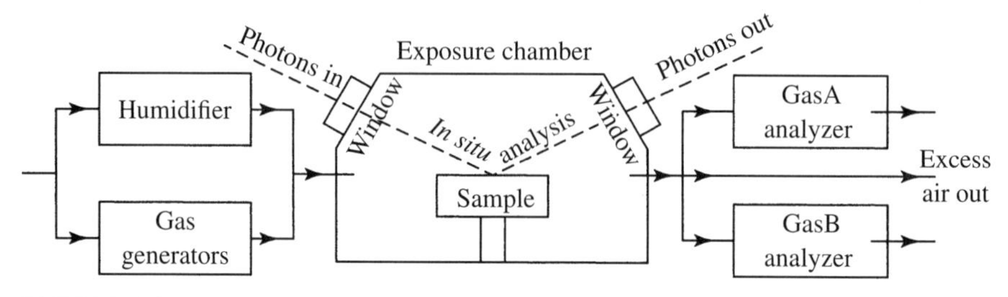

FIGURE 6.1 Schematic diagram of an experimental setup for exposures in controlled laboratory environments.

are summarized in Table 4.1 for a selection of important metals, alloys, and stones. The variations in susceptibility can be partly explained by the ability for a given metal to react with counterions as explained by the Lewis acid–base concept (Section 2.3.4). The resulting corrosion effect for a given corrodent-material system depends on the test severity or corrosivity of the atmosphere. Corrosivity depends not only on the presence and concentration of certain corrosive substances but also on other exposure parameters, the most important being relative humidity, temperature, and exposure time. These parameters are briefly discussed in the next section.

6.3 DESIGN CONSIDERATIONS

6.3.1 Exposure Chamber

The laboratory exposure is normally performed in an exposure chamber whereby the freshly prepared sample is bathed in a flowing atmosphere consisting of filtered air, Figure 6.1. The internal walls of the chamber and connecting pipes, valves, taps, gaskets, and seals in contact with the corrosive atmosphere should be made of an inert material such as glass or polytetrafluoroethylene. The chamber should be designed so that condensation can be avoided and well-defined flow conditions can be considered. It should have a volume of at least several cubic diameters to permit exposure of samples of a few square centimeter surface area without touching each other. Before humidification and addition of corrosive substances, the delivered air is filtered and purified from particles and gases by oil traps, particle filters, activated charcoal, or other devices.

6.3.2 Sample Preparation

The test sample, usually a flat specimen of a pure metal, is prepared prior to exposure by some mechanical or chemical treatment such as abrasion with SiC paper, diamond polishing, or immersion in a mild acid. This is usually followed by rinsing with demineralized water and perhaps alcohol. When testing a commercial product, the sample will normally only undergo some rinsing treatment. The purpose of the surface preparation is to clean the sample from contamination and to remove surface films in order to obtain reproducible conditions prior to exposure start. Most metals and alloys spontaneously form an oxide or oxyhydroxide immediately after surface preparation when exposed to the ambient environment (Sections 2.2.1 and 2.2.2), and this thin surface film often possesses corrosion protective properties that may influence the subsequent laboratory exposure. The properties of the film may age with time as may their corrosion protective ability. Partly this aging process may be due to changes of the surface film caused by hydration or dehydration of the film, partly due to contamination through adsorption of a very thin layer of carbonaceous material found on most materials exposed in ambient air. This layer, sometimes referred to as adventitious carbon, contains a variety of relatively short-chain hydrocarbons. As an example, a freshly polished copper surface has somewhat lower corrosion protective ability than the same surface after storage in dry, unpolluted conditions for a few weeks in ambient indoor conditions. Hence, for comparative purposes it is important always to prepare and store the sample in the same way; otherwise the resulting corrosion effect may be different from one exposure to the next.

6.3.3 Relative Humidity

The air is humidified, typically within the range of 50–95% relative humidity. As further discussed in Section 8.3, the sensitivity for different metals to relative humidity changes varies considerably. Iron or steel, for instance, is very sensitive to relative humidity changes, silver is not very sensitive, and copper, nickel, zinc, and cobalt exhibit intermediate sensitivities. Hence, the accepted variation in relative humidity during a laboratory exposure varies from one metal to another but should typically be kept within ±3% of the desired value. The humidified air should not include any water droplets or aerosols. From this it follows that the air never should pass through water during the humidification step but only above the water surface.

6.3.4 Temperature

The temperature in most laboratory exposures is kept constant usually in the range between 20° and 30°C. In this temperature range the influence of temperature on the corrosion rate is not too pronounced, and it is usually sufficient to keep the temperature to within ±1°C of the desired value.

In order to mimic outdoor or near-outdoor exposure conditions, some laboratory exposures are based on cyclic variations of relative humidity and temperature, as exemplified in the subsequent section. However, this makes the experimental procedures more complicated and increases the probability of not obtaining sufficient reproducibility between different experimental setups.

6.3.5 Exposure Time

The exposure time varies typically from a few days to a few weeks and depends on the aim of the test, test severity, and the corrosion protective ability of the exposed materials. When extrapolating corrosion data from laboratory exposures, it may be

TABLE 6.1 Commonly Studied Air Constituents in Controlled Laboratory Exposures and Techniques for Their Generation and Continuous Analysis

Air Constituent	Generating Technique	Technique for Continuous Monitoring
SO_2	Permeation tube	Fluorescence
NO_2	Permeation tube	Chemiluminescence
NH_3	Permeation tube	Chemiluminescence
O_3	UV radiation	UV photometry
H_2S	Permeation tube	Lead acetate densitometry or conversion to SO_2 and fluorescence measurements
Cl_2	Permeation tube	No suitable technique for continuous monitoring
HNO_3	Permeation tube	No suitable technique for continuous monitoring
$(NH_4)_2SO_2$	Solution atomizer, spraying	No suitable technique for continuous monitoring
NaCl	Thermophoretic deposition, spraying	No suitable technique for continuous monitoring

useful to establish actual corrosion effects (M) as a function of time (t). As discussed in Section 7.3, the generalized form can be written as

$$M = At^n \tag{6.1}$$

with n usually in the range between 0.5 and 1.0 and A a constant of proportionality.

6.3.6 Corrosive Substances

In the design of laboratory exposures, it is important to consider what concentration of gaseous corrosive species should be employed. Lower levels decrease the corrosion rate and increase the exposure time needed to measure significant corrosion effects. Much higher levels, on the other hand, may introduce physical or chemical processes that are not representative of field exposures. Gas concentrations in the range from a few tens to a few hundreds of ppbv have proven to give reasonable agreement with field data. This concentration range is in marked contrast to studies well before the 1980s, which hardly ever were performed with gas concentrations below 1 ppmv and often at concentrations several orders of magnitude higher. While these earlier studies, performed without the instrumental capabilities available today, generated insight into interesting physicochemical processes, in many cases the insight gained was of little relevance to actual field exposures.

Table 6.1 summarizes gaseous and particulate air constituents used in laboratory exposures, together with methods of generation and techniques to monitor concentrations. For several important constituents, no suitable continuous technique is available. In the case of gaseous hydrogen chloride (HCl), no adequate generation technique has been developed. Gaseous chlorine (Cl_2) is often used as a substitute, but it is a matter of considerable discussion whether such a substitution can be made without introducing unrealistic corrosion chemistry into the experiments. Among corrosive gases listed in Table 6.1, nitric acid (HNO_3) should be highlighted. Legislations in

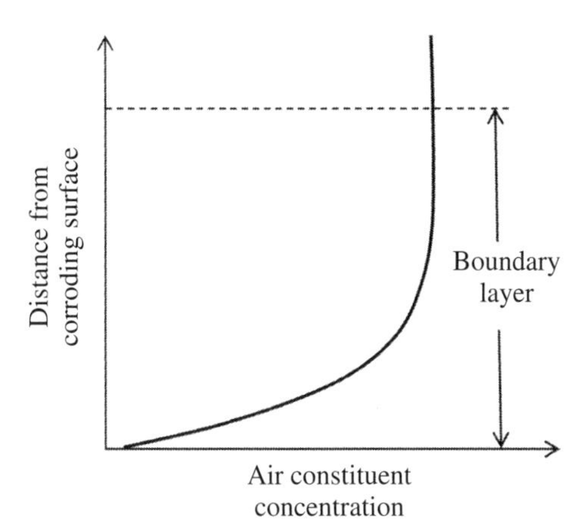

FIGURE 6.2 Schematic diagram of boundary layers caused by reaction of air constituent with the corroding surface.

many countries in the Western industrialized world have led to significant reductions in concentration of the acidifying gaseous pollutants SO_2. As a result, the relative corrosion effects of other pollutants have increased, including HNO_3.

Atmospheric corrosion rates are strongly dependent on the rates of transport of airborne corrosive substances through the gaseous regime to the corroding surface. This important factor is often overlooked in descriptions of laboratory exposures and is therefore discussed in some detail in Section 6.3.7. The transport rate of corrosive substances is determined by the local air velocity and the immediate configuration of airflow around the sample. In laminar flow, especially at low air speeds, diffusion is the dominant transport mechanism. In turbulent flow, the more usual and more realistic condition, convection generally dominates. As a result of the reaction of the airborne corrosive substance with the surface (generally a thin liquid layer atop the corroding material), there is a boundary region in the gaseous regime next to the liquid regime, in which the concentration of the air constituent is depleted, Figure 6.2.

6.3.7 Delivery of Corrosive Substances

The measured atmospheric corrosion rate is determined both by the delivery rate of constituents to the corroding surface and by the reaction rate of reactants with the surface. Hence, the rate of atmospheric corrosion is under mixed control, and efforts to correlate corrosion rates with concentrations of constituents must consider the possibility of limited transport of constituents to the surface. These are two extreme situations. Under well-stirred and highly turbulent airflow conditions, the limitation in transport rate is negligible and the corrosion rate will be solely determined by surface reaction rates. Conversely, surfaces that act as ideal absorbers experience no limitation in surface reaction rate and the deposition will be limited by transport processes to the surface. Figure 6.3 is an example of the influence of airflow velocity on the atmospheric corrosion rate of copper. Under fixed exposure conditions the

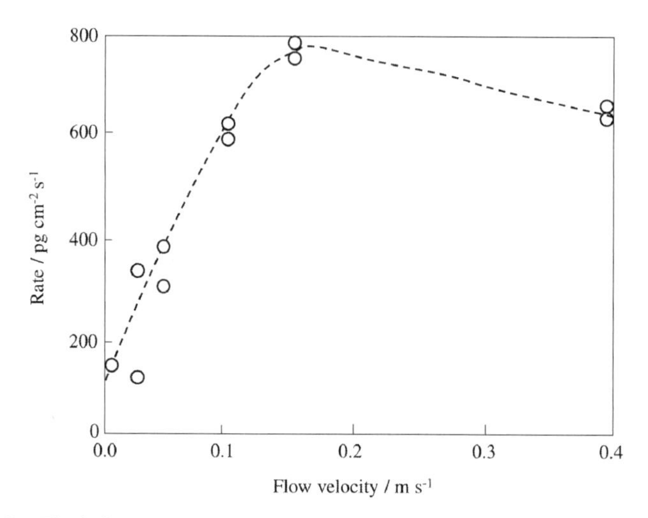

FIGURE 6.3 The influence of airflow velocity on the atmospheric corrosion rate of copper in a laboratory exposure with controlled aerodynamic conditions. The unfilled circles represent experimental data points. (Reproduced with permission from The Electrochemical Society; L. Volpe and P. Peterson, Mass-transport limitations in atmospheric corrosion, in *Proceedings of the 1st International Symposium on Corrosion of Electronic Materials and Devices*, ed. J.D. Sinclair, The Electrochemical Society, Pennington, pp. 22–39, 1991.)

corrosion rate in this particular example increases almost linearly with airflow velocity in the range up to $0.2\,m\,s^{-1}$, suggesting that the rate is limited by the transport of corrosive substances to the surface. Above $0.2\,m\,s^{-1}$ the corrosion rate levels off as a function of airflow, suggesting that surface reactions are rate limiting.

A useful interaction parameter in this context is the deposition velocity, defined in Section 4.2.9. Clearly this is a parameter that is affected by airflow rate and sample geometry. When designing an exposure in a controlled laboratory environment, it is important that the deposition velocity is similar to the situation one wishes to mimic.

6.3.8 Corrosion Effect Measurements

The corrosion effects are measured by different means depending on the actual application: mass increase, mass loss, change in visual appearance, change in electrical contact resistance of a thin film of the metal investigated, or evolving surface composition. In more sophisticated laboratory experiments, there are possibilities for measuring corrosion effects *in situ*. A set of analytical *in situ* techniques that has been used in the author's laboratory for exploring the interfacial regime in atmospheric corrosion is schematically depicted in Figure 6.4. Infrared reflection absorption spectroscopy (IRAS) integrated with the quartz crystal microbalance (QCM) provides overall information on the chemical composition of corrosion products and their amounts over the whole investigated surface. Confocal Raman spectroscopy (CRS) gives the lateral distribution of the corrosion products, which is important when relating the growth of corrosion products to the microstructure of the metal or

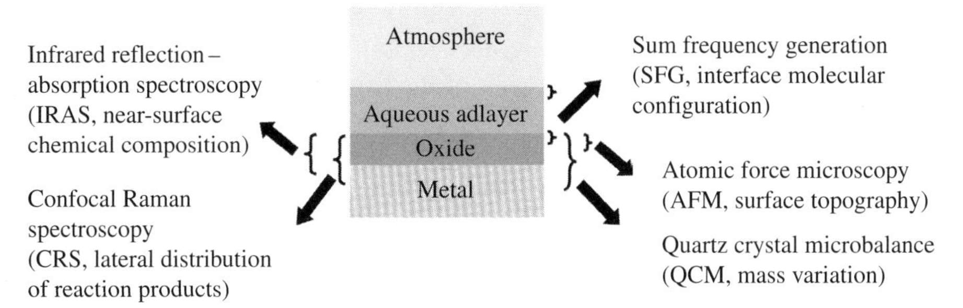

FIGURE 6.4 A set of analytical *in situ* techniques that has been used in the author's laboratory for exploring the interfacial regime in atmospheric corrosion. For further details, see J. Hedberg, S. Baldelli, C. Leygraf, and E. Tyrode, Molecular structural information of the atmospheric corrosion of zinc studied by vibrational spectroscopy techniques. Part I. Experimental approach, *Journal of the Electrochemical Society, 157*, C357–C362, 2010. Brief information regarding these and other techniques is given in Appendix A.

associating the growth corrosion products to anodic or cathodic processes. Atomic force microscopy (AFM) displays the topography of the corroded surface, while sum frequency generation (SFG) is a unique vibrational spectroscopy technique for exploring the chemistry of the interfaces (solid/liquid or liquid/atmosphere) involved in atmospheric corrosion. The actual gas consumption can also be monitored through measurements of the difference in gas concentration between the inlet and the outlet of the exposure chamber, Figure 6.1. A description of techniques for analyzing exposed surfaces is given in Appendix A.

6.4 EXAMPLES OF IMPORTANT LABORATORY EXPOSURES

Having described general design considerations, we next turn to a number of important examples of laboratory exposures for both scientific purposes and industrial applications. Pioneering developments have been reported over many years from Chalmers University of Technology in Gothenburg, Sweden (laboratory exposures studying the influence of primarily gaseous corrosive substances in combination with humidity), and from Lucent Technologies, Bell Labs, Murray Hill, NJ, United States (laboratory exposures studying the influence of particles in combination with humidity). The mechanistic insight generated from these laboratories and other have promoted the development of new laboratory exposures for various industrial applications, including coatings, microelectronics, and the automotive industry. A few examples are briefly described in this section.

6.4.1 Laboratory Exposures for Scientific Purposes

The major experimental advancement achieved at Chalmers was to allow laboratory exposures of metals in the presence of several gaseous corrosive substances in sub-ppmv concentrations, resembling those of moderately to heavily polluted outdoor environments. Temperature is carefully controlled throughout the experiment. One type of exposure, typically carried out over 24 h, involves ongoing analysis of gaseous corrosive substances at the exit of the exposure chamber. This permits the deposition rate or deposition velocity of each investigated corrodent to be estimated as a function of exposure time.

Another type of exposure permits the formation of corrosion products and the corrosion rate to be determined at different times of exposure up to typically a few weeks of. These studies, complemented with IRAS (see Appendix A) to provide *in situ* information on initial buildup of corrosion products, have resulted in insight into the corrosion mechanisms that govern the role of nitrogen dioxide (NO_2) and ozone (O_3) on the sulfur dioxide (SO_2)-induced atmospheric corrosion of zinc, copper, and other materials. Similar approaches have been adopted by several other laboratories.

The research group at Murray Hill has developed laboratory exposures in which submicron-sized particles of ammonium sulfate, $(NH_4)_2SO_4$, are aerosolized and deposited under dry conditions on the metal surface to be investigated. The deposited amounts range from less than 1 μg to several hundreds of μg over an area of 0.25 cm^2 (for comparison, the combined accumulation of particles on surfaces after 10 years of indoor exposure in the eastern half of the United States typically ranges between 5 and 10 μg cm^{-2}). A particular strength of this approach is the multitude of techniques used to follow the corrosion process after introduction of humidity into the exposure chamber, including scanning electron microscopy with energy dispersive X-ray analysis (*ex situ*), X-ray diffraction (*in situ*), Auger electron spectroscopy with ion sputtering (*ex situ*), infrared spectroscopy (*in situ*), and the Kelvin probe (*in situ*) (all of which are described in Appendix A). These studies have provided molecular insight into the interaction of $(NH_4)_2SO_4$ with zinc, copper, and aluminum at different relative humidities and temperatures. The results have aided in the interpretation of data from indoor or outdoor exposures, such as those shown in Figure 2.3 and discussed in Section 8.3.1.

6.4.2 Laboratory Exposures for Industrial Applications

For several decades, standardized laboratory exposures used as accelerated corrosion tests for industrial applications consisted of single-gas tests using high concentrations (>10,000 ppbv) of pollutants, mainly SO_2 or hydrogen sulfide (H_2S). Another accelerated test that has been used for a long time for evaluating the corrosion resistance of metallic materials and coatings is the so-called salt spray test (ASTM B117), originally published in 1939. Both types of tests, single gas or salt spray, tend to represent oversimplified corrosion conditions. Nevertheless, the salt spray test is still widely used for, for example, comparing the corrosion performance of various metals or coatings, with or without scratches, or for estimating the corrosion resistance in marine environments.

Overall, laboratory exposures for industrial applications represent a research area of utmost technical importance and numerous efforts have been made to establish tests with a high degree of predictability to mimic corrosion effects of different industrial products in a variety of environments. It is far beyond the scope herein to provide a comprehensive treatment of this important subject, since the design of the tests largely depend on the industrial application. However, two important industrial areas are treated in some detail in other chapters. Section 11.4 describes accelerated corrosion testing for electronic materials, and Section 12.5 accelerated corrosion testing for automotive applications. The former are so-called flowing mixed gas corrosion tests, which use a combination of H_2S, NO_2, Cl_2, or SO_2 and operate at constant humidity, temperature, and gas concentration conditions. The latter tests are based on cycling conditions in which the tested objects are allowed to go through phases of both dry and wet exposure conditions in which corrosive substances have been introduced. Both types of tests have been standardized, and examples are given in the list for "Further Reading."

6.5 CAN CORROSION PROCESSES IN THE FIELD BE REASONABLY SIMULATED BY LABORATORY EXPERIMENTS?

In considering the complex atmospheric environment, with its multitude of different air constituents and climatic factors, one may question whether corrosion processes operating in field environments can really be represented by processes operating under more simple laboratory exposures, with only a few parameters present. To derive an answer to this question, a study was performed in which the corrosion effects of nickel were compared after comparable exposures in the outdoor field and laboratory, respectively. Nickel was chosen because its outdoor field corrosion performance appears to be the simplest among the common metals. Its corrosion exhibits the best correlation with SO_2 concentration of all materials investigated in the UN Economic Commission for Europe (UN/ECE) exposure program. This program is a major international effort to evaluate the effect of sulfur pollutants in combination with nitrogen oxides (NO_x) on various materials. The linear relationship obtained between corrosion effect and SO_2 concentration after 1, 2, 4, and 8 years of exposure exhibits no time dependence, which implies that the corrosion rate, within experimental accuracy, is the same irrespective of exposure period. Nickel was also selected because the chemical composition of its corrosion products exhibits relatively little complexity. Sulfates are the dominant anions, with only minor participation of chlorides, nitrates, and carbonates.

Although SO_2 is the only explanatory variable in the corrosion relationship, this by no means suggests that SO_2 is the only air constituent responsible for the corrosion effect of nickel. In fact, laboratory exposures have shown that the corrosion rate of nickel caused by the combined action of SO_2 and NO_2 is approximately one order of magnitude faster than the corrosion rate of SO_2 alone (Section 8.3.7). Nonetheless, field variations in NO_2 are apparently small enough with respect to the nickel corrosion process that their inclusion in the empirical corrosion relationship is not mandatory.

Based on actual exposure conditions for nickel specimens inside a sheltering box used at 39 outdoor field exposure sites of the UN/ECE program, a representative set of (nonaccelerated) conditions for the laboratory exposures was defined as follows: temperature, 22.0 ± 0.5°C; relative humidity, $75 \pm 2\%$; airflow velocity, $8 \pm 0.5\ cm\ s^{-1}$;

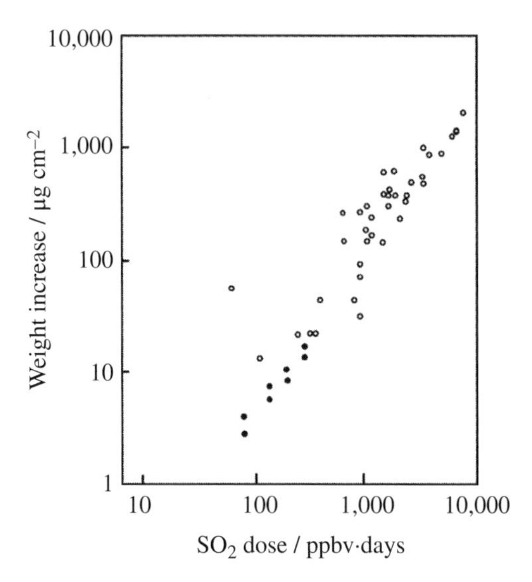

FIGURE 6.5 The weight increase of nickel as a function of SO_2 dose. The results are based on both laboratory exposures (filled circles) obtained after times up to 20 days and field exposures (unfilled circles) obtained after times up to 4 years. The conditions in the laboratory exposures represent the average of a majority of UN/ECE field sites. The field data points originate from sites with varying SO_2 concentration. (Reproduced with permission from The Electrochemical Society; S. Zakipour, J. Tidblad, and C. Leygraf, Atmospheric corrosion effects of SO_2, NO_2, and O_3. A comparison of laboratory and field exposed nickel, *Journal of the Electrochemical Society, 144*, 3513–3517, 1997.)

SO_2 concentration, 10 ± 1 ppbv; and NO_2 concentration, 30 ± 3 ppbv. The NO_2/SO_2 molar ratio of three represents an average value at a majority of field sites and the airflow velocity chosen results in similar SO_2 deposition velocities in the field and laboratory, respectively, between 0.3 and 0.4 cm s^{-1}.

Figure 6.5 displays the comparison between the corrosion effect data obtained from the outdoor field and laboratory exposures, respectively. The data is represented as weight increase versus SO_2 dose (the product of concentration and exposure time, ppbv·days). With increasing SO_2 dose there is clearly an increase in corrosion. As a result of shorter exposure times, the doses of the laboratory-exposed samples are lower than those of most of the field-exposed samples, but there is clear overlap in the data. When considering the many uncontrolled parameters that have influenced the field data points and the relatively few but controlled parameters that have influenced the laboratory data points, there is excellent agreement between the laboratory and field results.

Further evidence of similarities between field- and laboratory-exposed nickel samples is provided by comparing the surface composition of the corrosion products. By means of a combination of different analytical techniques, it has been shown that sulfate is the dominant anion in the corrosion products on both field- and laboratory-exposed samples, the most likely corrosion product being the hydrated nickel sulfate retgersite ($NiSO_4 \cdot 6H_2O$). Hence, the same mechanism, based on SO_2, NO_2, airflow velocity, and relative humidity as important corrosion parameters, appears to be operating during atmospheric corrosion of nickel under field and laboratory conditions.

The results shown so far for nickel exposed to humidified air with additions of SO_2 and NO_2 refer to efforts to simulate the corrosion effects of nickel outdoors. Similar efforts have also been performed in the author's laboratory to simulate indoor conditions. It turns out that zinc exposed to humidified air (typically in the range 80–90% relative humidity) to which around 100 ppbv of formic acid (HCOOH) or acetic acid (CH_3COOH) has been added results in similar corrosion effects as indoors. Both laboratory and indoor field exposures result in the formation of zinc carboxylates as the dominating group of corrosion products. Comparable amounts of zinc carboxylates after 2 days of laboratory exposure and after 1 year of representative indoor field exposures suggest an acceleration factor in the order of a few hundred times.

Each metal exhibits unique corrosion properties, and the conclusions obtained for nickel in Section 6.5 do not necessarily hold for other metals. In fact, the same investigation described earlier for nickel has also been performed on copper, with much less success. In this case neither the corrosion kinetics nor the chemical composition of corrosion products obtained in the field exposures could be reproduced by the laboratory exposure used for nickel. Thus, a more complex corrosion mechanism appears to occur for copper exposed in field sites compared with that in operation in laboratory exposures.

In order to exemplify the possibility to mimic different environments, Table 6.2 shows examples of laboratory exposures that have generated corrosion products similar to or identical to those seen in field exposures. Corrosion product agreement strongly suggests that the same corrosion processes are operating in the compared field and laboratory exposures.

TABLE 6.2 Characteristics of Laboratory Exposures That Have Generated Corrosion Products Similar to Those Observed in Certain Field Environments

Corrosive Species	Metal	Main Corrosion product	Type of Simulated Field Environment	Reference
SO_2, NO_2	Nickel	$NiSO_4 \cdot 6H_2O$	Rural, urban	Zakipour et al. (1997)
SO_2, NO_2	Copper	Cu_2O, $Cu_4SO_4(OH)_6$, $Cu_4SO_4(OH)_6 \cdot H_2O$	Urban	Tidblad and Leygraf (1995)
$(NH_4)_2SO_4$	Copper	Cu_2O, $Cu_4SO_4(OH)_6$, $Cu_4SO_4(OH)_6 \cdot H_2O$	Rural office	Lobnig et al. (1994)
Cl_2, NO_2, H_2S	Copper	Cu_2O, $Cu_2Cl(OH)_3$ $Cu_4SO_4(OH)_6 \cdot H_2O$	Indoor industry (pulp and paper)	Lenglet et al. (1995)
HNO_3	Copper	$Cu_2NO_3(OH)_3$, Cu_2O	Urban	Samie et al. (2007)
NaCl, O_3, UV	Silver	AgCl	Marine	Lin and Frankel (2013)
NaCl	Al–Zn coating	ZnO, $ZnAl_2O_4$, Al_2O_3 $Zn_6Al_2(OH)_{16}CO_3 \cdot 4H_2O$	Marine	Zhang et al. (2013)

In summary, it appears that certain combinations of corrosion stimulators at concentrations below 1 ppmv together with a proper choice of relative humidity and airflow rate can generate the corrosion products that have been observed in well-characterized field environments. Furthermore, the laboratory investigations on nickel with SO_2 and NO_2 as the only gaseous corrosive agents under nonaccelerated conditions have generated not only qualitative agreement (corrosion products) but also quantitative agreement (corrosion rates) between laboratory and field investigations.

Having identified the characteristics of a well-performed laboratory exposure, it is now time to describe the information that can be obtained on physicochemical processes at the corroding surface. In the next section, computational model studies are presented and compared to experimental results obtained in laboratory exposures involving SO_2 as the only gaseous corrosive agents.

6.6 COMPUTATIONAL MODEL STUDIES OF SO_2-INDUCED ATMOSPHERIC CORROSION OF COPPER

Several experimental studies based on IRAS (Appendix A) have been performed in which corrosion effects have been measured *in situ* on several metals in the presence of humidified air at constant temperature and a well-controlled level of SO_2. These studies have been followed by GILDES computer model investigations (see Chapter 3) designed to simulate the laboratory studies and thereby provide information on the most important physicochemical reactions that occur (see "Further Reading" at the end of this chapter for citations to the publications that provide details). We first describe this combined experimental and theoretical effort for copper exposed to some of gaseous corrosion stimulators.

6.6.1 Copper Corrosion Induced by SO_2

Short-term exposures of copper were performed in an exposure chamber that allowed the sample to be exposed to humidified air with additions of SO_2. From *in situ* IRAS spectra one can deduce which species are present at the surface and in what amounts. In humidified air with no SO_2 present, a film of Cu(I) oxide forms, the growth following a logarithmic rate law. Atop the Cu(I) oxide is an aqueous layer with a thickness that depends on the relative humidity. Based on these studies, the thickness of the aqueous layer for the base set of computer model studies was set at 5 nm, a thickness in which a major part of the aqueous adlayer exhibits bulk properties.

Upon the introduction of 210 ppbv of SO_2 in flowing air with 80% relative humidity, a new peak appeared due to the formation of adsorbed sulfite (SO_3^{2-}). This adsorbed species reacts with copper to form a copper sulfite. Figure 6.6 displays the experimental data expressed in terms of number of monolayers of copper sulfite as a function of exposure time. Sulfite was the only sulfur-containing compound detected. This is in contrast to naturally occurring sulfur-containing corrosion products, all of which contain copper sulfate in various forms, but generated in an environment in

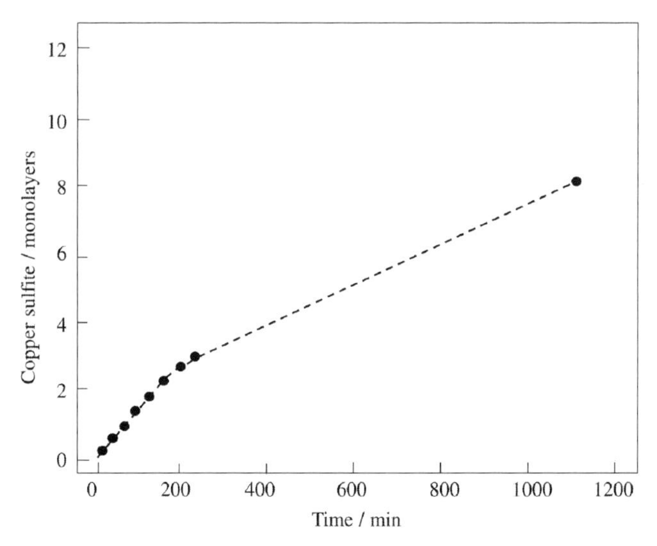

FIGURE 6.6 The experimentally determined amount of copper sulfite formed versus time of exposure. The filled circles represent quantified data points obtained at 80% relative humidity and 210 ppbv SO_2. The quantified data are presented in number of monolayers of copper sulfite, in which each monolayer is assumed to be 1.1 nmol cm^{-2}. The uncertainty in the quantification procedure is approximately 50%. (Experimental data from D. Persson and C. Leygraf, Initial interaction of sulfur dioxide with water covered metal surfaces: an in situ IRAS study, *Journal of the Electrochemical Society, 142*, 1459–1468, 1995. Adapted with permission from Elsevier; J. Tidblad and T. Graedel, GILDES model studies of aqueous chemistry. III. Initial SO_2-induced atmospheric corrosion of copper, *Corrosion Science, 38*, 2201–2224, 1996.)

which reactants that oxidize sulfur are abundant. Figure 6.6 shows that there is a relatively constant growth of copper sulfite during the first 200 min of exposure, after which the growth rate is reduced to a significantly lower value.

The complete set of physicochemical processes involved in the corrosion of copper under these conditions includes liquid-phase chemical processes, proton- and ligand-induced dissolution of copper, ion pairing, and precipitation and adhesion of corrosion products. The liquid-phase chemical processes include both equilibria and irreversible reactions. Altogether, 61 equilibrium reactions (gas–liquid reactions, hydration chemistry reactions, acid–base reactions, surface acid–base reactions, surface complex formation, and solubility equilibria) and 34 irreversible reactions (redox reactions involving oxygen and hydrogen, copper oxidation, reduction reactions, irreversible ligand- or proton-promoted dissolution) were employed in the computer model to simulate the corrosion of copper.

After assignment of equilibrium and rate constants (assumed equal to those of bulk water solutions) and subsequent computation, the dominant species and their major reactions were determined. Some of the results are summarized in Figure 6.7. The left part of this figure displays the dominant sulfur chemistry and the right part the dominant oxygen chemistry after 59 min of simulated exposure. The figure is

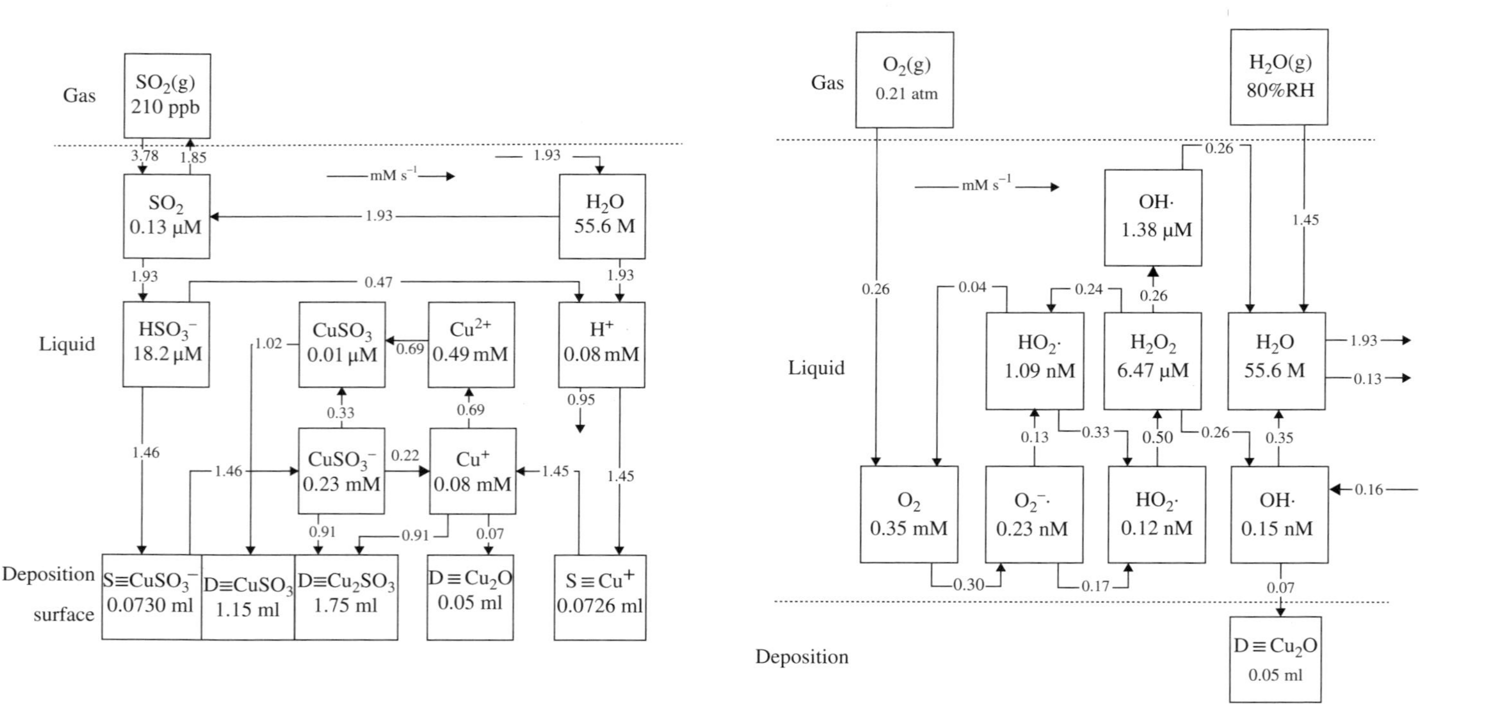

FIGURE 6.7 Significant species, their concentrations, and their principal reaction networks for GILDES model computations simulating the exposure of copper to 210 ppbv SO_2. Copper species (left panel) and oxygen species (right panel) are seen in the boxes, with their concentrations or amounts given in ppbv, M, or ml (monolayers). The rates between boxes are given in mM s^{-1} after 59 min of exposure of copper to the SO_2-containing atmosphere at 80% relative humidity. S≡X denotes a surface group extending from the copper(I)oxide into the aqueous layer and D≡X denotes the precipitated form of species X. (Reproduced with permission from Elsevier; J. Tidblad and T. Graedel, GILDES model studies of aqueous chemistry. III. Initial SO_2-induced atmospheric corrosion of copper, *Corrosion Science, 38*, 2201–2224, 1996.)

organized as a budget diagram, with the most important atmospheric species at the top, the dominant solution species in the middle, and deposited or surface species formed at the bottom.

Several interesting features of the results should be commented upon. The time scale for changes in species concentration (e.g., of bisulfate ions, HSO_3^-) is much slower than the time scale for the flow rates between the boxes. As a result, all flows between different boxes shown in Figure 6.7 are almost in balance. The exception is the SO_2 deposition as discussed in the following text. In fact, the flow rates and the concentration of species reach steady-state values within 1 min of simulated exposure. The exceptions are the amounts of deposited copper sulfite (assumed to be $CuSO_3$ and Cu_2SO_3) and Cu_2O, which slowly increase with exposure time.

Carbonate chemistry does not appear to play an important role. One reason is that the rate of ionization of dissolved carbon dioxide (CO_2) to form protons (H^+) and bicarbonate ions (HCO_3^-) is very slow. As a consequence, the amount of precipitated copper carbonate is insignificant. One of several important processes is the reaction of HSO_3^- with surface hydroxyl groups, Figure 2.7, to form $S{\equiv}CuSO_3^-$, which then dissolves as $CuSO_3^-$ (ligand-promoted dissolution). Another equally important process is the reaction of H^+ with surface hydroxyl groups to form $S{\equiv}Cu^+$, which also dissolves (proton-promoted dissolution). Note that the number of active surface sites, indicated by $S{\equiv}CuSO_3^-$ and $S{\equiv}Cu^+$, is of the order of 10% of the total number of surface sites. In agreement with mechanistic corrosion studies in the aqueous phase, this suggests that only part of a surface is actively participating in the corrosion process. Once dissolved, the Cu(I) ions can either precipitate, or be oxidized to Cu(II) prior to precipitation. The major routes for this to happen are indicated in Figure 6.7 (left panel). The description of redox chemistry is completed in the right panel of Figure 6.7, which shows the most significant reactions involving oxygen, hydrogen, and their reactive intermediates.

Among the reactions considered, no single one can be said to be controlling or rate limiting. All reactions and their products interact with each other. The only limiting factor is the net delivery of SO_2 to the surface, a process that depends on the SO_2 concentration, the deposition rate, and the escape rate of SO_2 from the liquid phase back to the gas phase.

Figure 6.8 shows the result of the calculations in terms of the total amount of copper sulfite formed as a function of exposure time. This amount has been derived from the calculated amounts of deposited $D{\equiv}CuSO_3$ and $D{\equiv}Cu_2SO_3$. The corresponding experimental results, based on IRAS measurements and previously seen in Figure 6.6, are also shown for comparison. Considering the uncertainty in the quantified reaction rate parameters of around 50%, the calculated and experimental results are in good agreement. This suggests that the computational model has captured most of the physicochemical processes that are important in the atmospheric corrosion of copper under actual exposure conditions. A number of sensitivity studies have been performed in which different parameters have been varied in the computational model and the results compared. For example, the line with the lower slope in Figure 6.8 has been obtained by reducing the proton- and ligand-induced dissolution rates by a factor of 10, this computational experiment

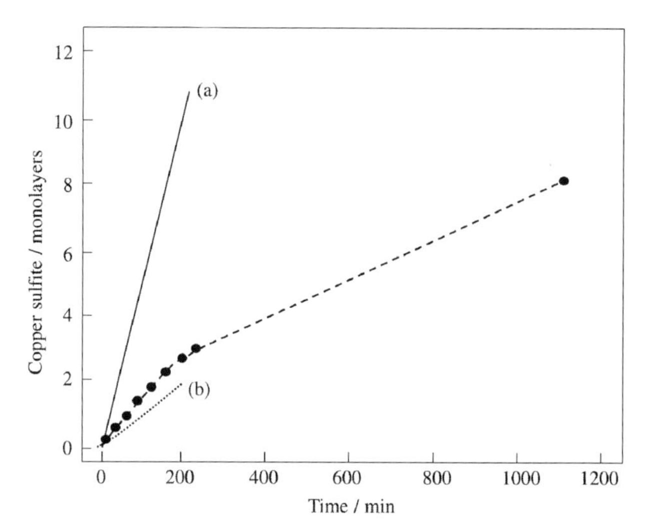

FIGURE 6.8 The amount of copper sulfite formed as a function of time of exposure in laboratory experiments and theoretical calculations. The filled circles represent experimental data from Figure 6.6. The straight lines represent calculated data, one with the base set of equilibrium and rate constants (a), the other with the proton- and ligand-induced dissolution rates each reduced by a factor of 10 (b). (Reproduced with permission from Elsevier; J. Tidblad and T. Graedel, GILDES model studies of aqueous chemistry. III. Initial SO_2-induced atmospheric corrosion of copper, *Corrosion Science, 38*, 2201–2224, 1996.)

being suggested by the fact that those rates are poorly constrained by existing data. The result implies that dissolution rates cannot be so low if reasonable results are to be obtained.

6.6.2 Copper Corrosion Induced by $SO_2 + O_3$ or $SO_2 + NO_2$

Whereas the experimental and GILDES computational studies shown before only involve SO_2 as the gaseous corrosion stimulator, efforts have also been undertaken to include NO_2 or O_3, in addition to SO_2. In this case, copper interacted with 200 ppbv SO_2 in combination with either NO_2 or O_3 at different concentrations. In the computational studies, the number of physicochemical processes increased considerably. In all, 61 equilibrium reactions (gas–liquid reactions, hydration chemistry reactions, acid–base reactions, surface acid–base reactions, surface complex formation, and solubility equilibria) and 34 irreversible reactions (redox reactions involving oxygen and hydrogen, copper oxidation, reduction reactions, irreversible ligand- or proton-promoted dissolution) were considered in the computer model to simulate the corrosion of copper.

In this case, the IRAS studies show that copper forms cuprite (Cu_2O) and various copper sulfates in both $SO_2 + NO_2$ and $SO_2 + O_3$ mixtures, also copper hydroxynitrate in $SO_2 + NO_2$ after an incubation time of 4 h. The GILDES model is able to correctly predict the formation of copper sulfate in both gas mixtures and copper hydroxynitrate

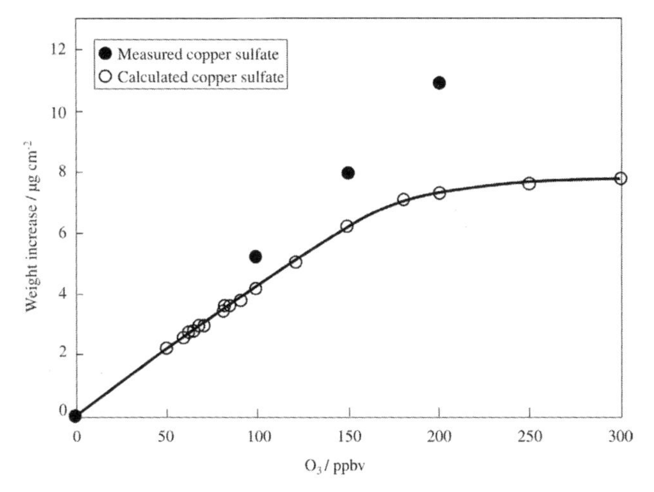

FIGURE 6.9 The amount of copper sulfate as a function of time of exposure in laboratory experiments (SO_2+O_3 and SO_2+NO_2 mixtures) and theoretical calculations. (Reproduced with permission from The Electrochemical Society; J. Tidblad, T. Aastrup, and C. Leygraf, GILDES model studies of aqueous chemistry. VI. Initial SO_2/O_3- and SO_2/NO_2-induced atmospheric corrosion of copper, *Journal of the Electrochemical Society, 152*, B178–B185, 2005.)

in the (SO_2+NO_2) mixture. GILDES is also able to predict the formation rate of copper sulfate in (SO_2+O_3) mixtures, but not the formation rate of copper sulfate and copper hydroxynitrate in (SO_2+NO_2) mixtures. As an example of successful prediction, Figure 6.9 displays the amount of copper sulfate as a function of time of exposure at 200 ppbv SO_2 and different O_3 concentrations in the range from 0 to 250 ppbv. It is evident that GILDES predicts the amount of copper sulfate in the (SO_2+O_3) mixture with high accuracy but is not able to perform a good prediction with the same accuracy in the (SO_2+NO_2) mixture. This shows that the mechanistic role of O_3 and NO_2 for the SO_2-induced atmospheric corrosion of copper—and probably of other metals—is different.

The description of important processes involved in the (SO_2+O_3)-induced atmospheric corrosion of copper is schematically depicted in Figure 6.10. They are all assumed in the GILDES model calculations and may serve as starting point for more complicated calculations involving other corrosive species as well.

6.6.3 Anodic and Cathodic Reactions in the GILDES Model

The processes that are represented in Figure 6.7 constitute a more elaborate way to describe atmospheric corrosion than does the traditional approach, which simply uses an anodic reaction (metal dissolution) and a cathodic reaction (oxygen (O_2) reduction). From Figure 6.7, left panel, it is evident that the metal dissolution process can be described by a series of physicochemical reactions that include the formation and subsequent detachment of surface species and their dissolution into the liquid layer. Similarly, the cathodic reaction can be divided into several elementary steps

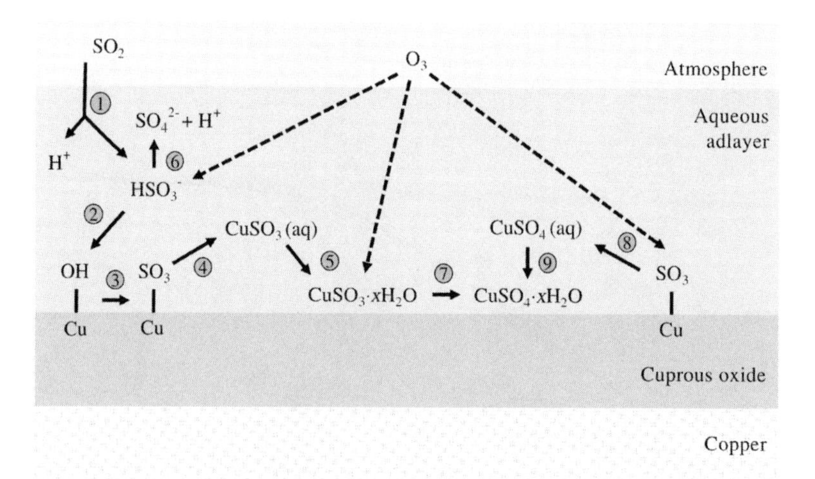

FIGURE 6.10 Schematic description of important processes involved in the (SO_2+O_3)-induced atmospheric corrosion of copper in the presence of a metal oxide film and a thin liquid layer. The main reactions seen are deduced from the GILDES model studies.

involving a number of reaction intermediates, such as the hydroperoxyl ion (HO_2^-) and hydrogen peroxide (H_2O_2). It is well known from independent electrochemical studies that both metal dissolution and O_2 reduction are processes that involve a number of elementary steps. Figure 6.7 illustrates this fact and is designed to reproduce the most important steps and intermediates based on relevant rate and equilibrium constants.

The GILDES model assumes that the oxidized and reduced species involved in the redox chemistry reactions have intimate contact with each other so that electrons liberated at sites where an oxidation (anodic) reaction occurs are transported without any resistance to sites where a reduction (cathodic) reaction occurs. This is most likely a valid assumption as long as the deposited layers formed are very thin (less than a few nanometers) and therefore the distance for electron travel is very short. With increased thickness of the deposited layer (more than a few nanometers) or the aqueous layer (more than a few tens of micrometers), the situation becomes more complex and kinetic constraints from processes such as electron and O_2 transport may become important.

6.7 SUMMARY

Laboratory exposures are mainly aimed at studying the influence of individual parameters under well-controlled exposure conditions or to mimic specific field exposures under accelerated conditions. For the latter to occur, it is important that the conditions used in the laboratory exposure trigger the same corrosion processes as in the field exposure.

In the design of laboratory exposures, it is important to consider sample preparation prior to exposure as well as relative humidity, temperature, exposure time, corrosive agents, and their delivery rate to the corroding surface during exposure. Most laboratory exposures performed so far, both for scientific purposes or for industrial applications, have been based on introduction of gases rather than on particles. Proper combinations of the most commonly used gaseous corrosive agents (H_2S, NO_2, Cl_2, and SO_2) in concentrations ranging from a few tens to a few hundreds of ppbv have proven to result in reasonable agreement with field data.

However, in view of the specific corrosion behavior of each material in a given corrosive environment, it is evident that there exists no universal laboratory exposure capable of simulating the corrosion processes of all materials and in any environment. If one uses an increased number of corrosion stimulators in the laboratory test, there is an increased probability of simulating actual corrosion processes. The increased complexity, however, also leads to increased difficulties in obtaining reproducible results among different laboratories. Two or three corrosion stimulators in a laboratory experiment, in addition to humidified air and temperature, are probably an optimum number if reliable and realistic results are to be obtained.

Future developments in this field will depend on continuing experimental and theoretical studies that provide molecular insight into the operating corrosion processes and on the introduction into these studies of the corrosion effects produced by atmospheric particles, especially those containing NaCl and $(NH_4)_2SO_4$.

FURTHER READING

Examples of Important Laboratory Exposures and Accelerated Tests

ASTM, ASTM B845-97(2013)e1, Standard Guide for Mixed Flowing Gas (MFG) Tests for Electrical Contacts, *American Society for Testing and Materials*, 2013.

D. Bengtsson Blücher, J.-E. Svensson, and L.-G. Johansson, Influence of ppb levels of SO_2 on the atmospheric corrosion of aluminum in the presence of NaCl, *Journal of the Electrochemical Society, 152*, B397–B404, 2005.

H. Gil and C. Leygraf, Quantitative in-situ analysis of initial atmospheric corrosion of copper induced by acetic acid, *Journal of the Electrochemical Society, 154*, C272–C278, 2007.

ISO, ISO/TR 16335:2013, Corrosion of Metals and Alloys—Corrosion Tests in Artificial Atmospheres—Guidelines for Selection of Accelerated Corrosion Tests for Product Qualification, 2013.

R.E. Lobnig, R.P. Frankenthal, D.J. Siconolfi, J.D. Sinclair, and M. Stratmann, Mechanism of atmospheric corrosion of copper in the presence of submicron ammonium sulfate particles at 300 and 373 K, *Journal of the Electrochemical Society, 141*, 2935–2941, 1994.

Examples of Laboratory Exposures for Possible Simulation of Field Exposures

M. Lenglet, J. Lopitaux, C. Leygraf, I. Odnevall, M. Carballeira, J.-C. Noualhaguet, J. Guinement, J. Gautier, and J. Boissel, Analysis of corrosion products formed on copper in $Cl_2/H_2S/NO_2$ exposure, *Journal of the Electrochemical Society, 142*, 3690–3696, 1995.

H. Lin and G.S. Frankel, Accelerated atmospheric corrosion testing of Ag, *Corrosion, 69*, 1060–1072, 2013.

R.E. Lobnig, D.J. Siconolfi, R.P. Frankenthal, M. Stratmann, and J.D. Sinclair, Atmospheric corrosion of aluminum, copper and zinc in the presence of ammonium sulfate particles, in *Proceedings of the 3rd International Symposium on Corrosion and Reliability of Electronic Materials and Devices*, eds. R.B. Commizzoli, R.P. Frankenthal, and J.D. Sinclair, The Electrochemical Society, Pennington, pp. 63–78, 1994.

F. Samie, J. Tidblad, V. Kucera, and C. Leygraf, Atmospheric corrosion effects of HNO_3—comparison of laboratory exposed copper, zinc and carbon steel, *Atmospheric Environment, 41*, 4888–4896, 2007.

J. Tidblad and C. Leygraf, Atmospheric corrosion effects of SO_2 and NO_2. A comparison of laboratory and field exposed copper, *Journal of the Electrochemical Society, 142*, 749–756, 1995.

S. Zakipour, J. Tidblad, and C. Leygraf, Atmospheric corrosion effects of SO_2, NO_2, and O_3. A comparison of laboratory and field exposed nickel, *Journal of the Electrochemical Society, 144*, 3513–3517, 1997.

X. Zhang, C. Leygraf, and I. Odnevall Wallinder, Atmospheric corrosion of Galfan coatings on steel in chloride-rich environments, *Corrosion Science, 73*, 62–71, 2013.

Computational Model Studies of SO_2-Induced Atmospheric Corrosion of Copper

J. Tidblad, T. Aastrup, and C. Leygraf, GILDES model studies of aqueous chemistry. VI. Initial SO_2/O_3- and SO_2/NO_2-induced atmospheric corrosion of copper, *Journal of the Electrochemical Society, 152*, B178–B185, 2005.

J. Tidblad and T. Graedel, GILDES model studies of aqueous chemistry. III. Initial SO_2-induced atmospheric corrosion of copper, *Corrosion Science, 38*, 2201–2224, 1996.

7

CORROSION IN INDOOR EXPOSURES

7.1 GENERAL CHARACTERISTICS OF INDOOR ENVIRONMENTS

While vast experience from outdoor exposures has been accumulated over most of the twentieth century, the history of indoor corrosion is much shorter. It goes back to the growing interest in corrosion of electronics during the 1960s and 1970s, followed by the increasing concern of acid deposition effects, in particular objects of cultural heritage and on equipment during storage or transport.

The characteristics of indoor environments may differ greatly from those outdoors. The same fundamental corrosion-related processes described in Chapters 2 and 3 operate in both environments, however. As a result of different environmental conditions, the corrosion rates are generally found to be lower indoors than outdoors. While the comparatively low indoor rates have little influence on the structural integrity of most materials, they may nevertheless have a significant influence on surface-related material properties, such as electrical performance of connectors or visual appearance of statues. This chapter deals with general aspects of indoor corrosion, whereas Chapter 11 describes one main application area of indoor corrosion: electronic materials.

Unlike what one might first suppose, indoor environments are far from "outdoor environments without the rain." Rather, they differ substantially with respect to a large number of environmental parameters. Some of the more important of these differences are discussed in the following.

7.1.1 Relative Humidity and Temperature

Water present in outdoor environments can interact with material surfaces through adsorbed moisture, condensed water, or direct precipitation. Indoors, on the other hand, the only important component in the formation of the wetted surface is adsorbed water.

Atmospheric Corrosion, Second Edition. Christofer Leygraf, Inger Odnevall Wallinder, Johan Tidblad and Thomas Graedel.

Exceptions are warm, humid climates with air-conditioned buildings, where water sometimes condenses on surfaces that are cooled below the dew point. Here, condensed water may promote mold and fungi growth and also indoor corrosion processes.

The indoor relative humidity and temperature depend on the degree of ventilation on one hand (a high degree of ventilation makes indoor values closer to outdoor values) and the degree of heating, thermal isolation, and air-conditioning on the other hand (which in general decrease indoor relative humidity and may increase indoor temperature). Representative indoor values of relative humidity may range from 15 to 85% with average values close to 50%.

Variations in relative humidity and temperature are much less dramatic indoors than out but may still be significant. Depending on the degree of sheltering, the variations may approach those seen outdoors, a circumstance that can greatly accelerate the corrosion rate. Often, the variations in temperature and relative humidity are equally or more important than their absolute levels.

7.1.2 Deposition Velocity

Wind velocities are normally lower indoors than outdoors by an order of magnitude or so, a situation that has a significant effect on the dry deposition velocity. In an indoor environment with perfectly still air, the transport of gases is determined by molecular diffusion. The resulting dry deposition velocity is governed by a combination of the rate of transport to the surface and the probability of transformation of the gas molecule after collision with the surface. As a result, the dry deposition in an indoor setting is a strong function of the relative humidity. Dry deposition velocities in still air vary with the characteristics of the gaseous molecule but have a lower limit around 10^{-3} cm s^{-1}. Because of temperature differences and heat transfer between different parts of indoor environments, however, there is always some natural convection. This results in typical indoor deposition velocities of gases ranging from 10^{-3} to 10^{-1} cm s^{-1}. Deposition velocities and other properties of indoor gases are given in Table 7.1.

Representative indoor deposition velocities of particles vary strongly with particle diameter, as discussed in Chapter 5. The estimated deposition velocity for both sulfate- and chloride-containing particles in the accumulation mode (diameter $<2\,\mu m$) is 0.005 cm s^{-1} and in the coarse mode (diameter $>2\,\mu m$) is 0.7 cm s^{-1}.

7.1.3 Atmospheric Gases

The indoor concentration of any given gaseous air constituent is a complex function of a number of sources and sinks. The sources include transport from outdoors or from other indoor rooms, direct indoor emission of gases, and production of gases through chemical reactions. The sinks include removal of gases by transport to other rooms, removal by indoor surfaces or chemical filtering devices, and loss by chemical reactions. In general, all processes are a function of time and of the concentration of each constituent.

TABLE 7.1 Characteristics of Selected Gaseous Indoor Air Constituents: Concentration Range (C, ppbv), Henry's Law Coefficient (H, M atm^{-1}, see Section 3.3), and Deposition Velocity (v_d, cm s^{-1}, see Section 4.2.9)

Constituent	C (ppbv)	H (M atm^{-1})	v_d (cm s^{-1})
O_3	3–30	1.8×10^{-2}	0.036
H_2O_2	5	2.4×10^{5}	0.07
SO_2	0.3–15	1.4	0.05
H_2S	0.1–0.7	1.5×10^{-1}	0.03
NO_2	1–30	7.0×10^{-3}	0.006
HNO_3	3	9.1×10^{4}	0.07
NH_3	15–50	1.0×10^{1}	0.05
HCl	0.05–0.20	2.0×10^{1}	0.04
Cl_2	0.001–0.005	6.2×10^{-2}	
HCHO	10	1.4×10^{4}	0.005
HCOOH	20	3.7×10^{3}	0.006

Data from C. Leygraf, Atmospheric corrosion, in *Corrosion Mechanisms in Theory and Practice*, ed. P. Marcus, CRC Press/Taylor & Francis Group, Boca Raton, pp. 669–704, 2012.

Chemical reactions between indoor air constituents, gases as well as particles, are important if they occur on a time scale shorter than the renewal time for indoor air. This condition is very often fulfilled, and chemical reactions therefore play a significant role indoors, as will be demonstrated in the following.

Many important chemical reactions are photolytic in nature, meaning that their rates are influenced by the presence of photons. Although lighting levels in general are much less indoors than outdoors, the influence of both artificial light and sunlight, which enters into the building through windows or skylights, may have an effect. Indoor surfaces, such as floors, ceilings, and walls, may also play important roles not only as absorbing surfaces but also because they permit heterogeneous reactions to take place at the solid/gas interface, in addition to homogeneous reactions in the gas phase. Our understanding of chemical reactions between indoor air constituents is gradually improving. Research has revealed that homogeneous chemistry can generate highly reactive species, such as the nitrate and hydroxyl radicals (given the appropriate precursors), which can have a strong influence on indoor corrosion processes. Heterogeneous chemistry may also result in highly reactive species.

Despite the complex indoor situation, successful mathematical models have been developed; they are capable of generating simulated gaseous pollutant concentrations that show good agreement with measured values. Figure 7.1 is an example of a result from such a study, in which a comparison is made between modeled and measured O_3 concentrations for a 2-day period in a museum in the Los Angeles metropolitan area. The outdoor concentration of O_3 during this period exceeded 120 ppbv and exhibited the characteristic daily cycle caused by variations in NO_x (i.e., mainly $NO+NO_2$) emissions and solar light (see Section 4.2.2). Indoor measured values show good agreement with the modeled base case, in which

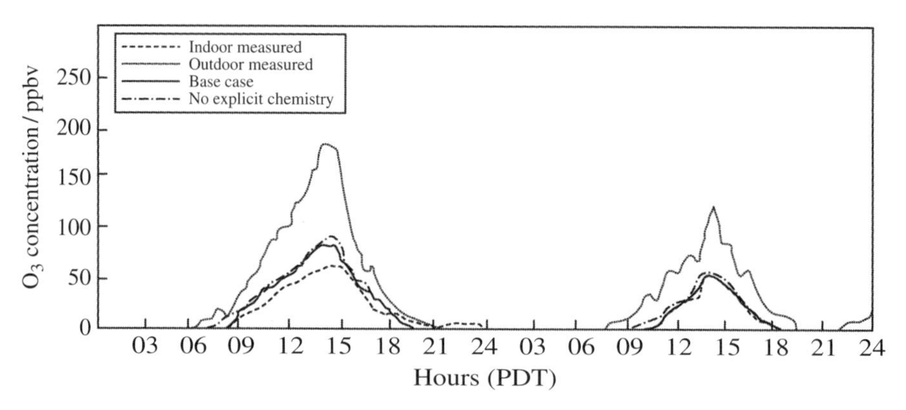

FIGURE 7.1 Comparison of modeled and measured O_3 concentrations for a 2-day period inside and outside a museum building. PDT, Pacific Daylight Time. (Reproduced with permission from American Chemical Society; W.W. Nazaroff and G.R. Cass, Mathematical modelling of chemically reactive pollutants in indoor air, *Environmental Science and Technology, 20*, 924–934, 1986.)

20 gaseous constituents were considered, as were 57 possible chemical or photochemical reactions involving these constituents. The modeled case involving no reactions between the constituents, but treating them as chemically independent, predicts slightly higher O_3 concentrations. The reason is that O_3 is partly consumed because of the reaction

$$O_3 + NO \rightarrow NO_2 + O_2 \tag{7.1}$$

This study emphasizes the significant role of homogeneous and heterogeneous chemical reactions and the importance of absorption effects on walls. The concentrations of several species (e.g., O_3) are perturbed when going from outdoor to indoor conditions. In addition, the model predicts significant indoor production of several nitrogen-containing species, including HNO_2, HNO_3, HNO_4, NO_3, and N_2O_5, as the result of chemical reactions among indoor constituents.

The effect of different pollutant removal devices is exemplified in Figure 7.2, which displays the SO_2 concentrations in various indoor locations of another museum environment. The figure shows that, depending on air treatment conditions, the SO_2 concentrations can vary over one order of magnitude.

Figure 7.3 shows indoor concentration values of important corrosive species compiled from a larger study of indoor corrosion effects at eight sites in the United States. The data could be plotted in lognormal probability plots, meaning that the pollutant data are approximately randomly distributed. The abscissa in such plots, the cumulative percentage, can be regarded as the probability that a given pollutant concentration is lower than the corresponding ordinate value. Figure 7.3 shows the large distribution of measured pollutant data, including dust, in the investigated indoor environments.

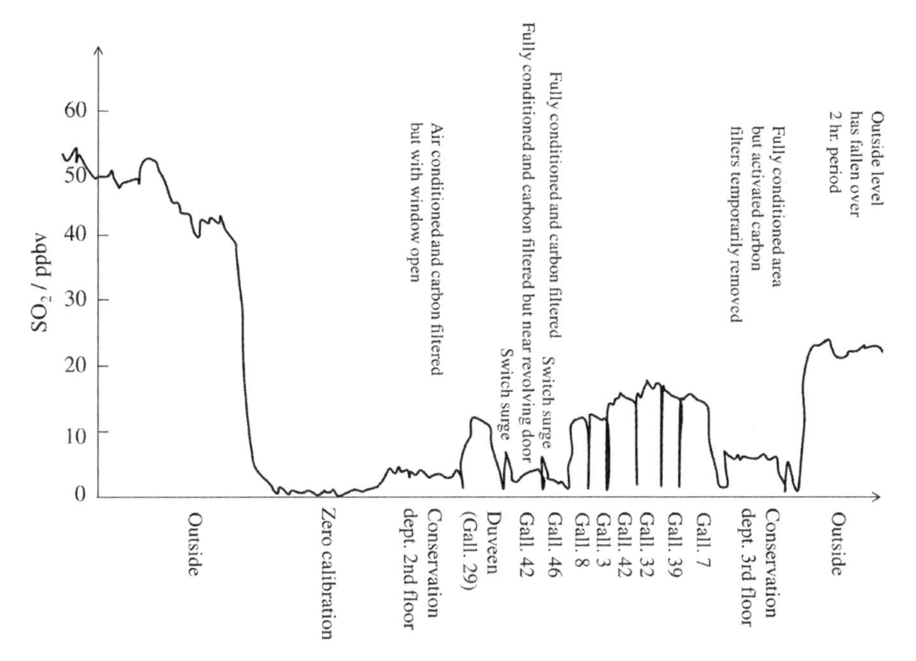

FIGURE 7.2 Influence of air conditions on SO_2 concentration in various indoor locations of the Tate Gallery, London. (Reproduced with permission from Maney Publishing; S. Hackney, The distribution of gaseous air pollution within museums, *Studies of Conservation, 29*, 105–116, 1984.)

When comparing mean indoor concentrations with outdoor concentrations, SO_2 and NO_2 values were about 2/3 of those outdoors, dust about 1/4, Cl_x ($HCl+Cl_2$) about 1/7, and S_x (reduced sulfur pollutants) about the same indoors as outdoors. NH_3 was the only species investigated that showed higher levels (about 1.5 times) indoors than outdoors. Indoor environments often exhibit higher NH_3 concentrations than outdoor settings, reflecting the fact that human occupants emit NH_3 as a consequence of routine metabolic processes. Other emission sources are cleaning agents, photocopiers, and people.

Although not commonly measured, many organic species, including aldehydes, organic acids, and volatile organic compounds (VOC), such as hexanes and benzenes, are known to exhibit higher concentrations indoors than outdoors. As with NH_3, the higher indoor concentrations reflect indoor sources: floor wax, cigarettes, deodorizers, uncured wood, particle board, etc.

Table 7.1 is an effort to summarize concentration ranges of gaseous air constituents in indoor environments together with their corresponding Henry's law coefficient (Section 3.3) and deposition velocity (Section 4.2.9). The concentrations vary, however, among different types of indoor locations. For instance, in indoor locations

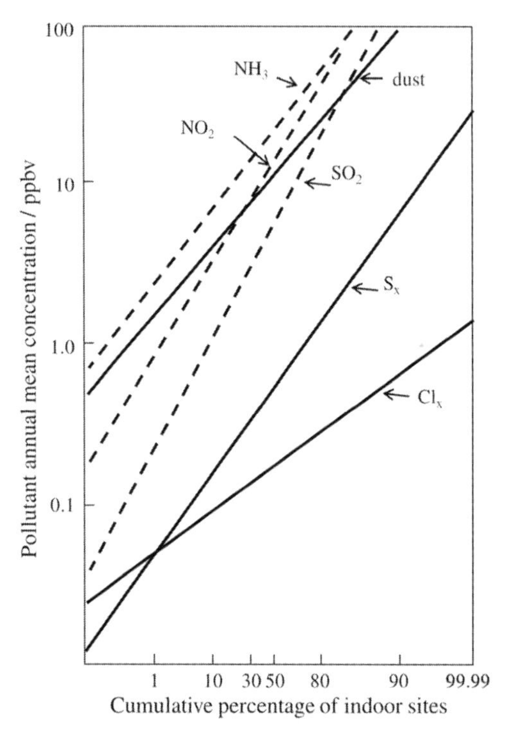

FIGURE 7.3 Mean pollutant concentration at indoor locations in the United States as a function of the cumulative percentage of the indoor sites. S_x represents reduced sulfur compounds, mainly H_2S and COS, and Cl_x represents Cl_2 and HCl. (Reproduced with permission from D.W. Rice, R.J. Cappell, P.B.P. Phipps, and P. Peterson, Indoor atmospheric corrosion of copper, silver, nickel, cobalt, and iron, in *Atmospheric Corrosion*, ed. W.H. Ailor, John Wiley & Sons, Inc., New York, pp. 651–666, 1982.)

at or close to pulp and paper plants, H_2S and SO_2 are dominating air constituents, whereas in museums and other exhibition rooms with wooden display cases, organic species are more influential.

7.1.4 Atmospheric Particles

Deposition of atmospheric particles to corroding surfaces is of significant importance not only outdoors but also indoors. As with gaseous constituents, the indoor concentration of particles is determined by a number of sources, including transport from outdoors or from indoor spaces or direct emission through smoke or other human activities. Sinks include removal by transport to other rooms or absorption to walls and particle filtering devices.

As discussed in Chapter 5, particles in the accumulation mode and coarse mode are very different chemically as well as physically. The concentration of coarse particles in office buildings is usually much lower than outdoors, because of the relatively high efficiency of coarse particle removal in air filtration systems and the relatively large deposition velocities of coarse particles. In homes, most of which have no air filtration, the concentration of coarse particles will generally be higher than in office buildings but lower than outdoors, depending on the degree of sealing. Industrial environments with processes such as cutting, grinding, or welding may also be characterized by locally high concentrations of coarse particles. The concentration of fine particles (accumulation mode) in offices and homes is usually in the range of 20–50% of their outdoor concentrations. Processing activities in manufacturing environments are occasionally significant sources of fine particles.

Indoor to outdoor concentration ratios in the presence of standard air filters and in the absence of indoor sources are typically 0.3 for particles in the accumulation mode and 0.05 for particles in the coarse mode. These ratios, however, should not be taken too literally. They may vary considerably with the removal efficiency of the air filters, particularly for particles in the accumulation mode.

7.2 THE INTERPLAY BETWEEN POLLUTANTS AND CORROSION RATES

It is evident from the previous section that the indoor environmental situation is as complex as that of outdoors and that indoor corrosion rates can be influenced by a large number of gases and particles. The ability to predict indoor corrosion rates based on information from only a few pollutant concentrations is very limited and cannot yet be performed with high accuracy. Nevertheless, we are able to measure the change in corrosion rate caused by a change in the indoor environment. This can be illustrated by experiments, based on simultaneous measurements of the mass response of gold, silver, and copper, to concentrations of H_2S, SO_2, O_3, and NO_2, together with those of temperature and relative humidity, in an electrical control room within a pulp and paper industry. The air outside the building is highly enriched in H_2S, with mean concentrations between 7 and 32 ppbv during four consecutive 3-month periods. The air in the control room is normally changed four times per hour with 50% new and 50% return air, dehumidified to between 30 and 40% relative humidity, thermostated to between 20 and 24°C, and conditioned by means of both chemical (active carbon) and particle filters (down to diameter 0.04 μm).

In order to test how the corrosion rates of metallic components in electronic equipment inside the room would be influenced by malfunctions of the air filtration system, the system was equipped with a bypass so that nonfiltered air could be admitted into the room during supervised malfunction simulations. This was accomplished by recording all parameters during normal operating conditions, after which the air was allowed to bypass the filter system and the response of the system followed in real time. After a few hours, each test was ended by closing the bypass and recording the parameters until normal, clean, indoor values were reached again.

Figure 7.4 displays the results from one malfunction simulation, which started after about 4820 h (more than 6 months) of normal conditions and lasted for 7 h. The figure shows the simultaneous mass response of gold, silver, and copper, respectively,

(a)

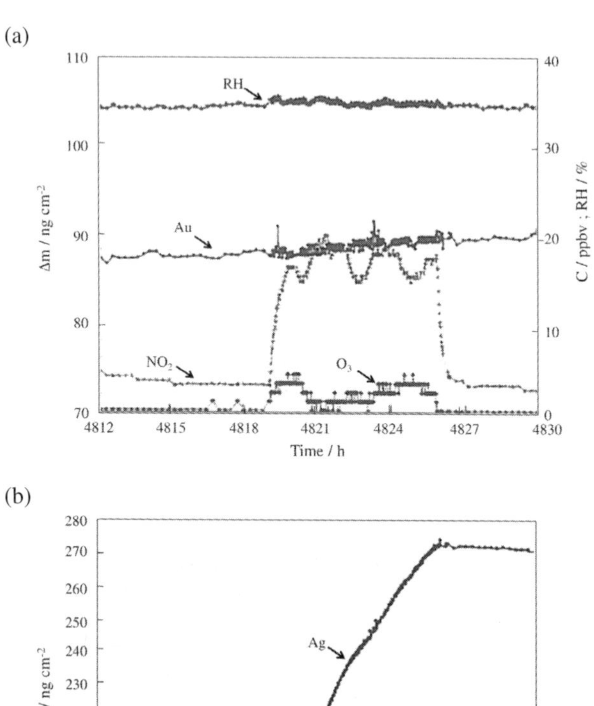

(b)

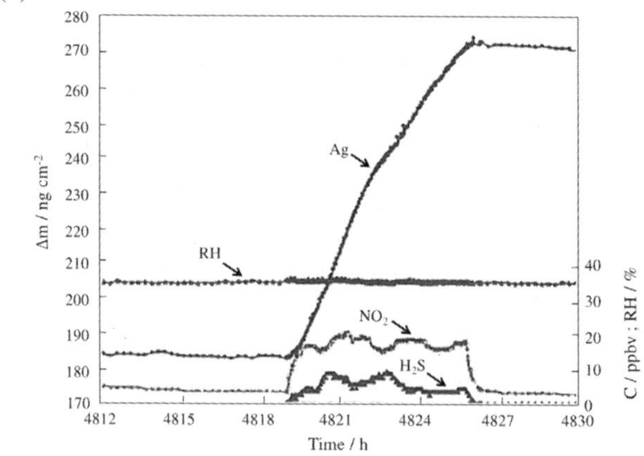

(c)

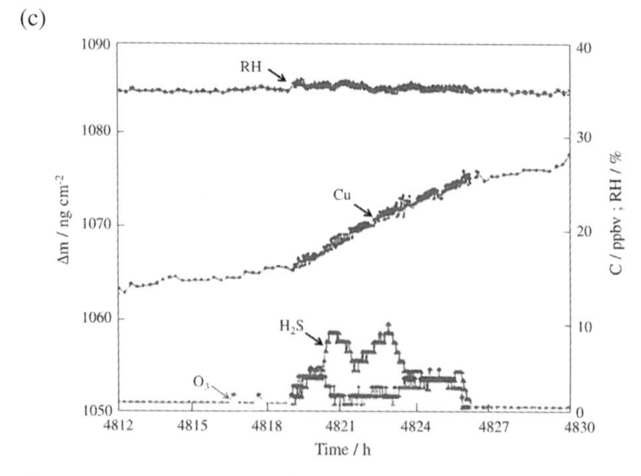

FIGURE 7.4 Total mass response, Δ*m*, of (a) gold, (b) silver, and (c) copper; concentrations of H_2S, NO_2, and O_3; and relative humidity (RH) as a function of time during a simulated malfunction of the air filter system in an electrical control room at a pulp and paper factory. Mass response data were obtained by the quartz crystal microbalance. Since the gaseous concentrations are the same for each metal, one gas is deleted from each figure to obtain better clarity. (Reproduced with permission from The Electrochemical Society; M. Forslund, J. Majoros, and C. Leygraf, A sensor system for high resolution in situ atmospheric corrosivity monitoring in field, *Journal of the Electrochemical Society, 144*, 2637–2642, 1997.)

and simultaneous variations in concentrations of H_2S, NO_2, and O_3 and of relative humidity. Temperature (almost constant at 25.0°C) and SO_2 concentration (below detection limit even during the malfunction) are not shown. The concentration of NO_2 was normally not zero in the room, despite the air filtration. This is probably due to the local formation of NO, generated by electrical discharges, and its subsequent oxidation in air to NO_2. During the simulated malfunction the NO_2 concentration increased from 3–4 ppbv to 14–18 ppbv. The O_3 concentration was normally below 1 ppbv but varied between 1 and 4 ppbv during the altered conditions, whereas the H_2S concentration, practically zero under normal conditions, varied between 2 and 9 ppbv during the simulated malfunction.

As judged from the mass response curves, the gold surface seemed to be practically unaffected by the increase in gas concentrations, whereas the silver and copper surfaces were clearly affected. Both metals showed very fast mass responses, for silver within a few minutes after the start of the malfunction. The total mass increase rate was also much faster for silver than for copper, about one order of magnitude under these conditions. Other malfunction simulations, with less enhanced concentrations of H_2S but with elevated levels of NO_2 and O_3, showed slower mass increase rates for silver.

The results are in good agreement with several laboratory exposures of silver performed with combinations of the three gases. They suggest that the fast mass response seen in Figure 7.4 is due to the rapid surface kinetics of silver sulfidation, a process that is mainly controlled by mass transfer of H_2S to the silver surface. At lower H_2S concentrations in laboratory exposures, and hence lower H_2S mass transfer rates, the silver sulfide formation rate is also slower. At lower NO_2 and O_3 concentrations, on the other hand, the surface kinetics of the sulfidation process is slower, as is the silver sulfidation process. The observations suggest a synergistic effect between H_2S and NO_2 or O_3. Subsequent surface analysis showed that the corrosion products formed on silver during simulated malfunctions largely contain sulfide and (to a lesser degree) chloride, most likely as Ag_2S and AgCl.

The lower mass increase rate of copper compared to silver, Figure 7.4, can be due to several reasons. Copper is known to be more sensitive to relative humidity than silver, and with low indoor levels of relative humidity (around 35%), the low rate is perhaps not surprising. Copper may also have formed a more protective layer of corrosion products, most likely a combination of Cu_2O and $Cu_2Cl(OH)_3$, as suggested by subsequent surface analysis of the exposed copper surface. Although their atmospheric concentrations were not directly measured, the results indicate that chlorine-containing corrosive species also may have played an important role in these exposures.

Overall, the present indoor exposures show that the corrosion effects of both silver and copper can be directly related to variations in indoor pollutant concentrations. Furthermore, the results and interpretations are consistent with insights into corrosion mechanisms gained from laboratory exposures.

Although the interplay between indoor corrodents and materials in general is difficult to describe quantitatively, the sensitivity of materials to corrode can be qualitatively specified, as was done in Table 4.1. The data in the table has been obtained mainly under laboratory or outdoor exposure conditions but may be applied to indoor conditions as well.

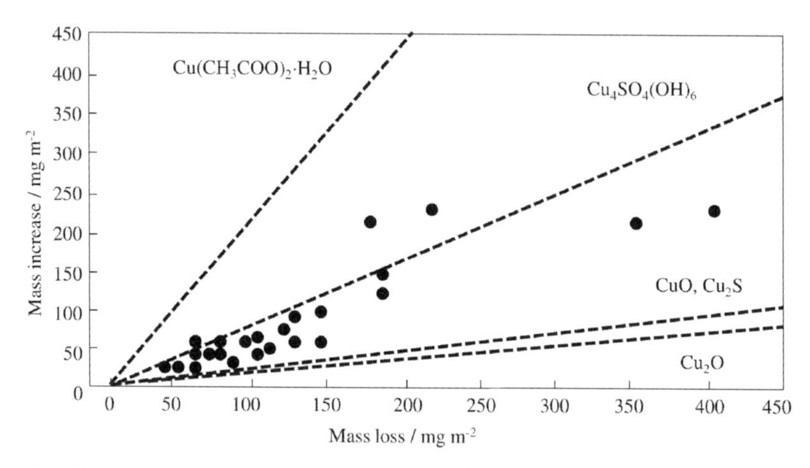

FIGURE 7.5 Mass increase as a function of mass loss data of copper samples exposed for 1 and 2 years at indoor sites in pulp mills, museums, and military stores. Dashed lines represent calculated ratios between mass increase and mass loss of different copper-containing phases. (Reproduced with permission from Royal Institute of Technology; E. Johansson, Corrosivity measurements in indoor atmospheric environments: a field study, Licentiate Thesis, Royal Institute of Technology, Stockholm, Sweden, 1998.)

7.3 CORROSION RATES

7.3.1 Measurements of Indoor Corrosion Rates

Measurements of indoor corrosion rates are by no means trivial to perform, and the low corrosion rates generally found indoors require highly mass-sensitive methods. Techniques that have been used for indoor corrosion rate studies include microgravimetry for measurements of sample mass loss or mass increase upon indoor exposure, resistance sensors for monitoring change in electrical resistivity across a metal foil during exposure, quartz crystal microbalance for measurements of mass change during exposure, and electrolytic cathodic reduction of corrosion products formed during exposure. Their mass change detection limit ranges from about 10 to 0.1 $mg\,m^{-2}$. This is quite adequate, as exposure of copper during 1 year in a highly benign indoor environment results in mass changes of around 50 $mg\,m^{-2}$. Appendix A provides a closer description of these and other techniques for evaluation of corrosion effects.

Electrolytic cathodic reduction of copper and silver specimens, mass increase, and mass loss are by far the most common corrosion rate measurements for samples exposed indoors. Because of inherent differences between the techniques, corrosion rate data obtained from different techniques may give different results. As an example of relatively good agreement, Figure 7.5 shows a plot of mass increase and mass loss data obtained from copper samples that were exposed for 1 and 2 years in a variety of indoor sites in pulp mills, museums, and military stores. The dashed lines represent calculated ratios between mass increase and mass loss of different copper-containing phases. Most of the data points are positioned between the ideal lines for frequently occurring crystalline phases, mainly $Cu_4SO_4(OH)_6$ and Cu_2O.

7.3.2 Corrosion Kinetics

As with exposures outdoors, the properties of the corrosion products formed largely determine the corrosion rate in any given indoor situation. As with outdoors, indoor corrosion products possess a corrosion protective ability of varying degrees, from which it follows that the corrosion effect (*M*) versus time (*t*) in its generalized form may be written as (see Section 8.3.6)

$$M = At^n \tag{7.2}$$

with A being a constant and n usually in the range from 0.5 (parabolic rate law) to 1.0 (linear rate law). Whereas outdoors an n-value close to 1 indicates low corrosion protection ability due to repeated dissolution–reprecipitation cycles of corrosion products, a linear corrosion rate indoors may instead be caused by mass flow-controlled corrosion kinetics. As concluded in Section 7.1.2, indoor deposition velocities of both gases and particles may be very low, especially at low relative humidity, from which follows that the kinetics of surface reactions often is faster than that of transport and deposition of corrosive species to the surface.

In very mild indoor environments, characterized by mostly dry conditions with the relative humidity hardly ever above 50%, the opportunity for obtaining a wetted surface in which deposited aerosols are dissolved is strongly limited. Detailed studies of the corrosion effect versus time in such mild conditions have shown that the time behavior can be approximated by a power law with n as low as 0.3. There are no physical–chemical processes, however, from which a power law with $n = 0.3$ can be deduced. Because the relative accuracy in the corrosion rate data is generally low, such data seems likely to be following a logarithmic or inverse logarithmic rate law, generally written as

$$M = B \log t \tag{7.3}$$

for which there is a physical meaning. The inverse logarithmic rate law, derived by Cabrera–Mott for initial oxidation growth on metals, assumes that the growth rate is limited by the diffusion of metal ions under a strong electric field along the thin film of oxide. It is possible that similar transport mechanisms may be operating within corrosion products formed in the most benign indoor environments.

Examples of varying indoor corrosion kinetics are seen in Figure 7.6, which shows the mass increase of nickel as a function of time during 2-year exposures in four benign indoor sites. With increasing corrosion rates the sites are:

1. A library room at the Royal Palace in Stockholm, where the relative humidity never exceeded 50%, the temperature was always between 17 and 25°C, and the pollutant concentrations were very low
2. A cross-connection room in a paper mill, where the relative humidity hardly ever exceeded 40%, the temperature was within 1° at 24°C, and slightly elevated concentrations of H_2S and NH_3 were present

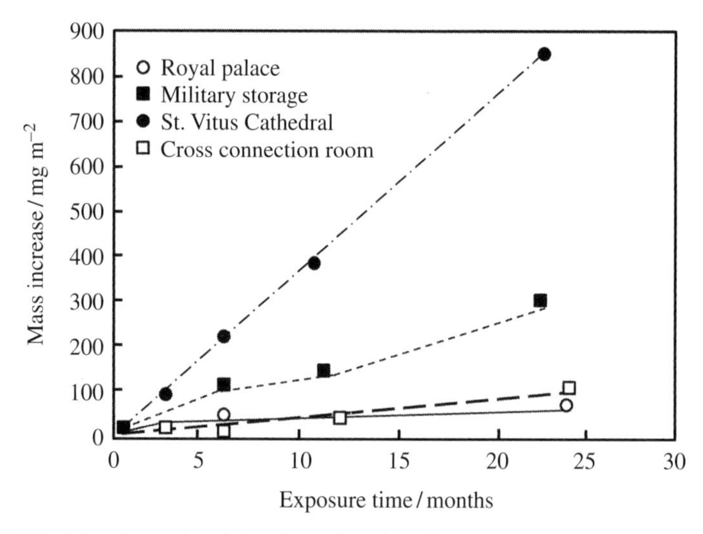

FIGURE 7.6 Mass increases of nickel as a function of exposure time in four indoor environments. (Reproduced with permission from E. Johansson, Swerea KIMAB, Stockholm, Sweden.)

3. A military equipment storage room with significant variations in relative humidity (from 50 to 90%, monthly averages) and in temperature (from −4 to +20°C, monthly averages) and with low pollutant concentrations
4. A church in an urban environment (St. Vitus Cathedral in Prague) with variations in relative humidity and temperature comparable to those in the military storage room but with elevated concentrations of both gases and particles

Corrosion kinetics close to linear are seen in all sites except the Royal Palace, where the mass increase is insignificant after the first 3 months of exposure. The results suggest that the linear behavior in St. Vitus cathedral and the military storage room is caused by insufficient protective ability of the corrosion products formed in the strongly varying indoor environment. The linear rate corrosion kinetics in the benign cross-connection room, on the other hand, is due to the limited mass flow rate of corrosive species to the nickel surface. The slow corrosion kinetics observed at the Royal Palace reflects the marked corrosion protective ability of the corrosion products formed in this highly benign environment.

7.3.3 Corrosion Rate Data

Giving the differences in concentrations of corrosive agents in different indoor environments, one expects corrosion rates to vary substantially. Figure 7.7 is a compilation of indoor corrosion rate data for five metals—iron, copper, cobalt, nickel, and silver—based on several indoor exposure programs. The corrosion rates, based

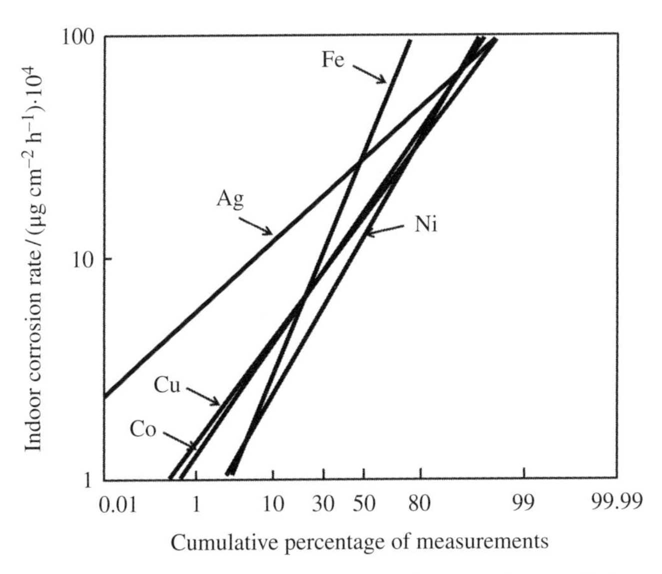

FIGURE 7.7 Mean indoor corrosion rates of five metals versus the cumulative percentage of measurements. (Reproduced with permission from D.W. Rice, R.J. Cappell, P.B.P. Phipps, and P. Peterson, Indoor atmospheric corrosion of copper, silver, nickel, cobalt, and iron, in *Atmospheric Corrosion*, ed. W.H. Ailor, John Wiley & Sons, Inc., New York, pp. 651–666, 1982.)

on mass increase measurements, could be plotted in lognormal probability plots similar to the indoor gas concentration data depicted in Figure 7.3. Several interesting observations can be made from the perspective of Figure 7.7.

Within a factor of three, the mean corrosion rates of all five metals is the same, with silver and iron corroding most and nickel corroding least. In benign indoor sites, corresponding to low cumulative percentage values in Figure 7.7, iron exhibited the lowest corrosion rates and silver the highest. In the most aggressive indoor sites, on the other hand, iron exhibited the highest corrosion rates.

A similar compilation was also made for outdoor corrosion rates of the same metals. In contrast to the situation indoors, outdoor mean values differed by a factor of 5000, with iron corroding most rapidly outdoors and silver least rapidly. Indoor and outdoor corrosion data, presented in lognormal probability plots, are compared for silver in Figure 7.8 and for iron in Figure 7.9. Silver is the only metal investigated that has higher corrosion rate mean values indoors than outdoors, Figure 7.8. It is likely that the principal reason is the independence of the silver corrosion rate on relative humidity. Another potential reason is the marked dependence of silver corrosion rate to the concentrations of reduced sulfur-containing species. As discussed in Sections 7.1.2 and 7.1.3, reduced sulfur concentrations indoors are about the same as outdoors, although their outdoor deposition velocities are higher.

The mean outdoor corrosion rate of iron is approximately 2000 times higher than the indoor rate, Figure 7.9. As laboratory studies have shown, the main reason is the

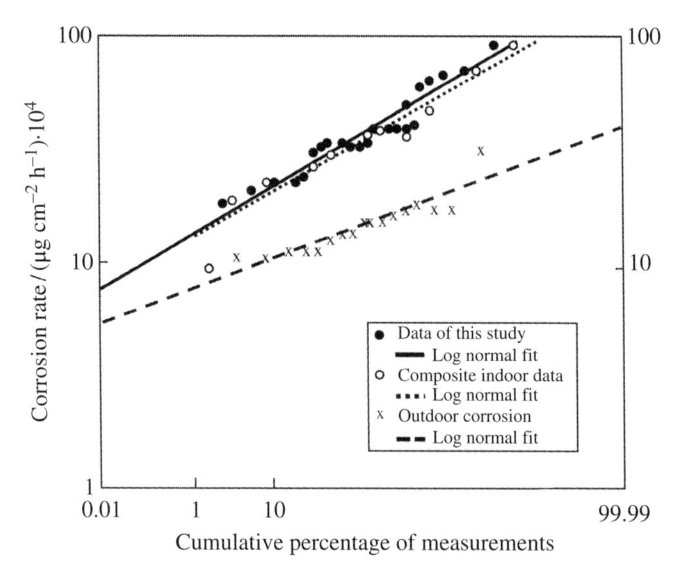

FIGURE 7.8 Mean indoor and outdoor corrosion rates of silver as a function of the cumulative percentage of measurements. Filled and unfilled circles are indoor corrosion data, whereas crosses are outdoor corrosion data. (Reproduced with permission from The Electrochemical Society; D.W. Rice, R.J. Cappell, W. Kinsolving, and J.J. Laskowski, Indoor corrosion of metals, *Journal of the Electrochemical Society, 127*, 891–901, 1980.)

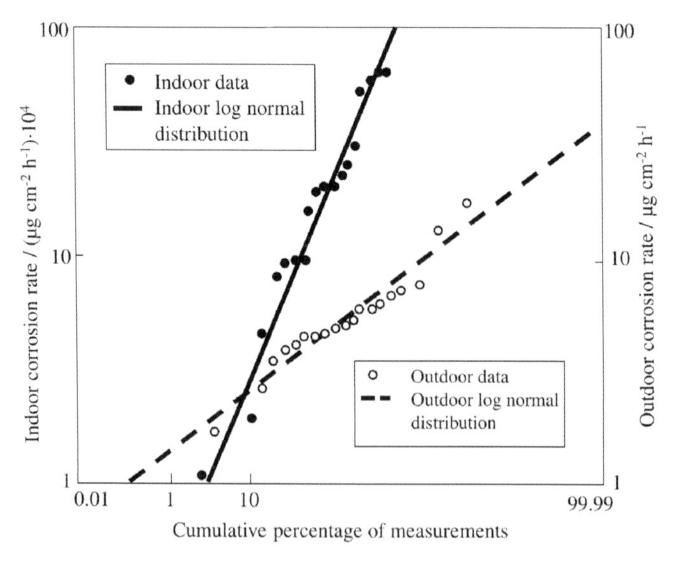

FIGURE 7.9 Mean indoor and outdoor corrosion rates of iron as a function of the cumulative percentage of measurements. (Reproduced with permission from The Electrochemical Society; D.W. Rice, R.J. Cappell, W. Kinsolving, and J.J. Laskowski, Indoor corrosion of metals, *Journal of the Electrochemical Society, 127*, 891–901, 1980.)

extremely high dependence of the iron corrosion rate on relative humidity. Outdoors, where sites with high relative humidity are common, iron will be most affected and silver least. Copper, nickel, cobalt, and most likely also zinc possess qualitatively similar sensitivities to relative humidity. The ratio between the outdoor and indoor corrosion rates of these four metals is also similar, of the order of 100. This ratio is partly explained by the higher relative humidity outdoors but partly also by the higher concentrations and deposition velocities of most corrosive species.

Overall, the results demonstrate the marked difference that exists between indoor and outdoor corrosion rates. They emphasize also that each metal has a unique corrosion behavior. Appendices C–J provide more detailed descriptions of the atmospheric corrosion of commonly used materials and their sensitivity to environmental parameters.

7.4 INDOOR CORROSION PRODUCTS

Since the chemistry of indoor atmospheric constituents often differs from corresponding outdoor constituents, one anticipates not only different corrosion rates but also different compositions of corrosion products formed during indoor exposures. Surface analysis studies of metals exposed to indoor environments reveal a rather complex mixture of corrosion products, including sulfates, chlorides, nitrates, sulfides, and carbonates of the corresponding metals. Phase determination of indoor corrosion products is often difficult to accomplish because of the thin layers formed and the frequent occurrence of amorphous phases, which makes X-ray diffraction for phase analysis difficult to use.

A detailed study of corrosion products, formed during indoor exposures of iron, copper, zinc, nickel, and silver, was completed by means of several complementary analytical techniques. The exposures were performed during 2 years in the same indoor environments as described in Section 7.3.2, the Royal Palace in Stockholm, a cross-connection room, a military storage room, and St. Vitus Cathedral in Prague. The results formed strong and unexpected evidence that the surface films on all metals, except silver, consist of corrosion products of metal carboxylates, such as metal acetate ($CH_3COO{-}Me$) and metal formate ($HCOO{-}Me$). Depending on the environment, the metal carboxylates were of different nature and with different proportions of, for example, metal acetate and metal formate. Other organic species with longer carbon chains may also have formed, although a precise identification was difficult to perform.

Sulfate (SO_4^{2-}), chloride (Cl^-), and ammonium (NH_4^+) ions were also detected on all metals. However, their relative surface abundance was smaller than that of the carboxylates, at least up to 2 years of exposure. These ions were all detected either in the deposited aerosols or in the corrosion products formed. With prolonged exposure, the surface films sometimes increased in sulfate content at the expense of the carboxylate content. The only metal that exhibited a different behavior was silver. On this metal Ag_2S was detected, and in addition also ammonium and sulfate, probably as ammonium sulfate ($(NH_4)_2SO_4$).

The organic compounds observed in indoor corrosion products are seen much less frequently in outdoor corrosion products. Somehow this observation must be related to the different nature between indoor and outdoor environmental characteristics. As discussed in Section 7.1.3, indoor air may contain increased concentrations of organic compounds, including unsaturated hydrocarbons. Upon emission, they can react with O_3 in a process involving several steps. The end product may be a variety of different aldehydes (RCHO) and ketones (RCOR) or more complicated organic species. These can undergo further oxidation, either in the atmosphere or after deposition in the liquid layer, resulting in carboxylic acids, such as formic acid (HCOOH) and acetic acid (CH_3COOH). To conclude, aldehydes, ketones, carboxylic acids, and other organic compounds as airborne constituents may all be involved in indoor atmospheric corrosion. This results in metal carboxylates as main compounds in corrosion products.

7.5 INDOOR ENVIRONMENTAL CLASSIFICATION

Based on indoor studies such as those shown herein, it has lately been possible to deduce standardized procedures for the classification of indoor corrosivities. In ISO 11844 the environments are classified into five categories: IC 1 (very low indoor), IC 2 (low indoor), IC 3 (medium indoor), IC 4 (high indoor), and IC 5 (very high indoor). The indoor corrosivity determination is based on measurements of the corrosion effect of four metals used as reference materials: carbon steel, zinc, copper, and silver. The corrosion effect can be deduced either through mass loss or through mass increase measurements after 1 year of exposure. If corrosion data is not available, the standard also provides guidelines for classifying the indoor corrosivity based on environmental parameters, such as humidity, temperature, and concentrations of corrosive gases or particles. It should be emphasized, however, that the indoor classification based on environmental parameters is associated with much more uncertainties than the corresponding outdoor classification. Outdoors a few parameters (e.g., SO_2, Cl^-, and relative humidity) may be sufficient to explain the corrosion effect of a given metal (see further in Chapter 8). Indoors the situation is much more complex because the corrosion effect usually cannot be explained on the basis of just a few parameters but rather seems to be the result of many nondominating parameters interacting simultaneously with the corroding metal surface.

The ISO 11844 standard is based on exposure of metals during 1 year. However, the environmental conditions may change during the exposure, and there is often a need for more continuous, real-time monitoring measurements. Figure 7.10 demonstrates how the environmental classification can be performed through continuous recording of the electrical resistance of 50 nm thick films of silver and copper. When the environment is changed, an immediate change in electrical resistance of both metals could be recorded, indicating a corresponding change from corrosivity category IC 2 to IC 1 in the case of the copper film.

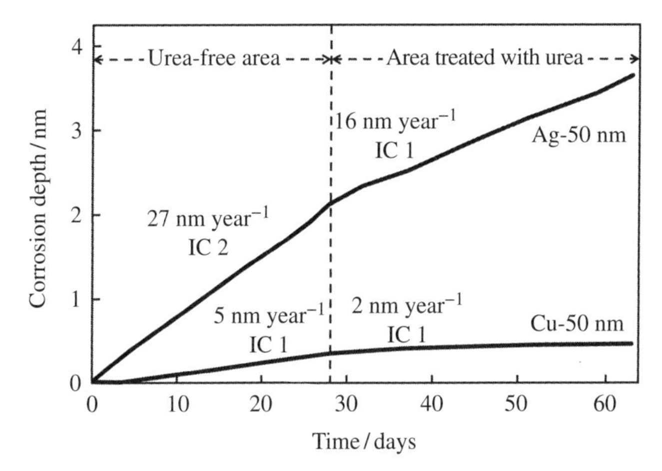

FIGURE 7.10 The corrosion effect of silver and copper during exposure in an archive building, measured as average corrosion depth and based on continuous recording of changes in electrical resistance of 50 nm thick metal foils. "Area treated with urea" refers to an environment in which the concrete of the building was treated with urea as an antifreeze agent, as opposed to the "urea-free area." The authors attribute the reduction in corrosion rate of both metals to the metal adsorption of urea released from the surrounding concrete and resulting in corrosion inhibition. (Reproduced with permission from Maney Publishing; T. Prosek, M. Kouril, M. Dubus, M. Taube, V. Hubert, B. Scheffel, Y. Degres, M. Jouannic, and D. Thierry, Real-time monitoring of indoor air corrosivity in cultural heritage institutions with metallic electrical resistance sensors, *Studies in Conservation, 58*, 117–128, 2013.)

7.6 AN EXAMPLE OF INDOOR CORROSION: METAL ARTIFACTS

Metal artifacts, more than most other objects, require extensive control of the relative humidity and low corrosive agent concentrations in museums or in museum storage rooms. Objects with metallic surfaces corrode during exposure, which results in a noticeable change in visual appearance due to formation of thin films of corrosion products. Artifacts made of silver, copper, or copper alloys are particularly sensitive in this respect. Due to their high reflectivity, these metals and alloys begin to exhibit visible discoloration at levels of corrosion, corresponding to weight increases as low as 20 mg m^{-2}. Even in highly benign indoor environments, such mass increases are frequently obtained within 1 year of exposure (Section 7.3.1). Thus, silver, copper, or copper alloy objects on display tarnish so quickly that their films of corrosion products have to be removed within 1 year, if the metallic luster of the objects is to be retained. This removal process usually involves unavoidable loss of material itself; the result is a regular thinning of the exterior walls of these irreplaceable objects.

Since many museums display cases and storage rooms contain nonmetallic as well as metallic objects, the selection of a suitable range of relative humidity is not straightforward. The lower limit is set by breaking of fibers and embrittlement damage of moisture-containing materials, which occurs at relative humidity levels of

TABLE 7.2 Choice of Relative Humidity Level in Museum Environments

65%	Acceptable for mixed collections in the humid tropics. Too high, however, to ensure stability of iron and chloride-containing bronzes. Air circulation very important
55%	Widely recommended for paintings, furniture, and wooden sculpture in Europe and satisfactory for mixed collections. May cause condensation and frosting difficulties in old buildings, especially in inland areas of Europe and the northern parts of North America
45–50%	A compromise for mixed collections and where condensation may be a problem. May well be the best levels for textiles and paper exposed to light
40–45%	Ideal for metal-only collections. Acceptable for museums in arid zones exhibiting local material

According to G.T. Thomson, The Museum Environment, 2nd edition, Butterworths, London, 1986.

less than about 40–45%. Mold growth and other biodeterioration processes, on the other hand, are highly probable at relative humidities above 65–70%. At such relative humidities, the atmospheric corrosion rate of iron and other metals can be significant, particularly if the surfaces of the objects have not been sufficiently cleaned but contain deposited hygroscopic salts. Experience has shown that clean and suitably coated iron objects and copper alloys with well-developed patinas will exist in reasonably good condition at 55%. Overall, therefore, $50 \pm 5\%$ relative humidity seems to be a good compromise for mixed collections of materials, as suggested by Table 7.2.

Depending on temperature changes, the relative humidity inside a display or a storage room may differ substantially from the surrounding environment. The interplay between temperature, absolute humidity (the water content per volume unit), and relative humidity can easily be displayed in a hygrometer chart. Increasing the ambient temperature (e.g., through spotlights) reduces the relative humidity. If the heating is turned off during night, the temperature decreases and the relative humidity increases. In this way, displays and storage rooms may experience dry–damp–dry cycles similar to outdoor situations, which have an accelerating corrosion effect (see further in Chapter 8). The effect is particularly serious during winter heating of dry air, such as in inland of northern areas of Europe and North America. In addition to daily cycling, dry–damp–dry cycling effects may also be operating on a weekly basis (e.g., caused by weekly services and the congregation giving off heat and temperature in churches) or on a yearly basis (historic buildings, which have been closed in winter and opened in the spring, when the cold interior surfaces are exposed to warm moist air resulting in condensation).

In regions of the Far East, the relative humidity outdoors for long periods is permanently above 65%. Since heating is rare in areas such as India, Sri Lanka, Malaysia, Indonesia, the Philippines, and parts of coastal China, the relative humidity indoors in those locations will also be above 65%. Hence, continuous mold and corrosion problems are expected in such environments, unless preventive measures are taken.

The influence of relative humidity on the rate of metal corrosion depends on the particular metal or alloy and on other factors, such as the cleanliness of the surface. As discussed in Section 7.3.3, the metal most sensitive to relative humidity is pure iron, followed by copper, nickel and zinc (all with comparable sensitivity), and silver, which is least affected. Brass alloys, having copper and zinc as main constituents, exhibit a sensitivity similar to that of their pure constituents. Tin bronzes and other tin-containing alloys may show a somewhat different behavior. Tin is a metal that forms a relatively resistant corrosion product and is not particularly sensitive to variations in relative humidity. Tin dioxide, SnO_2, is frequently found as an important part of the corrosion products formed on pure tin and on tin-containing alloys such as tin bronze or pewter. Although no systematic studies are at hand, the sensitivity of pewter and tin bronze to changes in relative humidity is expected to be between that of copper or zinc and that of silver.

When the metal is covered by hygroscopic salts, such as chlorides from soil or marine environments, moisture tends to be retained on the surface and the corrosion rate increased. Excavated objects may be particularly susceptible to such damage and may continue to corrode in the museum display or storage room if not properly cleaned. As further discussed in Chapter 8, the hygroscopic salts dissolve above their critical relative humidity, forming an aggressive aqueous layer that promotes the corrosion process. On iron and copper, clearly visible corrosion attacks can be seen. These are sometimes referred to as "sweating" or "weeping" of iron and "bronze disease," respectively.

In addition to its interplay with absolute and relative humidities, temperature has a more direct influence on physical, chemical, or electrochemical processes involved in metal corrosion. As discussed in Section 8.3, the temperature effect is twofold. On one hand, increased temperature increases the rate of most chemical processes, including corrosion. On the other hand, higher temperatures promote the drying of the liquid layer atop the metal. As a result, corrosion rates are relatively independent on temperature within the range of +10° to +25°C; hence, temperature is expected to have little direct influence on corrosion rates in most museum environments and storage rooms.

All air constituents that are known to corrode metals under outdoor or indoor conditions may influence the long-term survivability of metallic artifacts. They include sulfur dioxide (SO_2), nitrogen dioxide (NO_2), ammonia (NH_3), hydrogen sulfide (H_2S), hydrogen peroxide (H_2O_2), ozone (O_3), aldehydes, organic acids, and aerosol particles, mainly sodium chloride (NaCl) and ammonium sulfates ($(NH_4)_2SO_4$ or $(NH_4)HSO_4$). These constituents may influence nonmetallic artifacts as well. Table 7.3 lists some documented effects of air pollutants on metallic and nonmetallic artifacts.

Measurements of air constituents and of corrosivity have been performed in a number of museum environments. Two examples were given in this chapter. Figure 7.1 shows how the measured variations in O_3 concentration in a museum environment in the Los Angeles metropolitan area could be modeled by considering the influence of ventilation, filtration, heterogeneous removal, direct emission, and photolytic and thermal reactions during a 2-day period. Figure 7.2 shows the variations in SO_2 concentrations at different locations within a single museum, as well as the effects of different air treatment approaches. Such continuous measurements of air constituents in actual museum environments provide valuable experience on the effect of open or closed doors and windows, poor or good circulation, the effect of

TABLE 7.3 Selected Air Constituents of Concern for Cultural Artifacts

Pollutant	Effect
SO_2	Corrodes metals, alloys, and stones, especially in combination with oxidants. Attacks cellulose (paper, cotton, linen), dyes, pigments, paints, and photographic materials. Weakens leather.
NO_2	Promotes SO_2-induced corrosion of metals, alloys, and stones. Reduces strength of textiles. Attacks certain dyes.
NH_3	Promotes stress corrosion cracking of high-zinc brass.
H_2S	Tarnishes silver, copper, and copper alloys. Damages paint and dyes.
H_2O_2	Promotes SO_2-induced corrosion of metals, alloys, and stones. Discolors photographic prints.
O_3	Promotes SO_2-induced corrosion of metals and stones. Damages dyes, pigments, inks, and photographic materials. Cracks rubber.
Organic acids and aldehydes	Corrodes metals and alloys.
NaCl	Corrodes metals and alloys.
$(NH_4)_2SO_4$ and $(NH_4)HSO_4$	Corrodes metals and alloys.

outside weather conditions, and the effect of air-conditioning provided with or without chemical filtering. Figure 7.10 finally shows how continuous recording of the corrosivity of a museum environment can indicate changes that have occurred in the environment.

7.7 SUMMARY

Corrosion in uncontrolled indoor environments has important implications in such diverse application areas as electronics, equipment storage, and the display or preservation of cultural artifacts. Indoor corrosion of metals is of similar complexity as outdoors, although indoor environmental characteristics are significantly different. The variations in relative humidity are less dramatic indoors than outdoors, and the deposition velocities and concentrations of gas or particle corrosive agents are generally lower. The same physicochemical processes operate, including adsorption of water, deposition of gases and particles into the liquid layer, proton- and ligand-induced dissolution, ion pairing, and precipitation and growth of corrosion products. Precipitation–dissolution–reprecipitation cycles, common and very important processes outdoors, are less common indoors but are expected in locations with high outdoor–indoor exchange rates.

The indoor corrosion rate laws can be linear, parabolic, or intermediate between these two. In very mild indoor environments, the rate is slower than parabolic and can be approximated by either logarithmic or inverse logarithmic rate laws, as with room temperature oxidation. One consequence of the differences between indoor and outdoor characteristics is the relatively low indoor corrosion rate of many metals, including iron, copper, nickel, cobalt, and zinc. Mean indoor corrosion rate values turn out typically to be between two and three orders of magnitude lower than outdoors. The only known exception is silver, which exhibits similar indoor and outdoor corrosion rates. Another indoor–outdoor difference is in the chemical composition of corrosion products that are formed. Metal carboxylates are important constituents indoors but not outdoors.

FURTHER READING

General Characteristics of Indoor Environments

W.W. Nazaroff and G.R. Cass, Mathematical modelling of chemically reactive pollutants in indoor air, *Environmental Science and Technology, 20*, 924–934, 1986.

J.D. Sinclair, L.A. Psota-Kelty, C.J. Wechsler, and H.C. Shields, Measurement and modeling of airborne concentrations and indoor surface accumulation rates of ionic substances at Neenah, Wisconsin, *Atmospheric Environment, 24A*, 627–638, 1990.

C.J. Wechsler and H.C. Shields, Potential reactions among indoor pollutants, *Atmospheric Environment, 31*, 3487–3495, 1997.

Indoor Corrosion Rates and Corrosion Products

E. Johansson, Corrosivity measurements in indoor atmospheric environments: a field study, Licentiate Thesis, Royal Institute of Technology, Stockholm, Sweden, 1998.

T. Prosek, M. Kouril, M. Dubus, M. Taube, V. Hubert, B. Scheffel, Y. Degres, M. Jouannic, and D. Thierry, Real-time monitoring of indoor air corrosivity in cultural heritage institutions with metallic electrical resistance sensors, *Studies in Conservations, 58*, 117–128, 2013.

D.W. Rice, R.J. Cappell, P.B.P. Phipps, and P. Peterson, Indoor atmospheric corrosion of copper, silver, nickel, cobalt, and iron, in atmospheric corrosion, in *Atmospheric Corrosion*, ed. W.H. Ailor, John Wiley & Sons, Inc., New York, pp. 651–666, 1982.

Indoor Environmental Classification

ISO, ISO 11844-1:2006, Corrosion of Metals and Alloys—Classification of Low Corrosivity of Indoor Atmospheres. Part 1: Determination and Estimation of Indoor Corrosivity, 2006.

ISO, ISO 11844-2:2006, Corrosion of Metals and Alloys—Classification of Low Corrosivity of Indoor Atmospheres. Part 2: Determination of Corrosion Attack in Indoor Atmospheres, 2006.

ISO, ISO 11844-3:2006, Corrosion of Metals and Alloys—Classification of Low Corrosivity of Indoor Atmospheres. Part 3: Measurement of Environmental Parameters Affecting Indoor Corrosivity, 2006.

Metal Artifacts Indoors

G.T. Thomson, *The Museum Environment*, 2nd edition, Butterworths, London, 1986.

8

CORROSION IN OUTDOOR EXPOSURES

8.1 THE EFFECT OF EXPOSURE CONDITIONS

Outdoor exposures can be performed for many reasons. They allow assessments to be made of the corrosion behavior of a material in a given outdoor environment, as well as comparisons to be made of the corrosion performance of materials or of corrosion protective measures. They can also establish long-term changes in corrosivity as a result of changes in pollutants at a given outdoor environment. Information presented in previous chapters has made it obvious that there are a substantial number of atmospheric constituents and climatic parameters that may influence the corrosion behavior of a given material in any outdoor environment. This makes the actual corrosion situation quite complicated to describe, especially since the atmosphere varies in composition on a daily, seasonal, and yearly basis.

Most of our knowledge of the atmospheric corrosion behavior of different materials outdoors originates from numerous outdoor exposure programs, the first of which started early in the twentieth century in the United States and the United Kingdom. It was soon recognized that the corrosion rates varied considerably among different exposure sites. As an initial attempt at classification, the environments were designated as "rural," "urban," "marine," and "industrial." During most of the twentieth century, numerous exposure programs were implemented, aimed at identifying and possibly quantifying the most important environmental parameters from a corrosion perspective. Although generating little mechanistic insight, the large body of data generated from outdoor exposures has manifested the importance of at least the following parameters: the levels and variations of relative humidity and temperature, the deposition or concentration of sulfur dioxide (SO_2), the deposition of chlorides (Cl^-), and the effect of precipitation in general and acid rain in particular.

Atmospheric Corrosion, Second Edition. Christofer Leygraf, Inger Odnevall Wallinder, Johan Tidblad and Thomas Graedel.

Many other parameters are known to influence the corrosion rate as well, including the sunshine sheltering, wind conditions, specimen orientation, emission sources, and surface preparation. Hence, there is an obvious need to standardize the exposure situation so that comparisons can be made between the corrosion results from different outdoor exposures.

This chapter deals with general design considerations for outdoor exposure experiments. It also discusses important exposure parameters and their general influence on corrosion processes, describes what we can learn from outdoor exposures, and, finally, presents an important classification system for outdoor exposures that has emerged as a result of the numerous exposure programs.

8.2 DESIGN CONSIDERATIONS

8.2.1 Selection of Exposure Site and Instrumentation

The exposure site is usually selected so that it represents a general macroclimate, that is, uniform conditions of a specific rural, urban, marine, or industrial environment. Hence, point sources of emission should be avoided. The test site should be located so that the samples can be exposed in the direction of the highest corrosivity, normally facing south in the northern hemisphere and north in the southern hemisphere. In marine exposure sites, samples frequently face the ocean (the source for chloride aerosols), and in urban sites samples may face the road with highest traffic intensity.

For interpretation of corrosion data, it is important to gain as much information as possible about the environmental characteristics of the test site. For this reason the test site should ideally be located at or near facilities engaged in ongoing measurements of gaseous and/or particle measurements, performed, for instance, by environmental or health organizations.

The recommended instrumentation at any test site comprises a thermograph, hygrograph, or logging hygrometers for monitoring ambient temperature and relative humidity. The equipment should be installed in a louvered structure to act as a shield against precipitation. Other useful instrumentation includes equipment for measuring wind velocity and wind direction, solar radiation (duration-of-sunshine meters or pyrheliometers), frequency, duration and amount of precipitation (rain gauges), time of wetness (TOW, either through relative humidity and temperature data or through TOW sensors; see Section 8.3), chloride ion (Cl^-) deposition (through the wet candle method), and determination of sulfur dioxide (SO_2) concentration (through sulfation plates, the peroxide candle method or diffusive samplers). Integrated Cl^- and SO_2 sensors provide a minimal level of information, but their lack of time resolution can be a significant constraint to interpretation of the results.

Many test sites are located in remote places, and it is important to protect the test site from damage caused by the public and from growing plants reaching the specimens. Ideally, the test site should be fenced.

8.2.2 Specimens and Specimen Mounting

Many details can be specified when it comes to general requirements of atmospheric corrosion testing, and these are also thoroughly described in ISO 8565. A common specimen size is 100 by 150 mm and a thickness of 1–3 mm, although other shapes (fasteners, bolts, nuts, tension bars, etc.) can be used as well, provided that they can be accurately evaluated. Important is to identify each specimen through, for example, metal stamping, notching edges, or drilling holes.

Before exposure the specimens are cleaned to avoid surface contamination and commonly weighed by means of gravimetric analysis (see Appendix A) and then stored in a clean room with controlled temperature and relative humidity of 65% or less or for sensitive specimens in a desiccator or sealed in plastic bags with desiccant.

The specimens are usually mounted on racks fabricated from wood or galvanized steel, in order to keep the specimen in well-defined positions throughout the whole exposure as well as isolated from each other. In order to avoid galvanic effects between samples or between each sample and the rack (if made of a metallic material), the samples are cushioned with some electrically insulating material such as rubber or porcelain or fixed on Plexiglas fixtures. A typical rack is seen in Figure 8.1.

A recommended practice is to include a corrosion rate determination of reference samples, usually zinc and unalloyed carbon steel, sometimes also copper and aluminum (see Section 8.3.6).

Specimens are often also exposed under sheltered conditions by means of a roof or a louvered structure. The purpose is to prevent the wash-off effects of precipitation while allowing airborne corrosive agents and adsorbed or absorbed water to reach the specimen surface (see Section 8.3.4).

8.2.3 Specimen Removal and Corrosion Evaluation

On predetermined days the specimens are removed and transported to the laboratory for further evaluation. Common removal periods are 0.5, 1.0, 2.0, 4.0, 8.0, and 16.0 years after exposure start. After removal, the specimens may be stored under clean, dry conditions for a period of a month or so, although this period should be kept as short as possible.

By far the most common evaluation of corrosion is through mass loss measurements. This requires chemical stripping of the corrosion products and determination of the difference in specimen mass before and after exposure by means of gravimetric analysis (see Appendix A). Mass loss determination assumes relatively uniform corrosion attacks. Passivating metals or alloys such as aluminum and stainless steels may suffer from localized corrosion attack, such as pitting corrosion or crevice corrosion, in which case the pit density or number of pits should be measured by means of an optical microscope or a scanning electron microscope (see Appendix A).

For more mechanistic studies, the corrosion products may be analyzed by a wide variety of analytical techniques, most of which are briefly described in Appendix A.

FIGURE 8.1 A representative rack for outdoor exposure of specimens.

8.3 INFLUENCE OF EXPOSURE PARAMETERS

8.3.1 The Complex Interplay

What does the specimen surface actually "see" of the atmosphere during an outdoor exposure? From the kinetic theory of gases, we know that each second more than 10^{23} atmospheric molecules collide with each square centimeter of the surface. As pointed out in Chapter 3, most of these molecules do not participate at all in the corrosion process but rather leave the surface region as quickly as they entered it. A large number of molecules do participate, however, as do aerosol particles that are collected by the surface.

In order to illustrate the complex interaction between a specimen surface and the atmospheric environment, we start by analyzing a relatively simple outdoor exposure situation: the mass change of a gold surface (which cannot corrode) that is exposed to a sheltered outdoor environment. The mass change is obtained by using the quartz crystal microbalance (QCM; see Appendix A) technique, which allows mass changes of less than $10\,ng\,cm^{-2}$ to be monitored under *in situ* conditions. (For reference, a monolayer of water on a flat metal surface has a mass of $30\,ng\,cm^{-2}$.)

Figure 8.2 displays results from the exposure of the protected gold surface to an urban environment. The form of the curve is similar to that shown in Figure 2.3, but the exposure period is only 14 days. The figure also shows the simultaneous variations in relative humidity. It can be clearly seen that the mass change consists of two contributions: (1) an irreversible mass increase and (2) reversible mass changes that coincide with variations in relative humidity. By performing a number of similar studies and analyzing the environment in each case, it has been shown that the

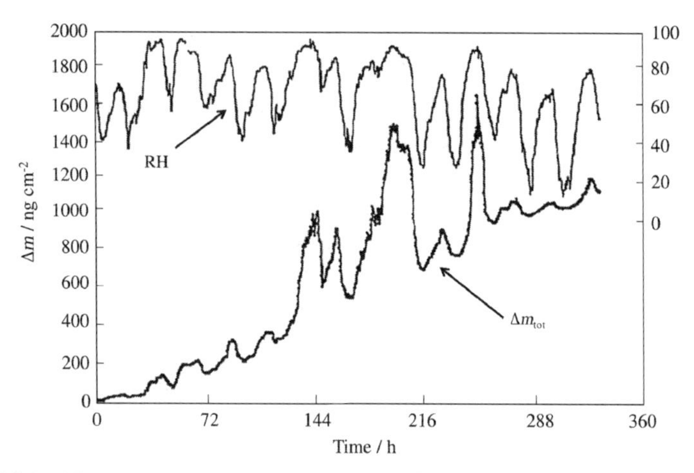

FIGURE 8.2 The total mass variation (Δm_{tot}, ng cm^{-2}) of a gold surface and the simultaneous variation in relative humidity (RH, %) as a function of time during a 14-day urban exposure. (Reproduced with permission from The Electrochemical Society; M. Forslund and C. Leygraf, Humidity sorption due to deposited aerosol particles studied *in situ* outdoors on gold surfaces, *Journal of the Electrochemical Society, 144,* 105–113, 1997.)

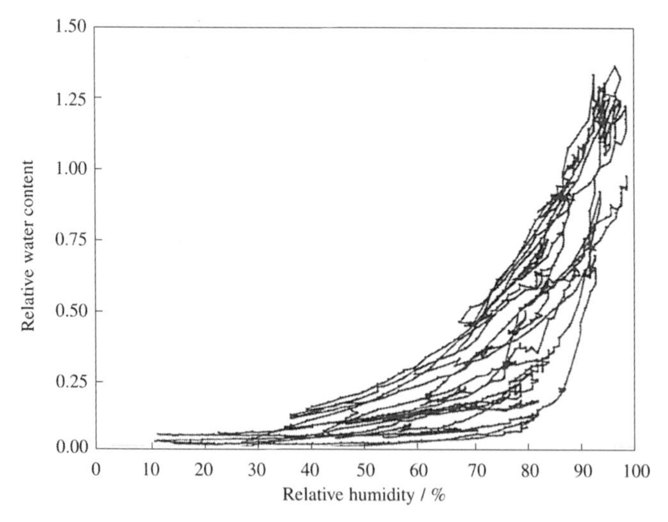

FIGURE 8.3 The relative water content, given as the mass of water divided by the mass of deposits, during 14 consecutive days exposure of gold in an urban environment. (Reproduced with permission from The Electrochemical Society; M. Forslund and C. Leygraf, Humidity sorption due to deposited aerosol particles studied *in situ* outdoors on gold surfaces, *Journal of the Electrochemical Society, 144,* 105–113, 1997.)

irreversible mass increase is mainly caused by airborne particles that are deposited and accumulated on the gold surface. The reversible mass variation, on the other hand, is due to water sorption and desorption. It is influenced by the properties and amounts of deposited particles and varies with the relative humidity. Surface analysis suggests that the deposited aerosols in this urban environment consist of a mixture of phases including $(NH_4)_2SO_4$, NH_4HSO_4, and NH_4Cl, all of which can be found in aerosol particles. In marine environments, NaCl, Na_2SO_4, and $MgCl_2$ occur.

Constituents of the aerosol particles, together with the ambient gases, dissolve in the liquid layer that forms on the gold surface and thereby modify the chemical composition of the layer. A very large number of chemical reactions may operate under field exposure conditions. Different reactions are anticipated in different types of environments, that is, a marine environment as opposed to an urban environment or an urban environment in, say, Los Angeles as opposed to an urban environment in Prague or Beijing. Each environment is unique and creates a specific chemistry in the liquid layer formed. Nonetheless, there are certain general observations that hold true despite the variability, as discussed in the following text.

8.3.2 Time of Wetness

Airborne gases and particles not only modify the chemistry of the liquid layer, but they also retain the liquid layer on the surface. The reason is the hygroscopic properties of many corrosion products and of deposited aerosol particles such as $(NH_4)_2SO_4$, $(NH_4)HSO_4$, and NH_4Cl. When the relative humidity reaches a certain critical value, the phase reaches its point of deliquescence, meaning that it will rapidly absorb water until a saturated solution is obtained. The critical relative humidity at 25°C for the ammonium compounds listed earlier is 79%, 39%, and 77%, respectively. While the mass for a single phase exhibits a sharp increase at the critical relative humidity, the corresponding mass decrease exhibits no critical relative humidity value. For a mixture of phases (a common occurrence), there is no sharp mass increase at any particular relative humidity but rather a continuously increasing curvature starting at a relative humidity lower than those of the single phases.

Measurements of surface wetting are of extreme importance in atmospheric corrosion and have been the subject of many studies based on devices that monitor the actual time during which the surface is wet. An important concept is the so-called TOW, regarded as the time during which a corrosion-stimulating liquid film exists on the surface. A widely adopted standardized definition of TOW is the time during which the temperature is above 0°C while the relative humidity exceeds or is equal to 80%. However, such a definition should not be taken too literally, since the actual time during which the surface is at least partially wetted and atmospheric corrosion can occur is much longer. This is illustrated in Figure 8.3, in which the relative water content on a gold surface, given as the mass of water divided by the mass of deposit, is plotted as a function of relative humidity during a large number of consecutive 24 h cycles. Each cycle is represented by one increase (during nighttime) and one decrease (during daytime) in water mass and relative humidity, respectively. From the figure it is evident that detectable amounts of water are present at a relative humidity far below 80%.

Despite uncertainties in the standardized definition and accuracy, TOW is nevertheless a useful parameter since under normal outdoor conditions most of the corrosion takes place under humid conditions roughly above 80% and during rain events. It can be estimated from relative humidity and temperature data, and it

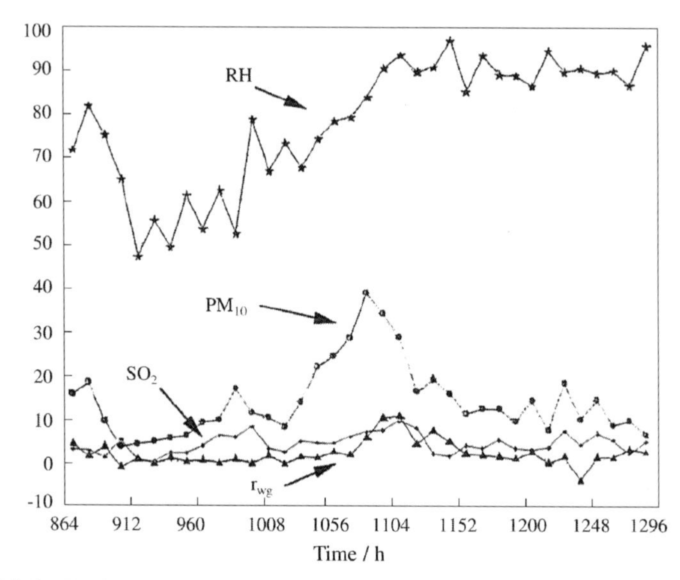

FIGURE 8.4 Twelve-hour mean values of mass increase rate of copper, r_{wg} (ng cm^{-2} h^{-1}), relative humidity (%), particle concentration (PM_{10}), particles with diameter less than 10 µm diameter (µg m^{-3}), and SO_2 concentration (µg m^{-3}). (Reproduced with permission from The Electrochemical Society; M. Forslund and C. Leygraf, *In situ* weight gain rates on copper during outdoor exposures, *Journal of the Electrochemical Society, 144,* 113–120, 1997.)

exhibits good correlation with outdoor corrosion rates (particularly of steel in many types of environments). The motivation of its use, however, is based on statistical grounds rather than on mechanistic grounds. In recent dose–response functions (see in the following text), the use of TOW has been abandoned in favor of the individual original parameters—temperature and relative humidity. One reason for this is that although TOW can be obtained for single locations with some effort—the calculation requires highly time-resolved values of temperature and relative humidity—the collection of data for calculating TOW for a large number of sites or grid cells, as a basis for mapping, is by no means easy.

The variation in surface wetness implies that the momentary corrosion rate also varies. Periods during which the surface is wetted and corrosion occurs are interrupted by periods during which the surface is dry and the corrosion is negligibly small. This is illustrated in Figure 8.4, which is an example of the combined action of several parameters (high relative humidity and high concentration of PM_{10} particles; see figure caption) in enabling the atmospheric corrosion process to occur during a specific period (between 1060 and 1120 h).

One might argue that a gold surface, located in a shelter and well protected from direct wind and precipitation, represents a situation that is far from a realistic outdoor exposure. Yet, from this carefully analyzed example and similar examples (see "Further Reading"), we can deduce a number of general conclusions concerning the properties of a wetted surface that we cannot possibly obtain from standard outdoor exposures such as those described in Section 8.2. First, the significance of the liquid layer with changing thickness and chemistry has been demonstrated. Second, the liquid layer's presence is retained because of corrosion products and aerosol particles with hygroscopic properties. Third, the time during which the surface is wet commonly exceeds the standardized definition of TOW. Fourth, when the thickness of the liquid layer decreases, the concentration of species increases, as do the conditions for enhanced corrosion rates. These conditions are enhanced further during periods of increased concentrations of corrosive substances in the atmosphere. Fifth, corrosion only occurs during limited periods of time and is interrupted by intervals with negligible corrosion rates.

8.3.3 Temperature

Based on insights gained in the many exposure programs, we now turn to the role of important exposure parameters and start with the possible influence of temperature. This issue is not straightforward. On one hand, increased temperature accelerates the rates of various chemical and physical processes, including chemical reactions, electrochemical reactions, and diffusion. On the other hand, increased temperature results in higher desorption rates of the liquid film, thereby reducing the time for the surface to be wetted. Increased temperature also reduces the solubility of various gases, including oxygen.

The overall effect of increasing temperature appears to be an increase in corrosion rate under conditions of permanent surface wetting, such as those obtained during precipitation. Under conditions of varying surface wetting, on the other hand, the corrosion rate increases with temperature up to a certain maximum value and thereafter decreases. As an example, the influence of temperature on carbon steel is displayed in Figure 8.5. Because of enhanced concentrations of dissolved atmosphere constituents, the freezing point of the liquid layer is normally reduced to far below 0°C, and significant corrosion rates can be seen at −5°C and even lower.

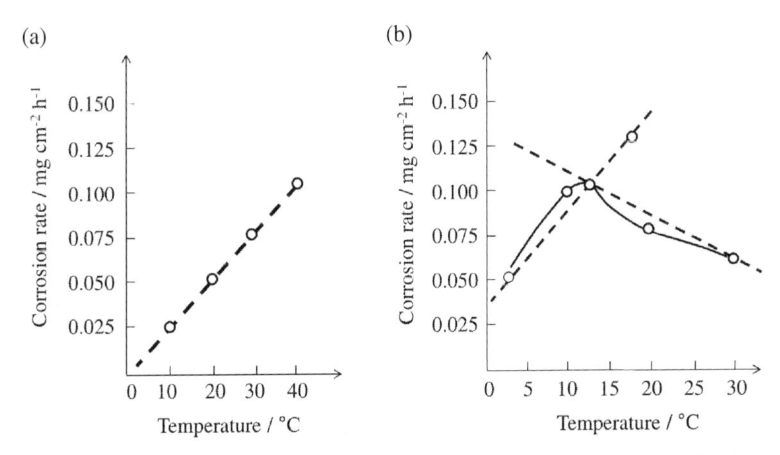

FIGURE 8.5 Influence of air temperature on the atmospheric corrosion rate of carbon steel: (a) during rain and (b) at drying out of a film of water at a constant relative humidity of 75%. (Reproduced with permission from Taylor & Francis; V. Kucera and E. Mattsson, Atmospheric corrosion, in *Corrosion Mechanisms,* ed. F. Mansfeld, Dekker, New York, pp. 211–284, 1987.)

At what approximate temperature is the corrosion rate expected to be maximum? When trying to explain the influence of climatic conditions on corrosion rate, most early field exposure programs have used the standardized definition of TOW (see Section 8.3.2) as the explanatory variable rather than to consider relative humidity and temperature separately. As a consequence, the direct influence of temperature is normally not discussed in analyses of corrosion data. Analyses from more recent exposure programs, however, have taken into account separate effects of temperature and relative humidity (see Section 8.4). From these analyses it has been shown that there is a positive correlation between corrosion and temperature up to around 10°C, depending on material, and a negative correlation above that temperature.

8.3.4 Rain Shelter

Sheltered exposure differs from unsheltered exposure in many ways. In the former, the surface is shielded from direct precipitation and solar radiation. The shelter also prevents coarse aerosol particles, such as windblown sea salt or soil dust, from reaching the corroding surface, while it allows interaction with smaller aerosol particles and gases. How these modifications of conditions influence the corrosion process depends on the actual exposure conditions and varies from one case to another.

The influence of precipitation is manyfold. On one hand, it can wash away corrosion stimulators that were accumulated through dry deposition, thereby decreasing the corrosion effect. On the other hand, it can add corrosion-stimulating ions (e.g., H^+ and SO_4^{2-}) to the liquid layer through wet deposition. Acidified precipitation may also partly dissolve the existing film of corrosion products and reduce the corrosion protective properties of the surface. The net effect of these opposing processes depends on exposure conditions and on the corroding material. This is illustrated in Table 8.1 for weathering steel and zinc. Weathering steel (see Section 13.5) is a low alloy steel developed for the purpose of being suitable in polluted environments (like Kopisty) but is only effective in unsheltered positions where it can be washed by rain regularly. Experience from a number of field exposures suggests that the wash-off effect of corrosive species increases with their concentration levels in the environment. Hence, in heavily polluted or chloride-contaminated sites, the corrosion rate is expected to be higher on sheltered samples than on unsheltered samples. In less polluted sites, the corrosion rate is expected to be higher on unsheltered samples. However, the efficiency of the wash-off effect of dry deposits not only depends on actual

TABLE 8.1 Corrosion Attack of Weathering Steel and Zinc after 8 Years of Exposure (1987–1995) in Unsheltered and Sheltered Position at Two Sites in the ICP Materials Exposure Program

	Weathering Steel (μm)		Zinc (μm)	
	Unsheltered	Sheltered	Unsheltered	Sheltered
Kopisty, Czech Republic	46	172	26	11
Stockholm, Sweden	51	30	11	3

precipitation conditions but also on the properties of the corrosion products. With increased aging, the corrosion products develop a more intricate morphology and a higher capacity to retain deposited species.

8.3.5 Sheltered and Unsheltered Outdoor Exposure: An Example for Zinc

The complex influence of a rain shelter can be illustrated by the behavior of zinc. Evidence presented in Section 8.3.2 has demonstrated that atmospheric corrosion outdoors often occurs during specific episodes characterized by cyclic humidity and increased pollutant conditions. In Section 9.2 we demonstrate how these alternating conditions gradually change the properties of the corrosion products formed. As a result, the corrosion products exhibit different principal phases during their evolution as well as a corrosion protective ability that slowly increases with exposure time. The first result is of significance for this section, and the second result for Section 8.4.4.

The corrosion products on zinc exposed in many marine environments exhibit the following sequence (for more details, see Section 9.2):

$$\begin{aligned}&Zn(OH)_2\,(\text{within seconds}) \rightarrow Zn_5(CO_3)_2(OH)_6\,(\text{hours}) \rightarrow \\ &\quad Zn_5(OH)_8Cl_2{\cdot}H_2O(\text{days}) \rightarrow NaZn_4Cl(OH)_6SO_4{\cdot}6H_2O(\text{weeks}).\end{aligned} \tag{8.1}$$

The latter two phases are formed largely because of the deposition of NaCl and Na_2SO_4 aerosols. These dissolve in the aqueous adlayer and gradually form zinc corrosion products through repeated dissolution–coordination–reprecipitation events (Section 2.4.2 and Chapter 9). This sequence is observed in both open and rain-sheltered exposure conditions but is much faster in unsheltered circumstances, possibly because the aerosols, due to their relatively coarse size, reach the corroding surface more easily under open than sheltered conditions. As a result, sheltering in the marine environment causes a delay in the initiation of the corrosion product sequence.

The corrosion product sequence of zinc in many rural and urban environments, on the other hand, is the following:

$$\begin{aligned}&Zn(OH)_2\,(\text{within seconds}) \rightarrow Zn_5(CO_3)_2(OH)_6\,(\text{hours}) \rightarrow \\ &\quad Zn_4SO_4(OH)_6{\cdot}4H_2O(s)(\text{days})\end{aligned} \tag{8.2}$$

In this case sheltering causes an acceleration of the corrosion product sequence: the hydroxysulfate, detectable after a few days of sheltered exposure, is not seen even after 1 year of unsheltered exposure. The reason is the delayed accumulation of sulfur-containing gases or particles on unsheltered zinc surfaces because of the wash-off effect of these species by precipitation.

8.3.6 Corrosion Rates

Atmospheric corrosion damage is usually measured in terms of mass loss per unit surface area or by average penetration depth, both values being obtained after the removal of corrosion products by chemical stripping (Section 8.2.3). The properties of the corrosion products determine to a large extent the corrosion rates in any given situation. In the case where no film of corrosion products is present or the corrosion film is porous and thus virtually without any protective ability, the corrosion rate will depend largely on the supply of corrosive agents and water to the surface. The corrosion damage (M) as a function of time (t) is then given by

$$M = At \tag{8.3}$$

where A is a constant termed the corrosion rate. This linear relationship is at odds with our earlier conclusion (Section 8.3.2) that atmospheric corrosion frequently occurs in episodes occurring on a relatively short time scale (days or weeks). For this linear relation to be valid, we have to observe the corrosion effects on a time scale of years rather than days or weeks and assume that the yearly mean concentration of corrosive agents does not change over that time period.

On the other hand, if we assume that the corrosion product layer is a highly protective barrier, then a necessary condition for atmospheric corrosion to occur is that ions from the corroding metal diffuse through the layer, react, precipitate, and increase the layer thickness further. In the case of diffusion control, the corrosion damage as a function of time is given by

$$M = At^{0.5} \tag{8.4}$$

In the previous section we mentioned that the total corrosion sequence often includes several phases resulting from repeated dry–wet–dry cycles over a time period of years that favor the formation of gradually more corrosion-resistant corrosion products. This is exemplified in Figure 8.6, which displays the corrosion sequence of copper, zinc, and nickel in a rural environment, the common observation for all metals being that the corrosion products possess successively higher protective ability with increased exposure time.

Due to the formation of gradually more resistant corrosion products through dissolution of the layer and subsequent ion pairing and reprecipitation, the corrosion rate can be expected to fall somewhere between the linear relationship

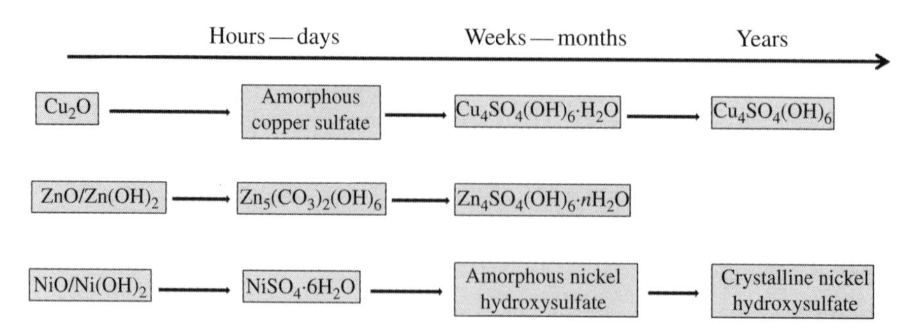

FIGURE 8.6 Corrosion sequence for copper, nickel, and zinc in a rural atmosphere (Aspvreten, Sweden) as a function of time of sheltered exposure.

(complete dissolution or poor protection) and the parabolic relationship (good protection). The generalized form is thus

$$M = At^n \tag{8.5}$$

This power law, with n usually falling between 0.5 and 1.0, turns out to be valid for many metals and alloys, including carbon and low alloy steels, galvanized steel, and zinc, copper, and copper alloys in a variety of different atmospheric environments, provided the relationship is determined for exposure over extended periods of time. The A factor in Equation 8.3 represents the initial corrosion rate, while the exponent n is a measure of the protective ability of the corrosion products formed. In very aggressive environments, n values close to 1 are normally observed. This reflects a situation in which the corrosion products possess meager protective properties and the corrosion rate is largely determined by supply of corrosive agents furnished to the surface. In benign environments that generate corrosion films having increasing protective properties, n is commonly observed to decrease to values of 0.6 or lower (see also Section 7.3, corrosion kinetics indoors).

To illustrate its validity, the power law equation has been applied to mass loss data of sheltered and unsheltered copper and bronze samples obtained from 1-, 2-, 4-, and 8-year exposures at 39 field test sites within the UNECE exposure program. The correlation coefficients are normally better than 0.99 for all exposure situations. Figure 8.7 shows copper and bronze mass loss data from one unsheltered rural

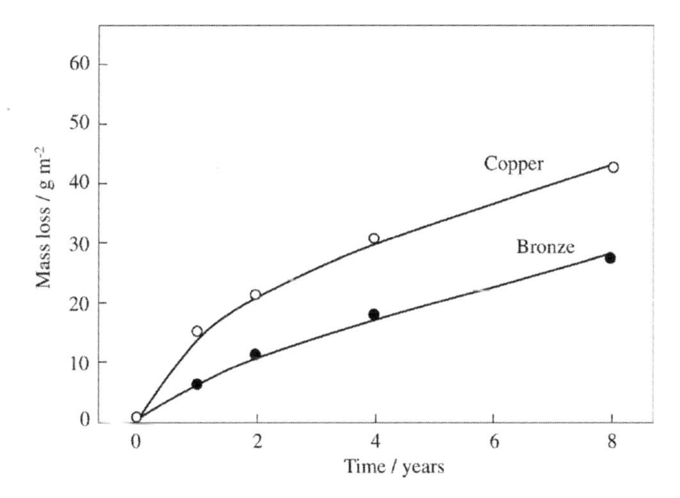

FIGURE 8.7 Comparison of copper mass loss data (unfilled circles) and bronze mass loss data (filled circles) as a function of time of exposure from one unsheltered rural exposure at Dorset, Canada, within the UNECE program. The circles are the experimental data and the solid line is a fit to the power law $M = At^n$. (Data from B. Stöckle, A. Krätschmer, M. Mach, and R. Snethlage, UN/ECE International co-operative programme on effects on materials, including historic and cultural monuments, *Report No. 23: Corrosion Attack on Copper and Cast Bronze. Evaluation after 8 Year of Exposure*, Bayerisches Landesamt für Denkmalpflege, Munich, Germany, 1998.)

exposure, together with the corresponding power law curves. The copper samples exhibit a higher initial corrosion rate ($A = 14.6\,g\,m^{-2}\,y^{-1}$) and higher protective ability ($n = 0.52$) than the bronze samples ($A = 6.4\,g\,m^{-2}\,y^{-1}$, $n = 0.71$). This can be explained by a high rate of initial formation of Cu_2O on copper, but not on bronze, and the formation of a more protective film on copper.

Corrosion rates for several metals are given in the corresponding appendices. However, corrosion rates defining the categories of corrosion (C1–C5 and CX) for the common construction materials—carbon steel, zinc, copper, and aluminum—deserve to be mentioned (Table 8.2). These categories are of widespread use for practical engineering purposes and often serve as a basis for defining need for appropriate corrosion protection systems. When defining these categories the levels of pollution are to be preferred over wide-ranged labels such as "rural," "urban,"

TABLE 8.2 Corrosivity Categories (ISO 9223) Showing Ranges of Corrosion Attack (First Year of Exposure) of Carbon Steel, Zinc, Copper, and Aluminum and Examples of Environments

Category	Corrosivity	Carbon (μm)	Zn (μm)	Cu (μm)	Al ($g\,m^{-2}$)	Example of Environments
C1	Very low	0–1.3	0–0.1	0–0.1	Negligible	Offices, schools, museums
C2	Low	1.3–25	0.1–0.7	0.1–0.6	<0.6	Storage, sport halls, low polluted rural areas
C3	Medium	25–50	0.7–2.1	0.6–1.3	0.6–2	Medium polluted inland areas or coastal areas with low deposition of chlorides
C4	High	50–80	2.1–4.2	1.3–2.6	2–5	High polluted inland areas or coastal areas without direct influence of spray of salt water
C5	Very high	80–200	4.2–8.4	2.6–5.6	5–10	Very high polluted inland areas or coastal areas with significant effect of chlorides
CX	Extreme	200–700	8.4–25	5.6–10	>10	Coastal areas with significant effect of chlorides in subtropical and tropical zones with very high time of wetness

"industrial," and "marine." For example, an urban location in a low polluted region can show much lower corrosion rates than a rural location in a high polluted region or country. This is further elaborated in Section 8.4 (dose–response functions).

8.3.7 Nitrogen Dioxide and Ozone

Since Vernon's original corrosion studies in the 1920s, SO_2 has been recognized as the most important gaseous air constituent. It is obvious, however, that SO_2 concentrations alone are not sufficient to explain the corrosion rates observed outdoors. This can easily be proven by a simple but very striking experiment in which a metal is exposed to humidified air to which SO_2 alone or in combination with an oxidizing constituent, such as nitrogen dioxide (NO_2), is added. Figure 8.8 shows results obtained for nickel, providing clear evidence that the corrosion rate caused by SO_2 under actual laboratory exposure conditions is one order of magnitude lower than that caused by the combination of SO_2 and NO_2. While SO_2 over the last decades has shown substantial reduction in many urban environments, at least in the Western hemisphere, the concentration of NO_2 in the same environments has been roughly constant. Hence, the NO_2/SO_2 molar ratio has increased with time, thus changing the mechanism for the oxidation of S(IV) to S(VI). Numerous laboratory experiments, using synthetic air with different concentrations of SO_2 and NO_2, have unambiguously proven that NO_2 increases the corrosion rates of many metals. Efforts to

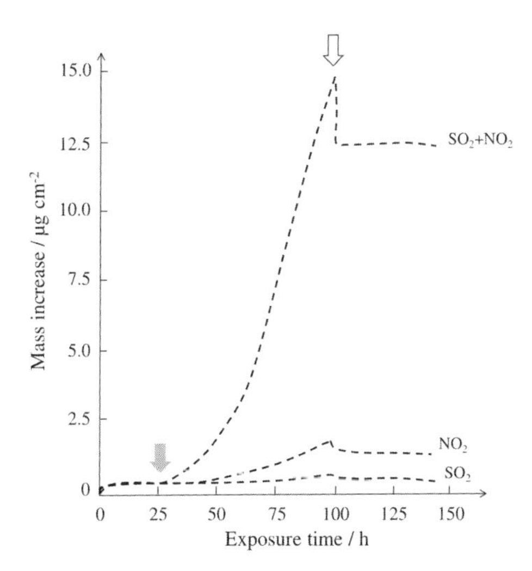

FIGURE 8.8 Mass increase of nickel as a function of exposure time in the presence of humidified air (75% relative humidity) to which 250 ppbv SO_2 and/or 300 ppbv NO_2 have been added. The filled and open arrows indicate introduction of gaseous pollutant and termination of pollutant and humidity, respectively. Results obtained by the quartz crystal microbalance. (Reproduced with permission from The Electrochemical Society; S. Zakipour, C. Leygraf, and G. Portnoff, Studies of corrosion kinetics on electric contact materials by means of quartz crystal microbalance and XPS, *Journal of the Electrochemical Society, 133,* 873–876, 1986.)

observe the influence of NO_2 in field exposures, however, have so far been unsuccessful. This paradox is probably related to the inherent difference between laboratory experiments, in which each gaseous constituent can be controlled separately, and field exposures, with their uncontrolled multipollutant exposure situation. As explained in Chapter 4, ozone (O_3) is formed during photolysis of NO_2 followed by reaction of NO with O_3. As a result, NO_2 is not the only oxidant; other oxidants, including O_3, are also operating. Since NO_2 and O_3 are intertwined, the important oxidizing role of NO_2 seen in the laboratory is hidden in the field because of the oxidizing effect of O_3 and perhaps other oxidizers and catalysts.

8.4 DOSE–RESPONSE FUNCTIONS

While earlier outdoor exposures classified the environments into categories, such as urban, rural desert, industrial, marine, industrial marine, and tropical, it was soon felt that a more quantitative prediction of site corrosivity was needed. Several international exposure programs were implemented during recent decades with the aim to more systematically obtain relationships between corrosion or deterioration rates and the levels or loads of pollutants in combination with climatic parameters. These relationships are frequently denoted dose–response functions even if not always the effect of pollution is expressed as a dose. They were developed based on statistical grounds, rather than on mechanistic grounds, but recent work has attempted to take into account known mechanistic grounds as well, and not only to optimize the functions from a data fitting point of view.

Two distinctly different purposes were in focus when developing these functions. The first was to predict a corrosivity category for worldwide rural, urban, and marine locations but excluding specific industrial pollution. Temperature, relative humidity, SO_2, and chloride deposition were measured but no other pollutants or pH of precipitation, resulting in dose–response functions for estimating corrosivity categories (Section 8.4.2). The second purpose was to predict and map corrosion rates with the aim to quantify effects of local and long-range transboundary air pollution. A wide range of climatic parameters and pollution parameters were measured, including pH of precipitation, at exposure sites with no significant marine influence, resulting in dose–response functions for the SO_2-dominating (Section 8.4.3) and multipollutant situations (Section 8.4.4). Thus, there is until today no international field exposure program that has systematically measured and quantified the simultaneous effect of SO_2, chlorides, and pH.

8.4.1 Determination and Estimation of Corrosion Rates

There are two different approaches of assessing the corrosivity of any environment. One approach is by exposing standard specimens of indicator materials such as steel, copper, zinc, and aluminum for 1 year and determining the mass loss. The other approach is based on dose–response functions as described in the following text. It is important to recognize that these two methods are not equivalent since the estimated level of uncertainty is very different. Therefore, within ISO, the first method is denoted "determination" with level of uncertainty in the range of 2–5%, while the second is denoted "estimation" with level of uncertainty in the range of 50%. At a distant future it may be possible to develop dose–response functions with better level of uncertainty, but with present knowledge it is always recommended to expose materials for a minimum of 1 year if it is important to assess the corrosivity of a location accurately.

8.4.2 Dose–Response Functions for Estimating Corrosivity Categories

Within the International Organization for Standardization (ISO), the system for classification of atmospheric corrosivity based on exposure of metallic materials is described in ISO 9223. As part of the revision of this standard, which originally was based on data only from the temperate region, the responsible technical committee initiated the international exposure program ISOCORRAG, which included more than 50 test sites in Europe, Argentina, Canada, Japan, New Zealand, and the United States. The measured parameters were SO_2 (concentration or deposition), chloride deposition, TOW, temperature, and relative humidity. The later MICAT project had a different aim, namely, to perform a corrosivity mapping in 12 Ibero-American countries and in Spain and Portugal. However, they copied the experimental protocol and measured parameters from ISOCORRAG, which made it possible to combine the databases from these two different exposure programs. This combined database, together with some additional data obtained with the same methodology from the colder parts of Russia, was then used to obtain new dose–response functions for carbon steel, zinc, copper, and aluminum. They all include the four parameters: SO_2, Cl^-, temperature, and relative humidity. Worth noting is that ISO 9223 allows for both concentration and deposition values of SO_2 when using these functions. Figure 8.9 illustrates the dose–response function for zinc in graphical form showing the complicated effect of temperature described in Section 8.3.3. In atmospheres dominated by SO_2, the corrosion rate increases with temperature up to a certain maximum value and thereafter decreases. In atmospheres dominated by chlorides, the decrease at higher temperature values is not observed, and the hygroscopic salt particles can retain moisture even at high temperatures. At high values of both temperature (tropical areas) and chloride deposition, corrosion can be extremely high, and this is the reason that the most recent version of ISO 9223 now includes an extra corrosivity category CX in addition to C1–C5.

Having determined or estimated a corrosivity class with measurements or dose–response functions, it is then possible to provide a prediction of the extent of corrosion damage of steel, copper, zinc, and aluminum during long-term exposures. Predicted data for long-term exposures can be made according to Equation 8.3 using the coefficients given in ISO 9224:2012.

The ISO classification system (ISO 9223–9226) provides adequate information for many practical purposes, including prediction of long-term corrosion behavior in different environments and the need for protective coatings. Experience from applying the ISO classification system has shown, however, that certain observations

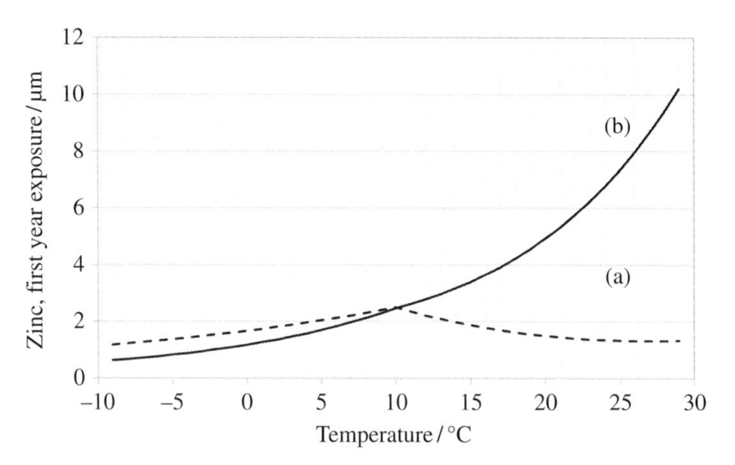

FIGURE 8.9 Calculated corrosion of zinc versus temperature based on the ISO 9223:2012 dose–response function for zinc and parameter values of (a) $SO_2 = 40\,\mu g\,m^{-3}$, $Cl = 3\,mg\,m^{-2}\,day^{-1}$; (b) $SO_2 = 1\,\mu g\,m^{-3}$, $Cl = 300\,mg\,m^{-2}\,day^{-1}$. (Adapted from A.A. Mikhalov, J. Tidblad, and V. Kucera, The classification system of ISO 9223 standard and the dose-response functions assessing the corrosivity of outdoor atmospheres, *Protection of Metals, 6,* 541–550, 2004.)

need further clarification. These include the frequent difference in corrosion rate between top side and bottom side of flat standard specimens and the possible role of other pollutants, as described in Section 8.4.4.

8.4.3 Dose–Response Functions for the SO_2-Dominating Situation

The international cooperative program on effects on materials including historic and cultural monuments (ICP Materials) is one of several programs and task forces dealing with effects of long-range transboundary air pollution. Other programs and task forces cover effects on forests, rivers, lakes, natural vegetation, crops, ecosystems, and health. ICP Materials performed the first exposures in 1987 at 39 test sites and is still ongoing with the latest exposure completed in 2012 at 22 test sites. One of the aims of ICP Materials is to perform a quantitative evaluation of the effects of multipollutants as well as of climate parameters on the atmospheric corrosion of materials in general, not only metals. A first set of dose–response functions was developed based on data obtained during the period 1987–1995 with exposure of materials after 1, 2, 4, and 8 years. All functions had more or less the same general form expressing corrosion in terms of dry and wet deposition functions (f_{dry} and f_{wet}) separately:

$$M = f_{dry}\left(T, \mathrm{RH}, \mathrm{SO_2}, t\right) + f_{wet}\left(\mathrm{Rain}, \left[\mathrm{H^+}\right], t\right) \tag{8.6}$$

where M is the mass loss, T is the temperature, RH is the relative humidity, SO_2 is the sulfur dioxide air concentration, t is time, Rain is the amount of precipitation, and $[H^+]$ is the hydrogen ion concentration of precipitation. Functions were developed for weathering steel, zinc, aluminum, copper, examples of bronze, limestone, sandstone, and paint-coated steel, as well as galvanized steel. At that point in time, SO_2 was still relatively high and was dominating among the gaseous pollutants in Europe, and therefore these set of functions were later denoted as the functions for the SO_2-dominating situation. An example of such an equation can be given for copper:

$$M = 0.0027\left[SO_2\right]^{0.32}\left[O_3\right]^{0.79} RH\cdot\exp\left\{f\left(T\right)\right\}t^{0.78} + 0.050 Rain\left[H^+\right]t^{0.89} \quad (8.7)$$

where M is in $g\,m^{-2}$, $[SO_2]$ and $[O_3]$ in $\mu g\,m^{-3}$, RH in %, $f(T)=0.083(T-10)$ when $T \leq 10$, otherwise $-0.032(T-10)$, T in °C, t in years, Rain in $mm\,year^{-1}$, and $[H^+]$ in $mg\,L^{-1}$. Note especially that the unit for $[H^+]$ is not the normal one used for this denomination ($[H^+]=10^{-pH}$); the relation between pH and $[H^+]$ is here $10^{-pH} \approx 10^{3-pH}$.

Equation 8.5 is the only exception for the dry deposition term in Equation 8.4 in that it includes the effect of O_3 in combination with SO_2 instead of only SO_2. Since this equation is obtained by a statistical analysis of the total corrosion attack, it is not possible to determine by this analysis why O_3 is significant for only copper, and not for other materials investigated, and to which extent O_3 contributes to the oxidation of SO_2 or to the formation of cuprite (Cu_2O).

8.4.4 Dose–Response Functions for the Multipollutant Situation

When SO_2 levels gradually decreased in Europe in the period 1960–2000, depending on location, it was realized that all of these functions were no longer sufficient for estimating corrosion and that other pollutants could contribute, such as nitric acid and particulate matter. A new exposure program within ICP Materials was launched in 1997 with exposures after 1, 2, and 4 years and with extensive environmental characterization within the EU project MULTI-ASSESS (2002–2005). New functions were developed based on the same concept as presented in Equation 8.4 but with additional additive terms for particulate matter (PM) and nitric acid (HNO_3). Not as many materials were exposed in this multipollutant exposure program, and new functions were developed only for carbon steel (additional PM term), zinc (additional HNO_3 term), bronze (additional PM term), and limestone (additional PM and HNO_3 terms). An example of a multipollutant equation is given for zinc:

$$R = 0.49 + 0.066\left[SO_2\right]^{0.22} e^{0.018RH+f(T)} + 0.0057 Rain\left[H^+\right] + 0.192\left[HNO_3\right] \quad (8.8)$$

where R is in μm (1-year exposure), $f(T)=0.062(T-10)$ when $T<10$°C, otherwise $f(T)=-0.021(T-10)$, and $[SO_2]$ and $[HNO_3]$ in $\mu g\,m^{-3}$.

The historical development with SO_2 concentrations increasing in Europe during the first part of the 2000s century and later decreasing SO_2 concentrations during the second part resulting in the multipollutant situation is characteristic for Europe. Other parts of the world are going through similar development paths, however at different pace and typically starting later in time. Taking Asia as an example, SO_2 concentrations are still very high and probably the most dominating source for atmospheric corrosion in nonmarine areas. When using any equation, however, outside the context of the original development a word of caution is in place as this is considered as an extrapolation.

8.5 SUMMARY

Corrosion in outdoor environments is complex, yet it has become possible to understand many aspects of these processes. The first is the importance of surface water, deposited under conditions of high relative humidity or rainfall and given up by evaporation under less humid conditions. The second is the sequence of steps, some slow, some faster, that mark the transition from initially exposed material to the final corrosion products. These steps and their importance will be described in more detail in Chapter 9.

Corrosion in outdoor exposures is influenced by many parameters, and standardized procedures have been developed to expose the samples under well-specified conditions. The rates of corrosion outdoors are reasonably well understood, at least qualitatively. The most important exposure parameters are the levels and variations of relative humidity and temperature, the time during which the surface is wetted, the concentration of gaseous and particle corrodents, and the presence of oxidants that promote the transformation of S(IV) to S(VI). The resulting corrosion rates depend on the protective nature of the corrosion products and the corrosivity of the environment.

Dose–response functions have been developed based on several international exposure programs. They provide adequate relationships between corrosion rates and levels or loads of atmospheric pollutants combined with climatic parameters.

FURTHER READING

Standardized Exposure Conditions

ISO, ISO 8565:2011, Corrosion of Metals and Alloys—Atmospheric Corrosion Testing—General Requirements, 2011.

ISO, ISO 9223:2012, Corrosion of Metals and Alloys—Corrosivity of Atmospheres—Classification, Determination and Estimation, 2012.

ISO, ISO 9224:2012, Corrosion of Metals and Alloys—Corrosivity of Atmospheres—Guiding Values for the Corrosivity Categories, 2012.

ISO, ISO 9225:2012, Corrosion of Metals and Alloys—Atmospheric Corrosion Testing—Measurement of Environmental Parameters Affecting Corrosivity of Atmospheres, 2012.

ISO, ISO 9226:2012, Corrosion of Metals and Alloys—Corrosivity of Atmospheres—Determination of Corrosion Rate of Standard Specimens for the Evaluation of Corrosivity, 2012.

The Complex Interplay and Time of Wetness

M. Forslund and C. Leygraf, Humidity sorption due to deposited aerosol particles studied *in situ* outdoors on gold surfaces, *Journal of the Electrochemical Society, 144,* 105–113, 1997.

M. Forslund and C. Leygraf, *In situ* weight gain rates on copper during outdoor exposures, *Journal of the Electrochemical Society, 144,* 113–120, 1997.

Influence of Temperature

J. Tidblad, A.A. Mikhailov, and V. Kucera, Model for the prediction of the time of wetness from average annual data on relative air humidity and air temperature, *Protection of Metals, 36*(6), 533–540, 2000.

Sheltered and Unsheltered Outdoor Exposure: An Example for Zinc

I. Odnevall and C. Leygraf, Formation of $NaZn_4Cl(OH)_6SO_4{\cdot}6H_2O$ in a marine atmosphere, *Corrosion Science, 34,* 1213–1229, 1993.

I. Odnevall and C. Leygraf, The formation of $Zn_4SO_4(OH)_6{\cdot}4H_2O$ in a rural atmosphere, *Corrosion Science, 36,* 1077–1091, 1994.

Useful information on various aspects, for example, influence of temperature, rain shelter, and corrosion rates can also be found in:

V. Kucera and E. Mattsson, Atmospheric corrosion, in *Corrosion Mechanisms*, ed. F. Mansfeld, Dekker, New York, pp. 211–284, 1987.

Dose–Response Functions

Functions for Classification of Corrosivity

A.A. Mikhailov, J. Tidblad, and V. Kucera, The classification system of ISO 9223 standard and the dose-response functions assessing the corrosivity of outdoor atmospheres, *Protection of Metals, 40*(6), 541–550, 2004.

Functions for the SO_2-Dominating Situation

J. Tidblad, V. Kucera, A.A. Mikhailov, J. Henriksen, K. Kreislova, T. Yates, B. Stöckle, and M. Schreiner, UN ECE ICP Materials. Dose-response functions on dry and wet acid deposition effects after 8 years of exposure, *Water, Air and Soil Pollution, 130*(1–4 III), 1457–1462, 2001.

Functions for the Multipollutant Situation

V. Kucera, J. Tidblad, K. Kreislova, D. Knotkova, M. Faller, D. Reiss, R. Snethlage, T. Yates, J. Henriksen, M. Schreiner, M. Melcher, M. Ferm, R.-A. Lefèvre, and J. Kobus, UN ECE ICP materials dose-response functions for the multi-pollutant situation, *Water, Air, and Soil Pollution: Focus, 7*(1–3), 249–258, 2007.

Major Outdoor Exposure Programs

ISOCORRAG

D. Knotkova, K. Kreislova, and S.W. Dean, *ISOCORRAG International Atmospheric Exposure Program: Summary of Results,* ASTM Data Series 71, ASTM International, West Conshohocken, 2010.

MICAT

M. Morcillo, Atmospheric corrosion in Ibero-America: the MICAT project, in *Atmospheric Corrosion,* ASTM STP 1239, eds. W.W. Kirk and H.H. Lawson, ASTM, Philadelphia, pp. 257–275, 1995.

ICP Materials (www.corr-institute.se/ICP-Materials)

J. Tidblad, V. Kucera, M. Ferm, K. Kreislova, S. Brüggerhoff, S. Doytchinov, A. Screpanti, T. Gröntoft, T. Yates, D. de la Fuente, O. Roots, T. Lombardo, S. Simon, M. Faller, L. Kwiatkowski, J. Kobus, C. Varotsos, C. Tzanis, L. Krage, M. Schreiner, M. Melcher, I. Grancharov, and N. Karmanova, Effects of air pollution on materials and cultural heritage: ICP materials celebrates 25 years of research, *International Journal of Corrosion, 2012*, 16, 2012.

9

ADVANCED STAGES OF CORROSION

9.1 INTRODUCTION

Corrosion products (constituents of the surface patina on many metals) that form on any metal object often consist of different compounds of varying characteristics that determine its overall corrosion behavior and surface appearance. Their formation and properties are consequences of complex interactions between the metal surface and the environment. Each metal behaves in a unique way with respect to the formation of corrosion products with compositions and properties that often reflect prevailing chemical conditions for a given environment. Hence, in an atmospheric environment with sulfur dioxide (SO_2) as the dominant corrosion stimulator, well-formed corrosion products most likely consist of different metal sulfates or hydroxysulfates or their combinations. Corrosion products rich in metal chlorides and hydroxychlorides will consequently form in chloride-rich environments. Combinations of sulfur- and chlorine-rich corrosion products have been shown to form at specific conditions. Since the barrier and solubility properties of these corrosion products differ, there is a need to understand the relationship between the formation of corrosion products and prevailing environmental and pollutant conditions at different atmospheric environments to, for example, assess and predict the service life or aesthetic appearance of a specific metal object.

In the following four sections, the corrosion product formation sequences of four commonly used construction metals—zinc, copper, carbon steel, and aluminum—will be described in some detail based on findings at outdoor conditions, partly supported by laboratory findings. This compilation will not include alloys, for which corrosion product sequences may be different depending on specific alloying constituents. The behavior of some alloys of these construction metals is though described in Chapter 13. Formation sequences are based on thorough analyses of corrosion

Atmospheric Corrosion, Second Edition. Christofer Leygraf, Inger Odnevall Wallinder, Johan Tidblad and Thomas Graedel.

products and their formation with time based on the use of several complementary analytical techniques, most of which are described in Appendix A. Information on environmental conditions are provided in terms of, for example, pollutant concentrations, particle deposition levels, relative humidity, temperature, etc. Detailed analyses of the formation of corrosion products from the first initial days of exposure up to several years of outdoor exposures in different environments permit a description of the evolution of corrosion products from the very earliest stages to fully developed constituents of the patina.

9.2 EVOLUTION OF CORROSION PRODUCTS ON ZINC

The zinc(II) ion is, according to the Lewis hard and soft acid–base (HSAB) principle (cf. Section 2.3.4), an intermediate Lewis acid from which follows that it is able to coordinate with different Lewis bases. As illustrated in Appendix J, carbonate, chloride, and sulfate are the most important Lewis bases that coordinate with zinc at outdoor atmospheric conditions.

A freshly prepared zinc surface is covered by a thin layer of zinc oxide (ZnO) with a thickness of the order of at least a few nanometer (nm) ($1\ nm = 10^{-9}\ m$). At humidified conditions a nm thick layer of zinc hydroxide ($Zn(OH)_2$) will form instantaneously. Zinc hydroxide exists in different amorphous and crystalline structures depending on the acidity or alkalinity of the aqueous film, which is determined by prevailing exposure conditions. Within a few hours of further exposure in any benign indoor or outdoor atmosphere, the hydroxylated zinc surface is commonly fully or partially transformed into a thin layer predominantly composed of a zinc hydroxycarbonate (hydrozincite, $Zn_5(CO_3)_2(OH)_6$). Since there is usually a time delay of at least several days between the preparation of a zinc surface and initiation of an actual outdoor exposure study, a thin layer of $Zn_5(CO_3)_2(OH)_6$ is normally present on the zinc surface prior to exposure. A similar layer is typically also present on as-received zinc surfaces.

Experimental studies of corrosion products formed on bare zinc metal or galvanized steel exposed to different atmospheric outdoor environments have revealed the existence of a variety of corrosion products. Appendix J lists the most commonly found, many of which also occur as minerals: zinc oxide (ZnO), zinc hydroxide ($Zn(OH)_2$), zinc hydroxycarbonate ($Zn_5(CO_3)_2(OH)_6$), hydrated zinc sulfates ($ZnSO_4 \cdot nH_2O$), zinc hydroxysulfate ($Zn_4SO_4(OH)_6 \cdot nH_2O$), and zinc hydroxychloride ($Zn_5(OH)_8Cl_2 \cdot H_2O$). In addition, two zinc chlorohydroxysulfates have been identified as stable and poorly soluble constituents of the zinc patina, $Zn_4Cl_2(OH)_4SO_4 \cdot 5H_2O$ and $NaZn_4Cl(OH)_6SO_4 \cdot 6H_2O$. In order to explore the possible relationships among these compounds, how they evolve as a function of time, and how their existence is influenced by different environmental characteristics, short- (days, weeks) and long-term (months, years) field exposures of bare zinc sheet were undertaken by Odnevall and Leygraf, Sweden, in rural, urban, industrial, and marine atmospheres. Identification of different corrosion products

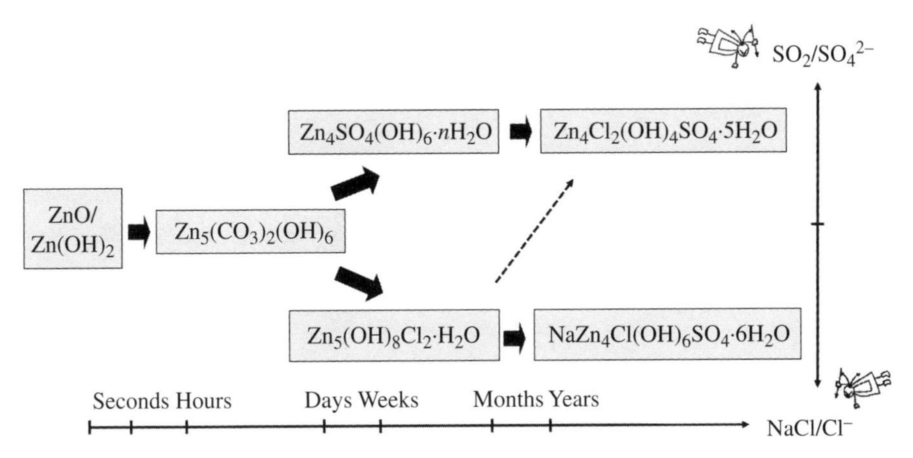

FIGURE 9.1 Schematic sequence for corrosion product formation on zinc sheet (or galvanized steel) at sheltered conditions in ambient atmospheres dominated by sulfur pollution (upper sequence) and chloride pollution (lower sequence). The dashed arrow indicates a possible transformation, shown in simulated field exposures. The time scale indicates the shortest exposure time period for a given corrosion product to be observed. The scheme is also representative for unsheltered conditions although with slower time scales for the formation of sulfate- and chloride-rich corrosion products. (Adopted with permission from I. Odnevall and C. Leygraf, Reaction sequences in atmospheric corrosion of zinc, in *Atmospheric Corrosion,* ASTM STP 1239, eds. W.W. Kirk and Herbert H. Lawson, ASTM, Philadelphia, pp. 215–229, 1995.)

was conducted using a multianalytical approach with quantitative X-ray microanalysis combined with scanning electron microscopy (SEM), infrared spectroscopy, and X-ray powder diffraction as main surface analytical tools (see Appendix A).

The corrosion product sequence on bare zinc sheet is firstly described for rain-sheltered conditions to avoid the influence of precipitation wash-off (runoff). The corrosion behavior of sheltered zinc is strongly influenced by the amount of sulfur (via SO_2 or dry-deposited sulfate aerosol particles) and chloride (e.g., as aerosol particles) that interact with the surface. Figure 9.1 displays schematically the overall reaction sequence for environments in which chloride interactions dominate over SO_2/sulfate interactions (the lower sequence), and vice versa (the upper sequence).

Both sequences include the prompt formation of ZnO and $Zn(OH)_2$ and the subsequent formation of $Zn_5(CO_3)_2(OH)_6$, as discussed previously. Depending on environmental conditions, these corrosion products can evolve and form relatively thick layers of corrosion products. $Zn_5(CO_3)_2(OH)_6$ is an example of a final end corrosion product on zinc in, for example, humid low polluted environments where it gradually evolves over time. The same compound is together with $Zn(OH)_2$, a main constituent of the white corrosion products that locally form and stain zinc sheet during wet storage, often referred to as white rust staining.

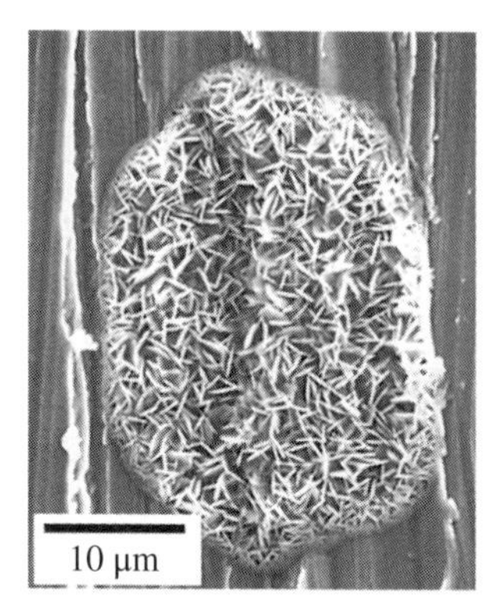

FIGURE 9.2 Scanning secondary electron images of typical corrosion products, locally formed as 20–30 μm sized islands on zinc sheet after 1 day (left) and 30 days (right) at marine sheltered exposure conditions. (Reproduced with permission from Elsevier; I. Odnevall and C. Leygraf, Formation of $NaZn_4Cl(OH)_6SO_4 \cdot 6H_2O$ in a marine atmosphere, *Corrosion Science, 34,* 1213–1229, 1993.)

9.2.1 Chloride-Dominated Sheltered Exposure

In marine and other chloride-dominated environments (e.g., traffic settings), chloride ions can deposit and interact with the zinc surface via dry or wet deposition processes of aerosol particles containing, for example, NaCl, $MgCl_2$, and $CaCl_2$ (cf. Section 12.3.3). The deposition of chloride ions is nonuniform and promotes thus spatially separated corrosion processes. Short- (days, weeks) and long-term (months, years) field exposure findings at a marine site in Sweden (Bohus-Malmön) will be used in the following to illustrate the evolution of corrosion products on zinc sheet at sheltered conditions. Small locally occurring islands of corrosion products were already observed within 1 day of exposure on the zinc surface that was predominantly covered by a thin layer of $Zn_5(CO_3)_2(OH)_6$ (hydrozincite). The islands consisted of thin platelets stacked together and had typical average sizes of 20–30 μm, Figure 9.2. Quantified X-ray microanalyses revealed a composition equal to $Zn_5(OH)_8Cl_2{\cdot}H_2O$ (simonkolleite). Local formation of these platelet islands increased with time and covered large surface areas after 30 days of exposure. An almost complete surface coverage was evident after 90 days of exposure.

The rapid formation of $Zn_5(OH)_8Cl_2{\cdot}H_2O$ at this early stage of marine exposure can be explained via detailed laboratory investigations that are able to reproduce the formation of the very same corrosion product. Locally deposited chlorides have at laboratory conditions been shown to accelerate the dissolution of zinc from the oxidized zinc surface, a process that is still poorly understood on a molecular level. The dissolution of zinc into Zn^{2+} ions is strongly localized and takes place at anodic sites of the zinc surface, whereas oxygen (O_2) reduction occurs at cathodic sites to produce hydroxyl ions (OH^-). To sustain charge neutrality positive ions (e.g., Zn^{2+}, Na^+) will migrate toward cathodic regions and negative ions (e.g., Cl^-, OH^-) toward anodic regions. The net result is the formation of $Zn_5(OH)_8Cl_2{\cdot}H_2O$ in the center of

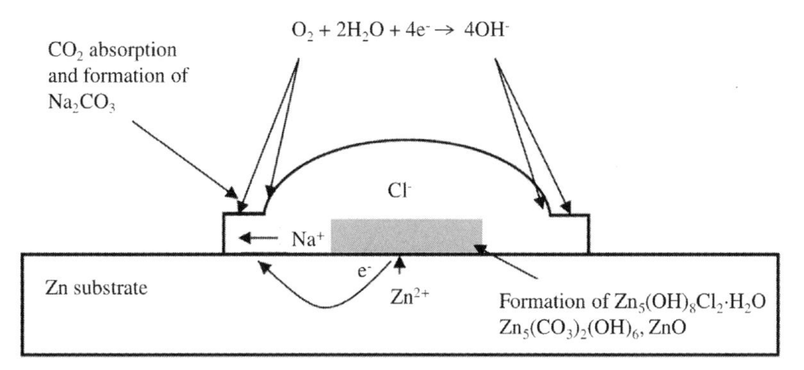

FIGURE 9.3 Schematic illustration of corrosion products locally formed on zinc at an early stage of chloride-dominated exposure conditions. (Adapted with permission from Z.Y. Chen, D. Persson, and C. Leygraf, Initial NaCl-particle induced atmospheric corrosion of zinc—effect of CO_2 and SO_2, *Corrosion Science, 50*(1), 111–123, 2008.)

the anodic region and growth and further formation of $Zn_5(CO_3)_2(OH)_6$, Figure 9.3, and possibly also of ZnO in cathodic regions. The presence of amorphous and crystalline ZnO adjacent to areas of $Zn_5(OH)_8Cl_2 \cdot H_2O$ has been shown both at laboratory and field conditions. As indicated by stability diagram calculations, $Zn_5(OH)_8Cl_2 \cdot H_2O$ is stable at an intermediate pH and relatively high chloride activity levels, whereas ZnO is stable at more alkaline pH and lower activity levels of chloride.

Exposure at the marine test site further resulted in the enrichment of sulfur locally within the $Zn_5(OH)_8Cl_2 \cdot H_2O$-dominated platelet-structured corrosion product layer. The chemical composition of these areas was after 30 days similar to sodium zinc chlorohydroxysulfate ($NaZn_4Cl(OH)_6SO_4 \cdot 6H_2O$), a compound that at the time for the investigation was novel within the discipline of atmospheric corrosion and that lacked a crystallographic structure. Its crystallographic structure was determined and later on given an official name, gordaite, and included in crystallographic databases by the International Mineralogical Association. The existence of both $Zn_5(OH)_8Cl_2 \cdot H_2O$ and $NaZn_4Cl(OH)_6SO_4 \cdot 6H_2O$ was confirmed by means of X-ray powder diffraction after longer exposure periods, and it became evident that the former gradually was transformed with time into the latter compound. $NaZn_4Cl(OH)_6SO_4 \cdot 6H_2O$ appeared as the predominating corrosion product after long-term (years) sheltered exposure conditions at this marine site.

How could this gradual transformation from one phase to another take place? Quantitative X-ray microanalyses of the chemical composition of a large number of randomly selected islands of corrosion products formed after different short time periods (days, weeks, months) of exposure revealed a gradual change in their chemical composition. This is illustrated in Figure 9.4 with a ternary diagram displaying the relative amounts of sodium, chlorine, and sulfur of individual islands after 2, 5, and 30 days of exposure. A point positioned in the chlorine corner represents an island with an atomic fraction of chlorine/sulfur/sodium equal to 1:0:0. The chlorine corner is equivalent to the composition of $Zn_5(OH)_8Cl_2 \cdot H_2O$, whereas the filled circle in the middle of the diagram represents a molar fraction equivalent to

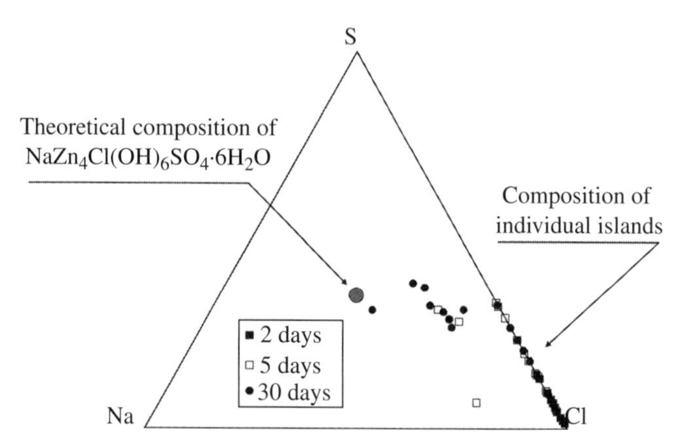

FIGURE 9.4 Relative atomic amount of sulfur, chlorine, and sodium of platelet-forming islands on zinc sheet after 2 (filled squares), 5 (unfilled squares), and 30 (filled circles) days of sheltered marine exposure. (Adapted with permission from I. Odnevall and C. Leygraf, Formation of $NaZn_4Cl(OH)_6SO_4 \cdot 6H_2O$ in a marine atmosphere, *Corrosion Science, 34,* 1213–1229, 1993.)

$NaZn_4Cl(OH)_6SO_4{\cdot}6H_2O$. The compositional range after 2 days of exposure suggests that the islands were mainly composed of $Zn_5(OH)_8Cl_2{\cdot}H_2O$ with some addition of sulfur. After 5 days the proportion of sulfur had increased and a few islands also contained sodium. The theoretical composition of $NaZn_4Cl(OH)_6SO_4{\cdot}6H_2O$ was for some islands approached after 30 days of exposure. The results demonstrate a gradual incorporation of first sulfur (as sulfate) into $Zn_5(OH)_8Cl_2{\cdot}H_2O$ and later of sodium. $NaZn_4Cl(OH)_6SO_4{\cdot}6H_2O$ continued to successively form and its existence was after 90 days unambiguously confirmed by means of X-ray powder diffraction.

Corrosion reactions have for decades been described by stoichiometric relationships. Hence, taking $NaZn_4Cl(OH)_6SO_4{\cdot}6H_2O$ as an example, the reaction could be written as

$$3ZnO(s) + Zn^{2}(aq) + SO_4^{2-}(aq) + Na^{+}(aq) + Cl^{-}(aq) + 9H_2O \rightarrow NaZn_4Cl(OH)_6SO_4{\cdot}6H_2O(s) \quad (9.1)$$

The utility of these relationships is that they only identify reactants and products and they do not specify the process that requires the determination of the many steps in the formation of a product or the sequential transformation of one compound, for example, $Zn_5(OH)_8Cl_2{\cdot}H_2O$, into another, for example, $NaZn_4Cl(OH)_6SO_4{\cdot}6H_2O$. There are, in theory, two possible routes to such a transformation, one in which the metal atoms are dissolved from the surface, react in the aqueous surface layer, and precipitate into the solid compound and another in which the metal atoms remain part of the solid phase but react with ions of the aqueous layer with which the surface comes into contact. As a general rule, the more chemically complex the final product, the more likely it is that most or all of the reaction steps leading to it occur at the solid–solution interface rather than in the aqueous phase.

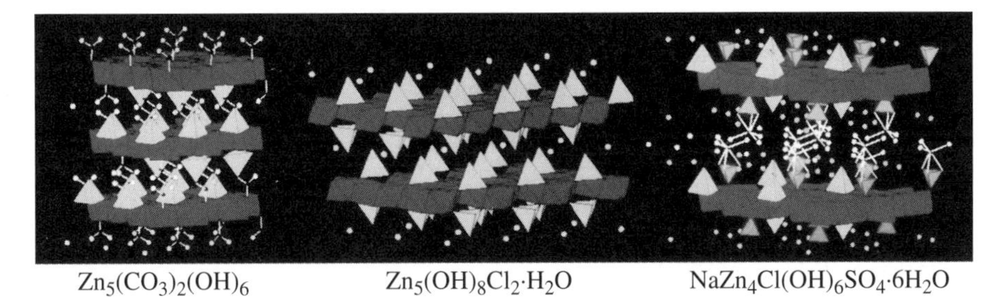

FIGURE 9.5 Crystallographic view of $Zn_5(CO_3)_2(OH)_6$, $Zn_5(OH)_8Cl_2{\cdot}H_2O$, and $NaZn_4Cl(OH)_6SO_4{\cdot}6H_2O$, all with layered-type sheet structures.

The similarity in crystallographic configuration of $Zn_5(OH)_8Cl_2{\cdot}H_2O$ and $NaZn_4Cl(OH)_6SO_4{\cdot}6H_2O$ is presented in Figure 9.5. Both compounds have a sheet structure with tetrahedrally and octahedrally coordinated zinc ions (Zn^{2+}). Each Zn^{2+} ion shares most of its bonding electrons with oxygen (O^{2-}) ions. In $Zn_5(OH)_8Cl_2{\cdot}H_2O$ the chlorine (Cl^-) ions are coordinated in the top corner of each tetrahedron, and the electrically neutral sheets are held together by weak O—H···Cl bonds. Similar coordination is valid in $NaZn_4Cl(OH)_6SO_4{\cdot}6H_2O$ that also have sulfate (SO_4^{2-}) groups connected to the octahedral sheet by corner sharing via one oxygen ion. Between the layers are sodium (Na^+) ions coordinated to three sulfate oxygen and three crystal water molecules. The structural resemblance between the phases is evident also for $Zn_5(CO_3)_2(OH)_6$ that reveals a similar layered structure, Figure 9.5 (left).

Formation of $NaZn_4Cl(OH)_6SO_4{\cdot}6H_2O$ from $Zn_5(OH)_8Cl_2{\cdot}H_2O$ requires that some bonds must be broken to enable the incorporation of sulfate groups and sodium ions into the structure and rearrange into the new crystal structure. This is possible at weak bonds such as O—H···Cl. As evident from the gradual change in composition, Figure 9.4, the replacement and rearrangement do not occur in a simple sequence.

Overall, the formation of corrosion products in chloride-dominated marine environments can be summarized via the following sequence:

$$\begin{aligned} &Zn(OH)_2 \text{ (within seconds)} \rightarrow Zn_5(CO_3)_2(OH)_6 \text{ (hours)} \rightarrow \\ &\quad Zn_5(OH)_8Cl_2{\cdot}H_2O \text{ (days)} \rightarrow NaZn_4Cl(OH)_6SO_4{\cdot}6H_2O \text{ (weeks)}. \end{aligned} \tag{9.2}$$

The sequence consists of a number of principal steps:

1. Hydroxylation of the zinc surface (very rapidly)
2. Formation of $Zn(OH)_2$ through ion pairing of Zn^{2+} ions and OH^- ions, the ions being produced by anodic and cathodic reactions, respectively (very rapidly)
3. Replacement of some of the OH^- ligands by carbonate (CO_3^{2-}) ligands, CO_3^{2-} originating from CO_2 dissolved from the air (fairly rapidly)
4. Rearrangement of the solid surface into a stable $Zn_5(CO_3)_2(OH)_6$ crystal structure, having a sheet structure similar to that of the phases to be formed later (relatively slow)

5. Replacement of CO_3^{2-} and OH^- ions at the $Zn_5(CO_3)_2(OH)_6$ surface/solution interface for Cl^- ions and OH^- ions, the Cl^- ions originating from sea-salt aerosols and OH^- ions probably from the oxygen reduction reaction (fairly rapidly)
6. Rearrangement of the solid surface into crystalline $Zn_5(OH)_8Cl_2{\cdot}H_2O$ with its characteristic sheet structure (relatively slow)
7. Partial replacement of Cl^- ions and OH^- ions at the $Zn_5(OH)_8Cl_2{\cdot}H_2O$ surface/solution interface with SO_4^{2-}, Na^+, and OH^- ions, the latter probably originating from the O_2 reduction reaction (relatively rapid)
8. Rearrangement of the solid surface into crystalline $NaZn_4Cl(OH)_6SO_4{\cdot}6H_2O$ with its characteristic sheetlike structure (slow)

These steps are promoted by repeated daily dry and wet cycles, described in Sections 8.3.1 and 8.3.2, and involve dissolution into the aqueous surface layer (in particular during periods of high acidification, such as in dew or during drying out), subsequent aqueous ion pairing, and precipitation of solid phases.

Findings from other more recent field exposures in chloride-rich environments (e.g., marine and traffic settings) confirm the proposed formation sequence of corrosion products on zinc.

Descriptions as detailed as this are not well developed for many corrosion reactions, and some of the sequences are much more complex than this example. Nevertheless, the goal in this book is to attempt to the degree possible to discuss corrosion processes from a *sequential* rather than a *stoichiometric* perspective. A complete scientific description of such corrosion sequences thus requires the following main steps:

- Identification of initial reactants and final products
- Identification of intermediate products
- Determination of the rates of transformation for each step of the sequence: reactants → intermediate products (several steps) → final products

9.2.2 Sulfate-Dominated Sheltered Exposure

Atmospheric conditions dominated by sulfur-containing air constituents are typical for most rural and urban locations and also for many industrial sites. The sequence of corrosion product formation at these settings is in the following primarily illustrated with field data from a rural site, Aspvreten, south of Stockholm, Sweden, as it clearly elucidates the influence of sulfate on the evolution of corrosion products on zinc and represents a scenario (though with less rapid processes) also for many urban and industrial conditions. Aerosol particles of $(NH_4)_2SO_4$ and relatively low gaseous levels of SO_2 and O_3 are the most important air constituents at the rural site, while chlorides are less important. The site is further characterized by high humidity levels, especially during morning hours. As previously discussed, a thin layer of ZnO, $Zn(OH)_2$, and $Zn_5(CO_3)_2(OH)_6$ was present at the surface already at exposure initiation.

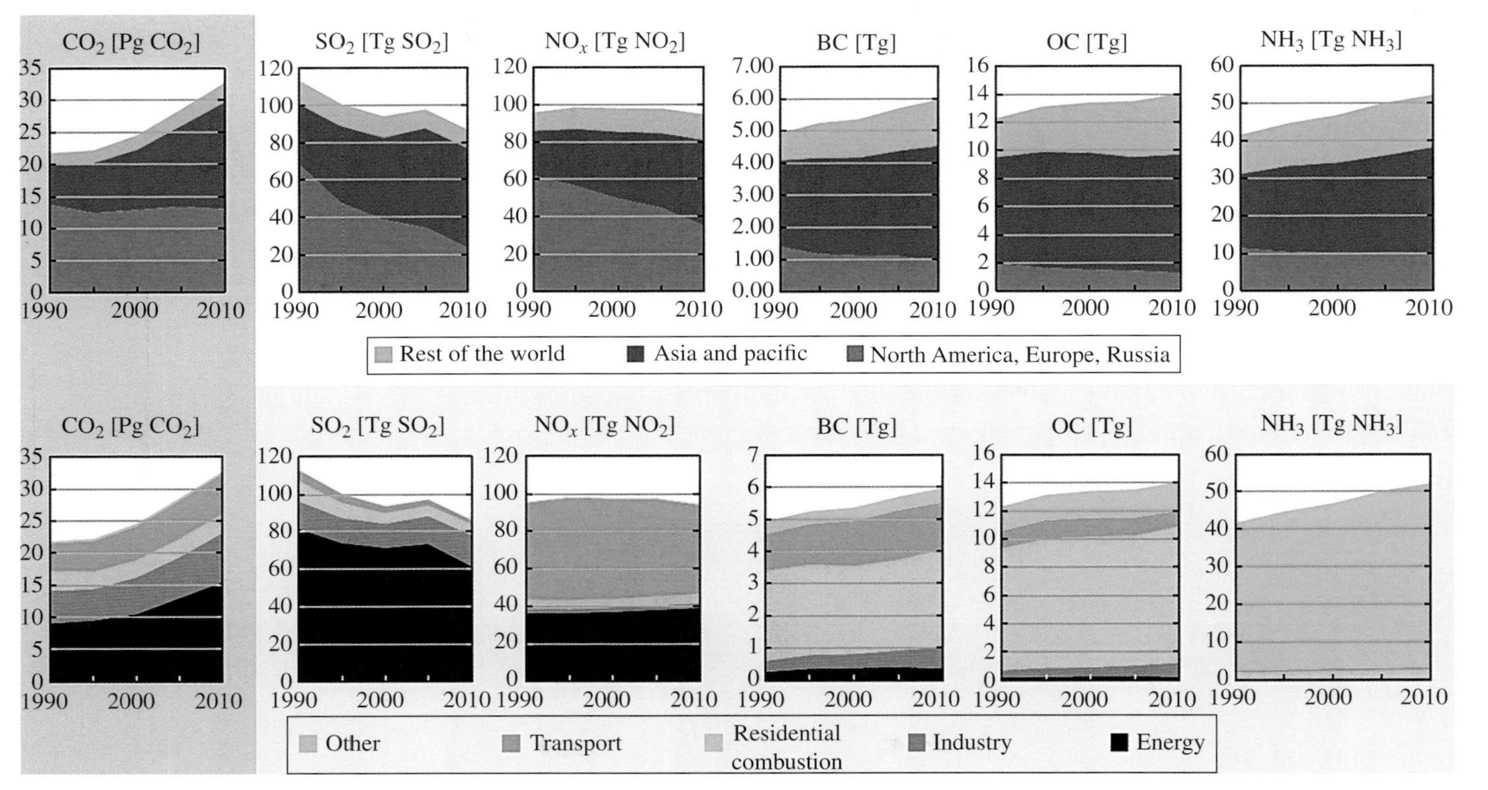

FIGURE 4.1 The evolution of anthropogenic emissions of CO_2 and key air pollutants, 1990–2010, by world regions (top) and source sectors (bottom), excluding international shipping and aviation. BC, black carbon; OC, organic carbon; Pg, petagram; Tg, teragram. (Reproduced with permission from Annual Reviews; M. Amann, Z. Klimont, and F. Wagner, Regional and global emissions of air pollutants: recent trends and future scenarios, *Annual Review of Environment and Resources, 38*, 31–55, 2013.)

Atmospheric Corrosion, Second Edition. Christofer Leygraf, Inger Odnevall Wallinder, Johan Tidblad and Thomas Graedel.

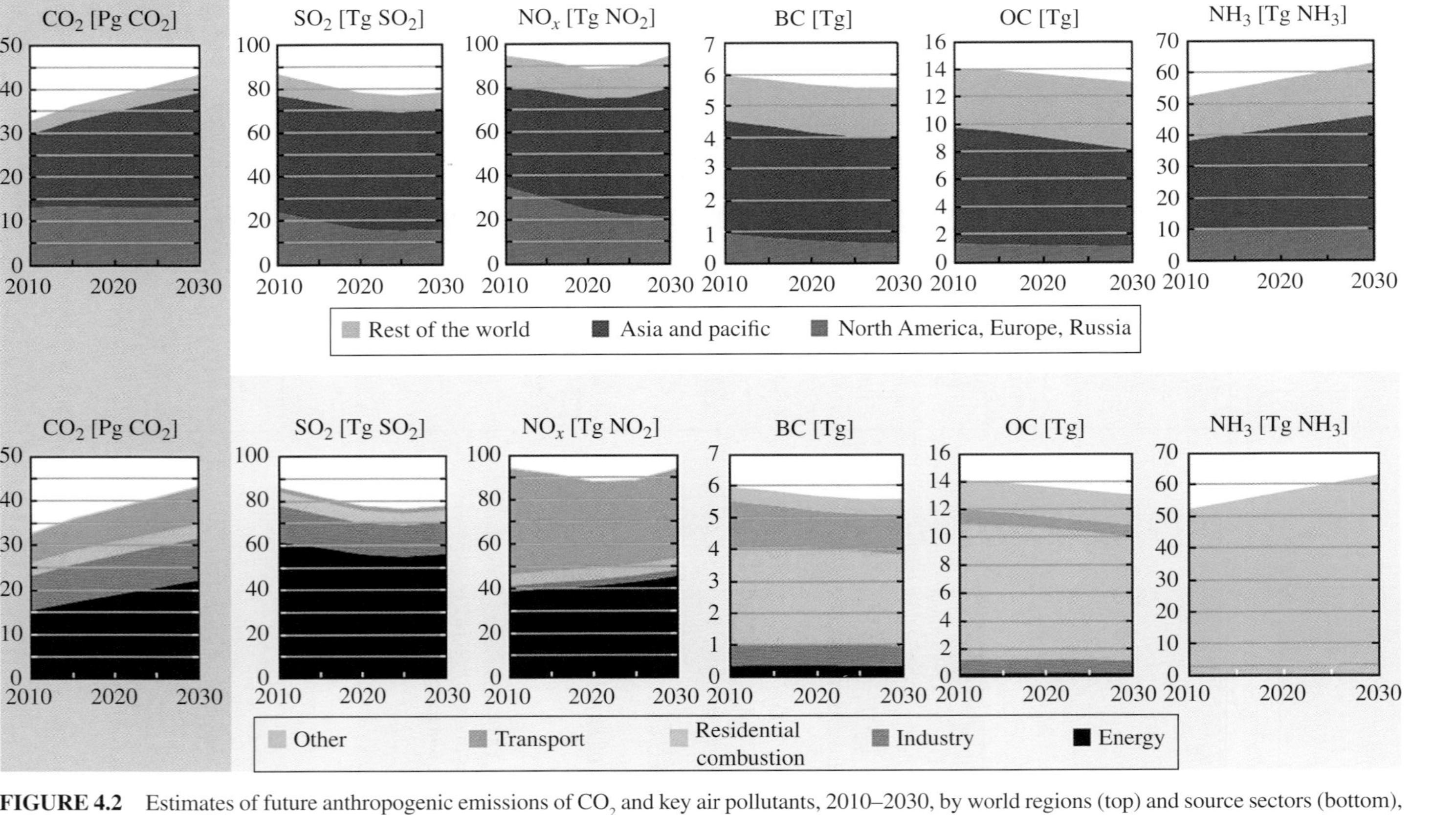

FIGURE 4.2 Estimates of future anthropogenic emissions of CO_2 and key air pollutants, 2010–2030, by world regions (top) and source sectors (bottom), excluding international shipping and aviation. BC, black carbon; OC, organic carbon; Pg, petagram; Tg, teragram. (Reproduced with permission from Annual Reviews; M. Amann, Z. Klimont, and F. Wagner, Regional and global emissions of air pollutants: recent trends and future scenarios, *Annual Review of Environment and Resources, 38*, 31–55, 2013.)

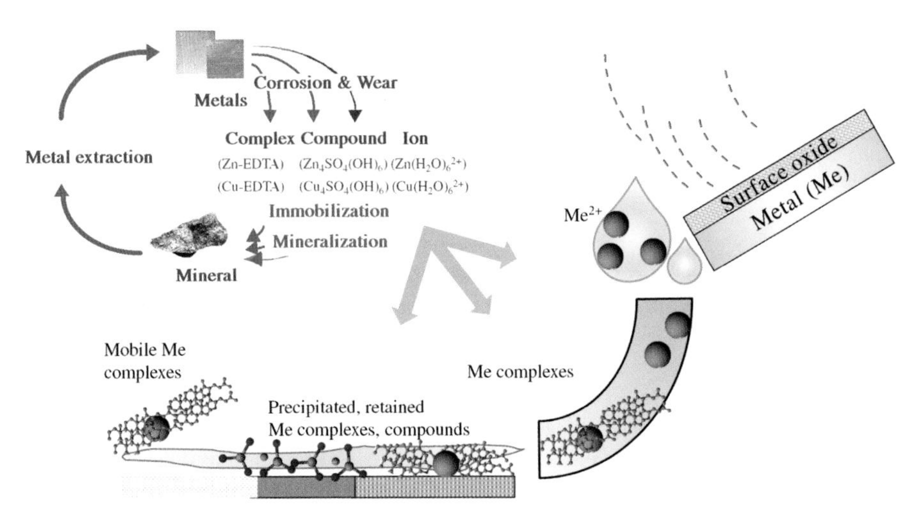

FIGURE 10.6 Illustration of the life cycle of and transition of extracted metals and illustration of changes in metal speciation, retention, and mobility of metals released from outdoor buildings in contact with metal retaining surfaces in the near vicinity of a building.

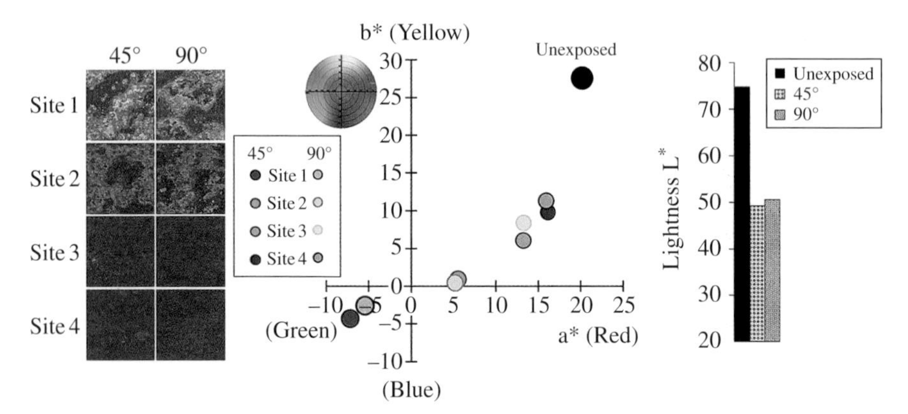

FIGURE 13.5 Spectrophotometric differences in visual appearance illustrated for copper sheet exposed at 45° and 90° from the horizontal at unsheltered conditions for 6 months at four different marine locations of increasing distance from the seashore (reduced deposition rates of chlorides from site 1 to 4; see also Figure 13.1c). Each color is represented by a point in the color space that is identified by three coordinates, a*, b*, and L*, with a* representing a color ranging from green to red, b* representing yellow to blue, and L* representing the brightness ranging from black to white. (I. Odnevall Wallinder, G. Herting, and S. Goidanich, unpublished data.)

During the first 30 days of exposure, the surface became successively covered by more $Zn_5(CO_3)_2(OH)_6$ forming on the surface. Platelet-forming islands rich in sulfur evolved in parallel and gradually increased in surface coverage. Surface analysis by X-ray photoelectron spectroscopy showed that significant amounts of $(NH_4)_2SO_4$ accumulated on the surface already during the first day of exposure and the amount of deposited sulfate increased on the surface after 3 and 5 days while the corresponding amount of ammonium (NH_4^+) decreased. This suggests that sulfate ions from $(NH_4)_2SO_4$ were interacting with the zinc surface. After one year of sheltered exposure, the platelet-forming corrosion products were more abundant than $Zn_5(CO_3)_2(OH)_6$ and unambiguously identified by X-ray powder diffraction as a zinc hydroxysulfate, $Zn_4SO_4(OH)_6{\cdot}4H_2O$. Zinc hydroxysulfates with different water content ($3 \leq n \leq 6$) have been observed at outdoor conditions. This formation sequence has been confirmed in other field exposures as well as at urban and industrial sites.

Lobnig and coworkers at Bell Laboratories in the United States studied the behavior of zinc in laboratory experiments involving humidity and aerosol particles containing $(NH_4)_2SO_4$. The end product in that study was a zinc hydroxysulfate, $Zn_4SO_4(OH)_6{\cdot}5H_2O$, very similar to the phase observed in the current field study. The processes discovered in the laboratory appear also to be valid at field conditions. Thus, the formation of corrosion products on zinc in the sulfate-dominated rural environment can be summarized by the following sequence:

$$Zn(OH)_2\,(\text{within seconds}) \rightarrow Zn_5(CO_3)_2(OH)_6\,(\text{hours}) \rightarrow \text{zinc hydroxysulfate } (Zn_4SO_4(OH)_6{\cdot}nH_2O)(\text{days}) \qquad (9.3)$$

The zinc corrosion product sequence observed in most rural, urban, and industrial environments can be divided into a number of principal steps. The first 4 steps, prior to outdoor exposure, are the same as in the chloride-dominated exposure and are followed by:

1. Replacement of CO_3^{2-} and OH^- ions at the $Zn_5(CO_3)_2(OH)_6$–solution interface with SO_4^{2-} ions and OH^- ions, the SO_4^{2-} ions originating mainly from aerosols and the OH^- ions probably from the O_2 reduction reaction (fairly rapid)
2. Rearrangement of the solid surface into crystalline $Zn_4SO_4(OH)_6{\cdot}nH_2O$ (relatively slow)

All steps (1–6) are promoted by repeated dry and wet exposure conditions.

Zinc hydroxysulfate ($Zn_4SO_4(OH)_6{\cdot}nH_2O$) is commonly observed on zinc surfaces after several decades of sheltered exposure and identified as an end product in many rural, urban, or industrial environments. However, if chloride deposition also is significant, but not dominant, $Zn_4SO_4(OH)_6{\cdot}nH_2O$ can incorporate chloride ions and eventually form zinc chlorohydroxysulfate ($Zn_4Cl_2(OH)_4SO_4{\cdot}5H_2O$), detected within days and weeks of sheltered urban and industrial exposure, Figure 9.1. This compound can be the end product in more aggressive industrial atmospheres characterized by high deposition rates of both SO_2/sulfate and chloride.

Corrosion products that predominantly form and evolve in chloride-dominated environments and at sulfur-dominated conditions on zinc include $Zn_5(CO_3)_2(OH)_6$, $Zn_5(OH)_8Cl_2·H_2O$, $NaZn_4Cl(OH)_6SO_4·6H_2O$, and $Zn_4SO_4(OH)_6·nH_2O$. All corrosion products bear structural resemblance in that they all have a layered-type structure with sheets consisting of both octahedrally and tetrahedrally coordinated zinc atoms. The main difference is the way in which the sheets are held together. In $Zn_5(CO_3)_2(OH)_6$ the sheets are held together by carbonate ions in a three-dimensional network, in $Zn_5(OH)_8Cl_2·H_2O$ the sheets are electrically neutral and held together by weak O—H···Cl bonds, and in $NaZn_4Cl(OH)_6SO_4·6H_2O$ sulfate groups and sodium contribute to the bonding between the sheets, Figure 9.5. Similar sheets exist also in $Zn_4SO_4(OH)_6·nH_2O$ bonded via sulfate groups connected on either side of the sheets through corner sharing.

The structural resemblance among the zinc-containing corrosion products facilitates the transformation from one compound to another through ion exchange or the growth of one compound from a preceding compound. This is an important reason why crystalline corrosion products are observed already after only a few days of outdoor exposure.

The structural resemblance between the phases also permits a gradual transformation from the chloride sequence to the sulfate sequence, as indicated in Figure 9.1. A prerequisite is that the environmental conditions change from an initial chloride-dominant deposition to subsequent sulfate-dominant deposition. In this case $Zn_5(OH)_8Cl_2·H_2O$ forms initially followed by $Zn_4Cl_2(OH)_4SO_4·5H_2O$. The opposite situation, namely, the initial formation of $Zn_4SO_4(OH)_6·nH_2O$ (in the SO_2/sulfate sequence) followed by the formation of $NaZn_4Cl(OH)_6SO_4·6H_2O$ (in the chloride sequence), has not so far been observed. However, the structural resemblance of the phases suggests that if the environmental conditions are favorable, such a transformation would be possible.

9.2.3 Unsheltered Exposure

Similar corrosion products as formed on sheltered zinc at outdoor atmospheric conditions have been observed on unsheltered zinc surfaces although their rate of formation was different due to the wash-off effect of deposited pollutants by means of the action of, for example, precipitation. On unsheltered zinc the formation of both $Zn_5(OH)_8Cl_2·H_2O$ and $NaZn_4Cl(OH)_6SO_4·6H_2O$ was delayed in the chloride-dominated environment, as was the formation of both $Zn_4SO_4(OH)_6·4H_2O$ and $Zn_4Cl_2(OH)_4SO_4·5H_2O$ in the SO_2/sulfate-dominated rural, urban, and industrial sites. The wash-off effect of precipitation hinders surface retention of chlorides and various sulfur-containing air constituents by dissolving and removing them from the surface at an early stage. This delays the interaction and/or nucleation of chloride and/or sulfate-containing corrosion products. Periods without wash-off events, though with daily cyclic temperature and humidity conditions and deposition of chlorides and sulfates, enable further patina evolution as described by the general corrosion product sequence.

9.3 EVOLUTION OF CORROSION PRODUCTS ON COPPER

Similar to the zinc(II) ion, the copper(II) ion is an intermediate Lewis acid according to the HSAB principle (cf. Section 2.3.4) and is hence able to coordinate with both soft and hard Lewis bases. However, as evident from Appendix E, sulfate and chloride are the most important Lewis bases that coordinate with copper at outdoor atmospheric conditions.

Copper(I) oxide, cuprite (Cu_2O), is the initial corrosion product that forms on copper at atmospheric conditions. Some studies report on the initial rapid formation of copper(II) oxide, tenorite (CuO). However, this compound is known to react very rapidly at ambient conditions with gaseous constituents, in particular SO_2, and transform into more stable compounds. Numerous studies of copper patina formation at atmospheric conditions report on the formation and growth of Cu_2O from an initial nanometer layer thickness to heterogeneous layers of locally varying thicknesses up to several tens of micrometers after years or decades of exposure. Its growth rate, thickness, and physical barrier properties depend on prevailing environmental conditions. Local or complete surface coverage and growth of Cu_2O rapidly change the copper surface appearance from an initially metallic luster into a dull brown or even blackish appearance already after a few days, weeks, or months, depending on exposure conditions.

In environments rich in SO_2/sulfates, chlorides, or their combinations, the patina gradually acquires a greenish or turquoise appearance related to the formation of sulfur- and/or chloride-rich corrosion products. These corrosion products form locally and grow atop the Cu_2O layer at a rate and with a chemical composition that depend on prevailing pollutant levels. Predominating corrosion products within this surface-adhering patina are relatively stable and of poor solubility. From this follows a patina that acts as a protective barrier of the copper object for long time periods (up to several decades) at many exposure conditions.

Appendix E contains a summary of corrosion products identified within the patina on copper at different atmospheric conditions. Besides Cu_2O (cuprite) are $Cu_4SO_4(OH)_6 \cdot H_2O$ (posnjakite), $Cu_4SO_4(OH)_6$ (brochantite), and $Cu_3SO_4(OH)_4$ (antlerite) predominantly identified in SO_2/sulfate-dominated environments, whereas $Cu_2Cl(OH)_3$ (atacamite) and/or its isomorphous compound paratacamite are common patina constituents on copper in chloride-dominated environments. Recent findings have also identified CuCl (nantokite) within the patina at marine conditions with high chloride deposition rates. Another copper hydroxysulfate, $Cu_{2.5}SO_4(OH)_3 \cdot 2H_2O$, has been reported to form during initial stages of copper patina formation. Although not an official crystallographic name, this compound is for convenience denoted strandbergite, after its discoverer.

9.3.1 Sulfate-Dominated Sheltered Exposure

A multianalytical short- and long-term field exposure investigation, performed by Odnevall and Leygraf in the 1990s of sheltered copper in a rural and an urban environment, revealed that the initial formation of corrosion products includes at least

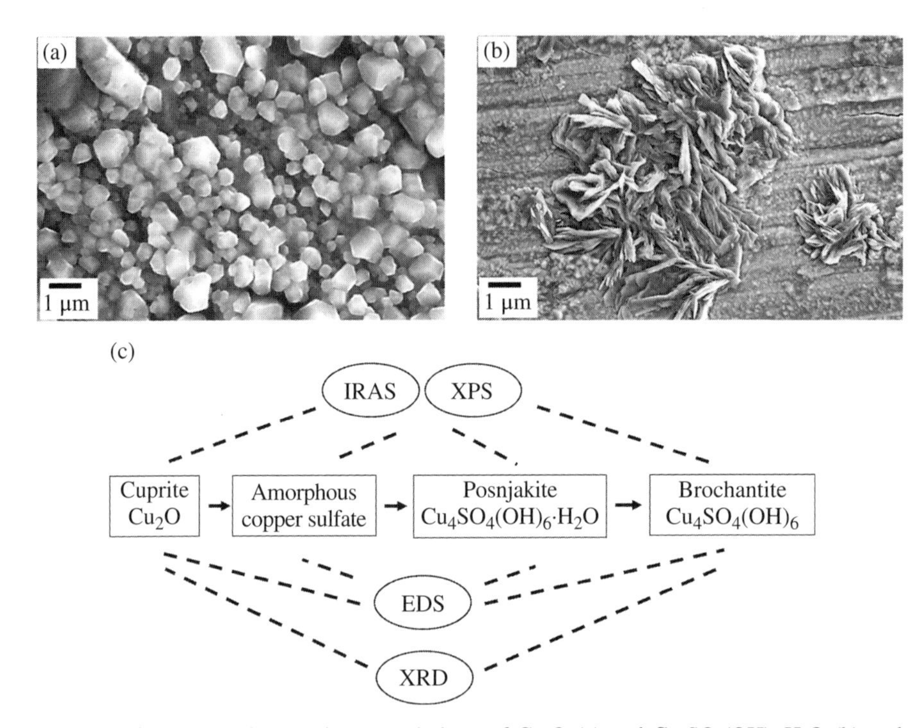

FIGURE 9.6 Corrosion product morphology of Cu_2O (a) and $Cu_4SO_4(OH)_6{\cdot}H_2O$ (b) and their formation sequence (c) on sheltered copper in a rural atmosphere together with the analytical techniques used for their identification. (Reproduced with permission from The Electrochemical Society; I. Odnevall and C. Leygraf, Atmospheric corrosion of copper in a rural atmosphere, *Journal of the Electrochemical Society, 142*(11), 3682–3689, 1995.)

four different stages. Cuprite (Cu_2O) was formed on the surface already within the first few hours of exposure and continued to grow throughout the entire 2-year period, gradually changing the surface appearance, as discussed previously, Figure 9.6a. The initial rate of formation was seasonal dependent and primarily related to prevailing humidity levels during the first month of exposure. These seasonal effects were however not influencing the long-term behavior.

A noncrystalline sulfate-rich corrosion product was locally formed during the first few days and weeks of exposure atop the Cu_2O layer that within a few months evolved into a crystalline basic copper sulfate, $Cu_4SO_4(OH)_6{\cdot}H_2O$ (posnjakite). This corrosion product revealed a sheetlike morphology, illustrated in Figure 9.6b, and was identified as a precursor for the formation of $Cu_4SO_4(OH)_6$ (brochantite), observed at the outer surface of the patina after 1 year of exposure. Figure 9.6c summarizes the sequence of the first three corrosion products formed on copper at the rural site, as well as the analytical techniques employed that permitted the detection of the different compounds. This reaction sequence has been shown to be representative for rural sites dominated by sulfate deposition and with no significant influence of chloride deposition and also for low polluted SO_2/sulfate-dominated urban sites.

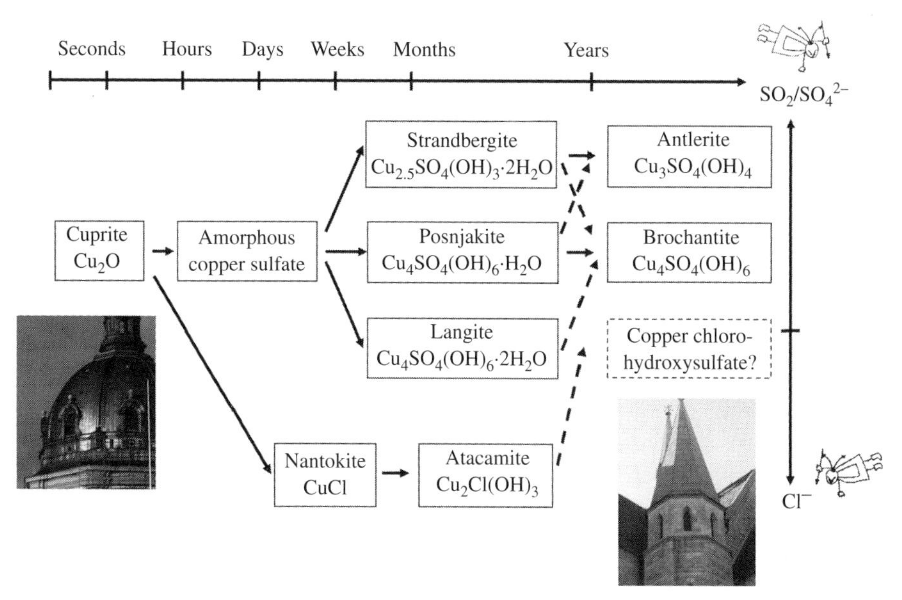

FIGURE 9.7 Evolution scheme of corrosion products on copper at sheltered conditions in atmospheres of varying extent of SO_2/sulfate surface interaction (top sequences) and chloride deposition (bottom sequence). The central sequence is representative for many rural and benign urban environments and the lower sequence for marine exposure conditions. The time scale indicates approximately the shortest time period after which a certain corrosion product has been observed. The scheme is also representative for unsheltered conditions although the formation time scales are different (typically slower for formation of sulfate- and chloride-rich corrosion products). (Adapted with permission from A. Krätschmer, I. Odnevall Wallinder, and C. Leygraf, The evolution of outdoor copper patina, *Corrosion Science, 44*(3), 425–450, 2002.)

Under a stronger influence of sulfur- or chlorine-containing atmospheric species, the reaction sequence becomes more complex and involves also other patina constituents.

Figure 9.7 summarizes the current knowledge of the evolution of different corrosion products formed on copper at sheltered atmospheric exposure conditions in SO_2/sulfate-dominated environments and in chloride-dominated conditions and their combinations. Presented sequences are based on observations from a large number of short- and long-term outdoor investigations and, in particular, on quantitative X-ray diffraction (XRD) analysis developed for copper patina evolved at rural and urban sites by Krätschmer and Stöckle, Germany. Their procedure has enabled the determination of absolute amounts of different crystalline corrosion products within the patina.

The central sequence of Figure 9.7 is the same as shown in Figure 9.6, with the gradual formation of Cu_2O, amorphous copper sulfate, $Cu_4SO_4(OH)_6 \cdot H_2O$, and $Cu_4SO_4(OH)_6$. In more acidified environments, characterized by higher levels of SO_2 and sulfate-containing aerosols, an alternative reaction route proceeds through $Cu_{2.5}SO_4(OH)_3 \cdot 2H_2O$ (strandbergite), which seems to be a precursor to $Cu_3SO_4(OH)_4$ (antlerite). These latter corrosion products are frequently observed within the copper patina in highly polluted environments, $Cu_3SO_4(OH)_4$ sometimes also in rural environments. $Cu_4SO_4(OH)_6{\cdot}2H_2O$ (langite) is another precursor that occasionally is observed together with $Cu_4SO_4(OH)_6{\cdot}H_2O$, especially in less polluted sites characterized by somewhat more humid and wet conditions.

The transition from cuprite (Cu_2O) to crystalline copper hydroxysulfate is thought to proceed via an amorphous copper sulfate, Figures 9.6 and 9.7 as a consequence of the marked difference in crystal structure between Cu_2O, with its high-symmetry cubic structure, and the copper hydroxysulfates, with their characteristic sheet structure (see following text). However, the amorphous copper sulfate compound is difficult to unambiguously identify due to its noncrystallinity and occurrence during early stages of patina formation.

Figure 9.8 shows examples of quantitative results that support the reaction sequence of Figure 9.7. Figure 9.8a displays the accumulated mass of different compounds formed during sheltered conditions at a rural site (Wells Cathedral, United Kingdom) as part of the UN/ECE exposure program. The results are in good agreement with the sequence

$$\begin{aligned}&Cu_2O(\text{cuprite}) \rightarrow Cu_4SO_4(OH)_6{\cdot}H_2O(\text{posnjakite}) \rightarrow \\ &\quad Cu_4SO_4(OH)_6(\text{brochantite})\end{aligned} \tag{9.4}$$

The fact that $Cu_4SO_4(OH)_6{\cdot}H_2O$ is a precursor to $Cu_4SO_4(OH)_6$ can be seen from the time dependence where the accumulated amount of $Cu_4SO_4(OH)_6{\cdot}H_2O$ reaches its maximum during the early years, whereas $Cu_4SO_4(OH)_6$ continues to increase in mass. Cu_2O is seen to grow throughout the whole exposure period.

Figure 9.8b shows the corresponding results from a highly polluted urban site (Prague, Czech Republic). It provides evidence for the sequence

$$\begin{aligned}&Cu_2O(\text{cuprite}) \rightarrow Cu_{2.5}SO_4(OH)_3{\cdot}2H_2O(\text{strandbergite}) \rightarrow \\ &\quad Cu_3SO_4(OH)_4(\text{antlerite})\end{aligned} \tag{9.5}$$

$Cu_{2.5}SO_4(OH)_3{\cdot}2H_2O$ seems to be a precursor to $Cu_3SO_4(OH)_4$, as judged from its time dependence with a maximum after a few years. Small amounts of $Cu_4SO_4(OH)_6{\cdot}H_2O$ and $Cu_4SO_4(OH)_6$ are also present in the patina. This suggests that the $Cu_2O \rightarrow Cu_4SO_4(OH)_6{\cdot}H_2O \rightarrow Cu_4SO_4(OH)_6$ sequence may occur simultaneously at other surface locations. Indeed, the copper patina often exhibits a heterogeneous appearance, with a patina that laterally diverges in composition.

Thus, a diversity of copper hydroxysulfates can form in the thin aqueous adlayer in contact with Cu_2O. The relative concentrations of cupric, sulfate, and hydroxyl ions will determine the copper hydroxysulfate most likely to precipitate. However, the rate of formation of phases is also influenced by kinetic constraints such as the oxidation rate of cuprous to cupric ions. It is interesting to note that the copper hydroxysulfates with the higher ratio between sulfate ions and hydroxyl ions ($Cu_{2.5}SO_4(OH)_3{\cdot}2H_2O$ with a SO_4^{2-}/OH^- ratio of 1:3 and $Cu_3SO_4(OH)_4$ with a SO_4^{2-}/OH^- ratio of 1:4) are

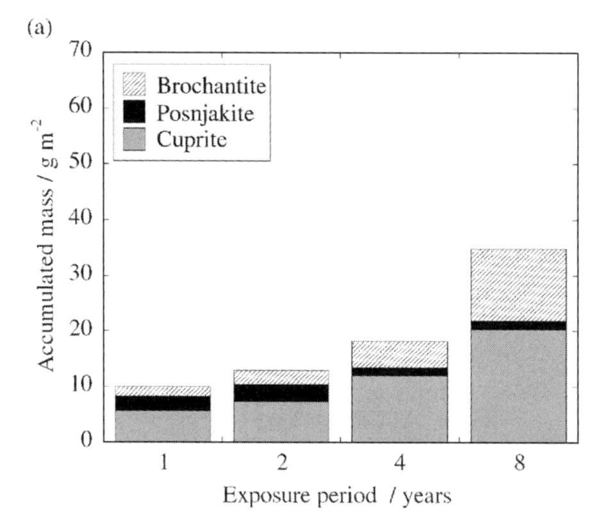

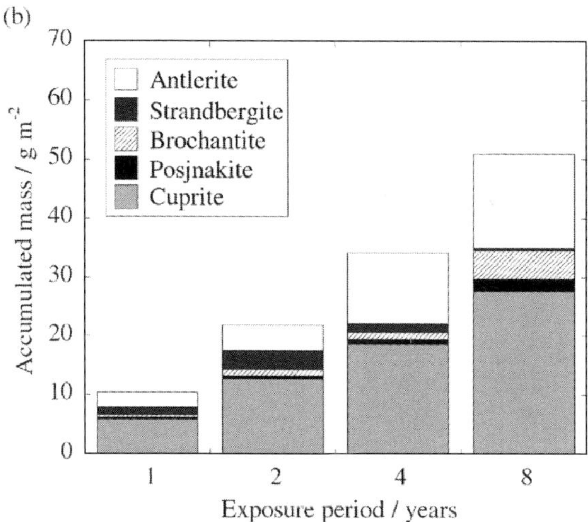

FIGURE 9.8 Accumulated mass versus exposure time for different crystalline corrosion products formed on copper during 8 years of sheltered exposure in a rural (Wells Cathedral, United Kingdom) (a) and an urban atmosphere (Prague, the Czech Republic) (b) determined by quantitative X-ray diffraction. (The technique does not allow detection of the amorphous copper sulfate, though almost certainly present after shorter time periods.) (Adapted with permission from A. Krätschmer, I. Odnevall Wallinder, and C. Leygraf, The evolution of outdoor copper patina, *Corrosion Science, 44*(3), 425–450, 2002.)

observed in more SO_2/sulfate-polluted environments, whereas the remaining compounds with the lower ratios ($Cu_4SO_4(OH)_6 \cdot H_2O$, $Cu_4SO_4(OH)_6 \cdot 2H_2O$, and $Cu_4SO_4(OH)_6$, all with a SO_4^{2-}/OH^- ratio of 1:6) are observed at less sulfur-polluted conditions. The gradual formation of these relatively complex compounds proceeds by the same schematic framework as for zinc presented in Section 9.2, in which

FIGURE 9.9 Crystal structure of $Cu_4SO_4(OH)_6{\cdot}H_2O$ (posnjakite) viewed along [001]. The structure consists of corrugated sheets of distorted (CuO_6) octahedra. Sulfate groups are connected to one side of the octahedral by corner sharing. The sheets are connected via hydrogen bonding (not indicated). (Reproduced with permission from Elsevier; A. Krätschmer, I. Odnevall Wallinder, and C. Leygraf, The evolution of outdoor copper patina, *Corrosion Science, 44*(3), 425–450, 2002.)

the cyclic environmental outdoor conditions result in many consecutive steps of surface reactions, each one being comprised of dissolution, ion pairing, and precipitation.

An interesting aspect of these results is that all five copper hydroxysulfates compiled in Figure 9.7 bear a structural resemblance. They all consist of corrugated sheets of distorted (CuO_6) octahedra, as illustrated for $Cu_4SO_4(OH)_6{\cdot}H_2O$ in Figure 9.9. Adjacent sheets are held together by sulfate groups, hydrogen bonds, and weak Cu—O bonds. The structural resemblance suggests that one phase can relatively easily be transformed into another, either through an ion exchange mechanism in which one phase (e.g., $Cu_4SO_4(OH)_6{\cdot}H_2O$) is partly replaced by another (e.g., $Cu_4SO_4(OH)_6$) or through precipitation of one phase onto another. The resemblance also suggests that the sequence of formation can change from the more sulfur-polluted sequence to the less sulfur-polluted sequence or vice versa. Hence, as indicated by the dashed arrows in Figure 9.2, it cannot be ruled out that $Cu_4SO_4(OH)_6{\cdot}H_2O$ (posnjakite) can act as precursor to $Cu_3SO_4(OH)_4$ (antlerite), whereas $Cu_{2.5}SO_4(OH)_3{\cdot}2H_2O$ (strandbergite) or $Cu_4SO_4(OH)_6{\cdot}2H_2O$ (langite) can act as precursors to $Cu_4SO_4(OH)_6$ (brochantite). The most frequently observed sequences, however, are the ones shown by solid arrows.

Figure 9.8b presents evidence that the main sequence of corrosion product formation can change as the environmental conditions change. During the first 4 years of exposure, the principal sequence was $Cu_2O \rightarrow Cu_{2.5}SO_4(OH)_3{\cdot}2H_2O \rightarrow Cu_3SO_4(OH)_4$, which is the expected route in this highly polluted exposure site (Prague, Czech Republic). A substantial reduction in annual SO_2 concentration was evident during the 8-year exposure, from $78\,\mu g\,m^{-3}$ during the first year to $32\,\mu g\,m^{-3}$ at exposure termination. As a result, the relative amount of $Cu_3SO_4(OH)_4$ was reduced between the fourth and eighth year of exposure, whereas the relative amount of $Cu_4SO_4(OH)_6$ increased. This illustrates hence a gradual transition from the "high-pollution sequence" to the "low-pollution sequence."

9.3.2 Chloride-Dominated Sheltered Exposure

Figure 9.7 also displays a sequence (bottom) that is frequently observed in chloride-containing environments such as marine sites or urban or industrial sites with high concentrations of chloride-containing aerosols. This route includes the formation of $Cu_2Cl(OH)_3$ (atacamite or paratacamite; see previous text) as end product and proceeds via CuCl (nantokite) as a precursor. CuCl is formed through the dissolution of Cu_2O to produce cuprous ions and the reaction of the latter with chloride ions. Once formed, CuCl can act as a seed crystal for the formation of $Cu_2Cl(OH)_3$ through many subsequent dissolution–ion pairing–precipitation steps. This sequence has recently been confirmed by Odnevall Wallinder and coworkers to take place on unsheltered copper exposed during short- (weeks, months) and long-term (up to 5 years) periods at marine field conditions in Brest, France. The same sequence is expected to take place at sheltered conditions, possibly with a higher formation rate. High deposition levels of chlorides onto the Cu_2O surface oxide layer resulted in the local formation of CuCl via reactions between chlorides and cuprous ions in the aqueous layer. This compound is relatively unstable and reacts easily with water and oxygen forming $Cu_2Cl(OH)_3$, a dimorph compound of significantly larger molar mass, and hence volume. CuCl was locally observed and identified within the patina in between the inner layer of Cu_2O and the outer layer of $Cu_2Cl(OH)_3$. Its local existence is illustrated in Figure 9.10 via confocal Raman imaging of a cross section of the copper patina after 3 years of exposure, findings supported by IRAS, XRD, and SEM observations.

Presented formation scenarios of copper products in different environmental conditions described in Figure 9.7 also postulate the formation of a copper chloro-hydroxysulfate. Its existence has though never been confirmed in natural outdoor environments. In laboratory exposures, however, a copper chlorohydroxysulfate of unknown crystal structure has been detected.

FIGURE 9.10 Confocal Raman microscopy (CRM) characterization of a cross section (SEM image, left) of copper exposed at unsheltered conditions for 3 years at a marine environment. The patina reveals an inner layer of Cu_2O and an outer layer of $Cu_2Cl(OH)_3$, separated by the local presence of CuCl. (Reproduced with permission from Elsevier; X. Zhang, I. Odnevall Wallinder, and C. Leygraf, Mechanistic studies of corrosion product flaking on copper and copper-based alloys in marine environments, *Corrosion Science, 85,* 15–25, 2014.)

9.3.3 Unsheltered Exposure

The corrosion sequence discussed previously applies to copper in sheltered exposure conditions. By comparing observed corrosion products within the patina on unsheltered and sheltered copper at all 39 UN/ECE sites (predominantly rural and urban sites), it was clear that a less number of corrosion products were observed in the patina on unsheltered copper compared with sheltered copper. As discussed in Section 8.3.4, the reason is the wash-off effect of precipitation that both hinders the surface retention of various deposited air constituents and dissolves water-soluble compounds. Figure 9.11 displays a representative example of the influence of precipitation on copper patina composition. It is based on the results from parallel exposures of copper at sheltered (a) and unsheltered (b) conditions in a rural site (Clatteringshaws Loch, Scotland—UN/ECE site). Several differences were observed. The cuprite (Cu_2O) formation rate was approximately constant at sheltered conditions but decreased with time at unsheltered conditions. The formation rate of other patina constituents was also much slower at unsheltered conditions, reflecting the fact that precipitation partly hinders the nucleation of copper patina and partly transports some part of the copper patina (discussed in Chapter 10) from the surface. In addition, neither langite ($Cu_4SO_4(OH)_6{\cdot}2H_2O$) nor atacamite ($Cu_2Cl(OH)_3$) is observed in the unsheltered patina, whereas they are clearly visible in the sheltered patina, with effects related to surface removal of deposited chlorides and of initial soluble corrosion products such as CuCl from the surface via the action of rainfall. The most frequently detected corrosion products within the patina on unsheltered copper at the rural and urban sites of low influence of chloride deposition within the UN/ECE project were Cu_2O (cuprite), $Cu_4SO_4(OH)_6{\cdot}H_2O$ (posnjakite), and $Cu_4SO_4(OH)_6$ (brochantite). All other corrosion products of the reaction sequences, including $Cu_4SO_4(OH)_6{\cdot}2H_2O$ (langite), $Cu_{2.5}SO_4(OH)_3{\cdot}2H_2O$ (strandbergite), $Cu_3SO_4(OH)_4$ (antlerite), CuCl (nantokite), and $Cu_2Cl(OH)_3$ (atacamite), were either not present or only present in trace amounts.

However, $Cu_2Cl(OH)_3$ has been identified as a main constituent of the patina, with traces of CuCl within the patina on copper exposed at unsheltered conditions in chloride-dominated environments after short- (days, weeks) and long-term exposures (up to 5 years).

From the fact that unsheltered copper patina has a lower mass or thickness than sheltered copper patina, it does not immediately follow that unsheltered copper corrodes with a slower rate than sheltered copper. In fact, determined corrosion rates (measured as mass loss rates) of unsheltered copper in the UN/ECE exposure program were almost always higher compared with sheltered copper. More soluble parts of the copper patina dissolve at unsheltered conditions, a process that seems to accelerate the corrosion rate. Differences in corrosion behavior between unsheltered and sheltered samples have been discussed in Section 8.3.5 and the influence of metal runoff in Chapter 10.

9.3.4 Other Patina Constituents

Several other known copper minerals, including CuO (tenorite), $Cu_2NO_3(OH)_3$ (gerhardite), and $Cu_2CO_3(OH)_2$ (malachite), have only occasionally been observed as copper patina constituents and are therefore not included in Figure 9.7. CuO is known

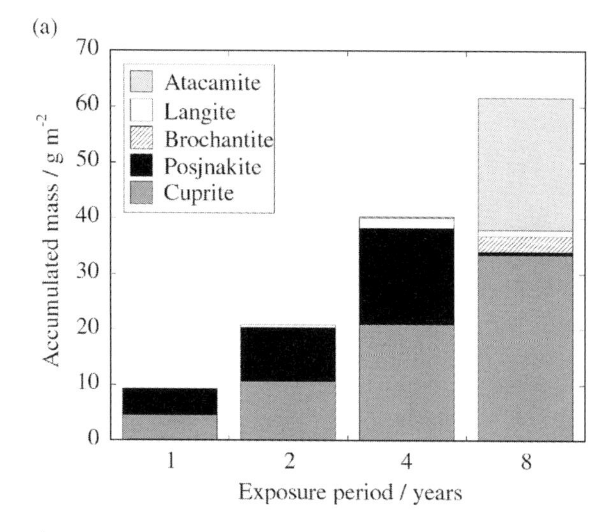

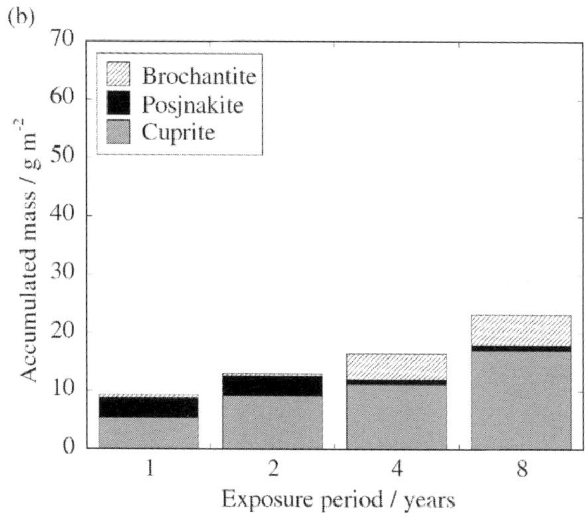

FIGURE 9.11 Accumulated mass versus exposure time for different crystalline corrosion products formed on copper during 8 years of exposure in a rural atmosphere (Clatteringshaws Loch, United Kingdom) at sheltered (a) and unsheltered conditions (b) determined by quantitative X-ray diffraction. (Adapted with permission from A. Krätschmer, I. Odnevall Wallinder, and C. Leygraf, The evolution of outdoor copper patina, *Corrosion Science, 44*(3), 425–450, 2002.)

to react very rapidly with gaseous constituents and is thereby transformed into more stable compounds. $Cu_2NO_3(OH)_3$ is occasionally formed on copper surfaces near electrical discharges. Very early studies state $Cu_2CO_3(OH)_2$ as a patina constituent of, for example, ancient statues and buildings. However, more recent studies during the last several decades do not support these observations. $Cu_2CO_3(OH)_2$ is predominantly present in aqueous environments such as tap water.

As with the more common corrosion sequence, these products reflect the chemical interplay between the metal itself and the environment to which it is exposed.

9.4 EVOLUTION OF CORROSION PRODUCTS ON CARBON STEEL

The outermost surface of freshly prepared iron becomes at dry conditions quickly covered by a typically 2–5 nm thick oxide consisting of an inner layer of Fe_3O_4 and an outer layer of Fe_2O_3. This oxide has good barrier properties in very clean air even at high humidity levels.

Rust layer formation on carbon steel is one of the most thoroughly studied systems in atmospheric corrosion (for more details, see Appendix F). Iron in carbon steel dissolves into the liquid layer as Fe^{2+} and is subsequently oxidized to Fe^{3+}. The ferrous ion is classified as an intermediate acid and the ferric ion as a hard Lewis acid (Section 2.3.4). The ability of the ferric ion to coordinate with different ions seems to be predominantly limited to oxygen- and hydrogen-containing species as well as chlorides and sulfates (all hard Lewis acids) involved in the formation of rust layers. Hence, the most common early corrosion products on iron are "green rusts," mixtures of ferrous and ferric oxyhydroxides. These compounds are of transitory nature and evolve into more stable constituents of the rust layer with prolonged exposure. Naturally formed rust layers on carbon steel can contain about a dozen different oxides or hydroxides. Various forms of ferric oxyhydroxide, FeOOH, including α-FeOOH (goethite), γ-FeOOH (lepidocrocite), Fe_3O_4 (magnetite), β-FeOOH (akagenite), and several amorphous phases are the most stable constituents of the rust layer at different environmental conditions.

Detailed analyses by Misawa and coworkers, Japan, of rust layers resulting from unsheltered exposures of up to 400 years have revealed that the proportions between the FeOOH phases alter with exposure time. As shown in Figure 9.12, three distinct regions can be identified. In the first, characteristic of rust layers of less than a few years of exposure, γ-FeOOH and the amorphous phase are of roughly equivalent abundance. In the second, after several years, the amorphous phase dominates; in the third region, characteristic of rust layers several decades old, α-FeOOH is the dominant phase. Sections 9.2 and 9.3 demonstrated that the buildup of corrosion products under outdoor conditions is governed by consecutive dry–wet–dry cycles, which favor the formation of less soluble compounds at the expense of more soluble compounds. The gradual evolution of the carbon steel rust layer is an excellent demonstration of this general conclusion since the solubility of α-FeOOH is one order of magnitude lower than that of γ-FeOOH.

The importance of cyclic exposure conditions for rust layer formation has been further demonstrated by Stratmann and coworkers, Germany, utilizing a number of different experimental techniques. It would be beyond the scope of this book to provide a comprehensive description of the results. Instead, a short summary will be presented to emphasize the variations in properties of the rust layer on iron during a dry–wet–dry cycle, similar to what a corroding steel structure experiences during a 24 h cycle.

A key feature of the results is the interplay between the rust properties and the kinetics of the O_2 reduction reaction. The rust layers investigated were formed by allowing a 10 μm thick iron layer atop of a gold substrate to react with highly humidified air to which NaCl and SO_2 were added. Within 14 days, the iron layer was completely transformed into a porous rust layer consisting of a mixture of α-FeOOH and

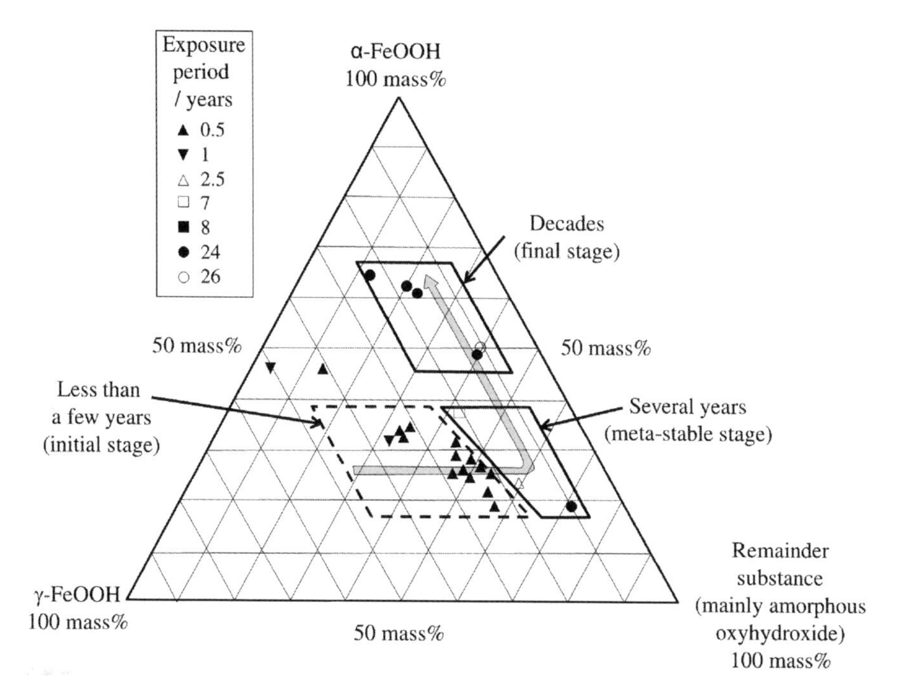

FIGURE 9.12 Change in relative proportions of lepidocrocite (γ-FeOOH), amorphous ferric oxyhydroxide, and goethite (α-FeOOH) in rust layers formed on carbon steel in industrial and rural regions as a function of exposure time. (Adapted with permission from Elsevier; M. Yamashita, H. Miyuki, Y. Matsuda, H. Nagano, and T. Misawa, The long term growth of the protective rust layer formed on weathering steel by atmospheric corrosion during a quarter of a century, *Corrosion Science, 36*, 283–299, 1994.)

γ-FeOOH, that is, the same crystalline compounds as observed outdoors. The average thickness of the rust layer was 70 μm. Iron was present in oxidation state +3 and no iron was detected in its metallic state. The changes in the physical properties of the rust layer during dry–wet–dry cycles were simulated under controlled conditions by immersing the gold-supported rust scale in a neutral electrolyte solution and varying the electrode potential from a more anodic potential (corresponding to dry atmospheric exposure conditions) to a more cathodic potential (corresponding to wet atmospheric exposure conditions) and back to a more anodic potential. The changes in the rust layer and in the kinetics of the anodic and cathodic reactions were followed by several analytical approaches that need not be detailed here.

The anodic and cathodic reactions of importance in the atmospheric corrosion of iron are as follows:

$$Fe \rightarrow Fe^{2+} + 2e^{-} \quad (\text{iron dissolution}) \tag{9.6}$$

$$1/2O_2 + H^{+} + 2e^{-} \rightarrow OH^{-} \quad (\text{oxygen reduction}) \tag{9.7}$$

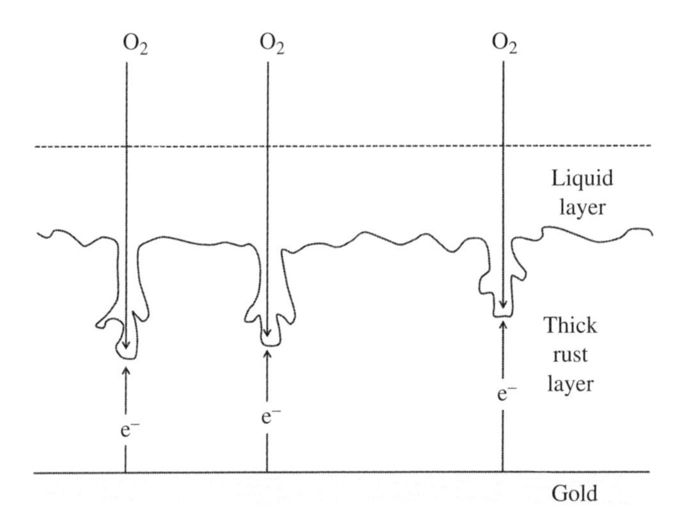

FIGURE 9.13 Schematic description of important processes involved in the oxygen reduction reaction during dry–wet–dry cycling of a thick, porous rust layer and a liquid layer of varying thickness. Under these conditions the electron transport from anodic to cathodic sites and the transport to the cathodic sites are rate-limiting processes whose importance varies during the cycle. (Reproduced with permission from John Wiley & Sons; M. Stratmann, The atmospheric corrosion of iron: a discussion of the physico-chemical fundamentals of this omnipresent corrosion process, *Berichte der Bunsengesellschaft/Physical Chemistry Chemical Physics, 94,* 626–639, 1990.)

When going from dry to wet conditions, corresponding to a decrease in electrode potential, the rust layer is partly changed in that the oxidation state of some of the Fe^{3+} ions in the γ-FeOOH lattice is reduced to Fe^{2+} through an additional cathodic reaction in the rust layer:

$$Fe^{3+} + 2e^- \rightarrow Fe^{2+} \tag{9.8}$$

This process is expected to be important when the O_2 reduction rate is limited.

The experiments demonstrate that the O_2 reduction under actual conditions occurs predominantly within the rust scale rather than at the interface between the metal and the liquid layer. From this follows that O_2 has to be transported within the pores of the rust layer to the site of its reduction. Simultaneously, the electrons required for the same reduction have to be transported from the anodic reaction sites through other portions of the rust scale, Figure 9.13.

As a result of the increased concentration of Fe^{2+} ions in the rust layer, the electronic properties of the layer are changed so that charge transfer reactions, such as O_2 reduction, are enhanced. During further reduction of the electrode potential, the γ-FeOOH layer is partly transformed into magnetite (Fe_3O_4). This increases the local electronic conductivity of the rust layer, since Fe_3O_4 has conductivity properties close to those of metals, and thereby the transport rate of electrons to the cathodic site is increased.

Overall, the change from dry to wet conditions results in reduction of the rust layer and enhanced rates of oxygen reduction. In the beginning of this change, the metal dissolution rate is balanced by the rate of reduction of Fe^{3+} to Fe^{2+}. During later stages of the change from dry to wet conditions, the metal dissolution rate is balanced by the O_2 reduction rate. Thus, the corrosion rate increases during this part of the cycle.

During the stage when the rust layer is wetted and covered by a thick liquid layer, the charge transfer for the O_2 reduction is so fast that the diffusion of O_2 from the gas phase through the pores of various sizes in the liquid layer to the cathodic reaction site now becomes rate determining. With increased reduction of the rust layer, the rate of the O_2 reduction is further increased because of an increased effective area for cathodic sites.

The processes described previously change at the beginning of the drying stage. The thickness of the liquid layer decreases, thereby increasing the O_2 diffusion rate to the cathodic sites. As a result, the corrosion rate increases further during the first part of the drying cycle. But the drying also results in an increase in electrode potential toward more anodic conditions and a parallel reoxidation of Fe^{2+} to Fe^{3+}. This reduces both the effective area for cathodic sites and the electronic conductivity of the rust layer. Consequently, the O_2 reduction rate is reduced, as is the corrosion rate.

Stratmann's study has shown that the corrosion rate and the properties of the rust layer undergo significant variations during a dry–wet–dry cycle. The complex interplay among corrosion product properties (including phase composition, electronic conductivity, and porosity), electrochemical reaction rates, and corrosion rates is expected to operate not only in the case of iron and carbon steel but with other metals and alloys as well.

Exposures of mild steel at marine conditions of relatively high levels of chloride deposition have revealed the additional presence of substantial amounts of β-FeOOH (akaganeite) within the rust layer after long-term (years) exposure periods (between 20 and 40 wt%). Its structure ($Fe(O,OH)_6$ octahedra) is stabilized by chloride ions in channels and has been shown to contain up to 6% chloride. Its presence in the rust layer after long, rather than short, time periods is supported by its complex crystal structure that may take time to evolve.

High chloride deposition levels at atmospheric conditions enable the formation of β-FeOOH in favor of α-FeOOH, and vice versa at lower chloride levels. Laboratory findings propose that the formation of β-FeOOH is possible at a Cl^-/OH^- concentration ratio ≥8, the formation of α-FeOOH at a ratio ≤6, and the formation of both compounds at ratios between 6.5 and 7.5. The findings are supported by observations made at field conditions.

The formation and existence of β-FeOOH in the rust layer of iron and mild steel at marine conditions are, for example, supported by findings of Asami and Kikuchi, Japan, who examined the rust layer of mild steel after 17 years of sheltered exposure at marine/industrial conditions by means of TEM/EDS and XRD. The rust layer consisted predominantly of homogeneously distributed α-FeOOH, β-FeOOH, and Fe_3O_4/γ-Fe_2O_3 (magnetite/maghemite) primarily present in the outer layer, though also observed within the layer, and of initially formed amorphous rust components (hydrated Fe(II)–Fe(III) oxyhydroxides) present at the interface between the rust layer and the steel substrate. These amorphous compounds have been proposed to gradually transform with time into a densely packed layer of α-FeOOH. The proportion of β-FeOOH compared with Fe_3O_4/γ-Fe_2O_3 was higher for thick porous sections of the rust layer into which chlorides easily could penetrate, whereas the opposite situation was evident for thin more dense (and protective) sections of the layer. Hence a negative correlation was observed between the formation of β-FeOOH on one hand and Fe_3O_4/γ-Fe_2O_3 on the other hand. No correlation existed between the rust layer thickness and the presence of γ-FeOOH. These compositional observations are supported by, for example, findings from long-term field studies (>10 years) in Spain and within the Ibero-American 6-year field investigation coordinated by Morcillo and coworkers, Spain, and from short-term field studies up to 24 months performed by Ma et al., China, who examined rust formation at sites of different chloride deposition.

It should be stressed that also soluble corrosion products such as chloride- (e.g., $FeCl_2$, $FeCl_2 \cdot nH_2O$ ($n=4, 6$)) and sulfate-rich corrosion products (e.g., $FeSO_4 \cdot nH_2O$ ($n=1, 5, 7$)) can be present, for example, sulfate nests (as shown at field conditions) within the rust layer, which can play an important role in the corrosion process. However, as they often are highly or very highly soluble, their amounts are typically low and their detection and assignment hence difficult.

9.5 EVOLUTION OF CORROSION PRODUCTS ON ALUMINUM

Freshly prepared aluminum forms spontaneously a 2–3 nm thick amorphous surface oxide of γ-Al_2O_3 (both amorphous and crystalline if formed at temperatures >500°C) at ambient conditions. This oxide is stable in neutral solutions but dissolves slowly in dilute acid solutions. The high oxygen affinity of aluminum ensures its passive properties. In contact with the environment, the outermost surface oxide will become hydrated and converted into an often highly porous aluminum oxyhydroxide, γ-AlOOH (boehmite). Other oxides and hydroxides that commonly evolve on aluminum at non-contaminated atmospheric conditions include $Al(OH)_3$ or $Al_2O_3 \cdot 3H_2O$ (bayerite) and/or other hydrated oxides and hydroxides. Further information is provided in Appendix C.

Al^{3+} is similar to Fe^{3+}, a hard Lewis acid that prefers to coordinate with hard Lewis bases, in particular oxygen- and hydrogen-containing species, and to different extent also with chlorides and sulfates.

Acidic aqueous conditions or interactions of pollutants such as sulfate result in the dissolution and consecutive thinning of the aluminum oxide or local destruction of the passive film in the presence of chlorides. Dissolved aluminum will combine with sulfate to form sulfate-rich corrosion products. Different kinds of hydrated aluminum sulfates, for example, $Al_2(SO_4)_3 \cdot nH_2O$ ($n=4, 5$), and poorly soluble aluminum hydroxysulfates, $Al_4SO_4(OH)_{10}$ (basaluminite), have been reported to gradually form with time on aluminum exposed in different sulfur (SO_2 or SO_4^{2-})-containing aerosols including rural, urban, industrial, and marine environments worldwide (India, China, North America, South America, Europe). Hydrated amorphous aluminum sulfate

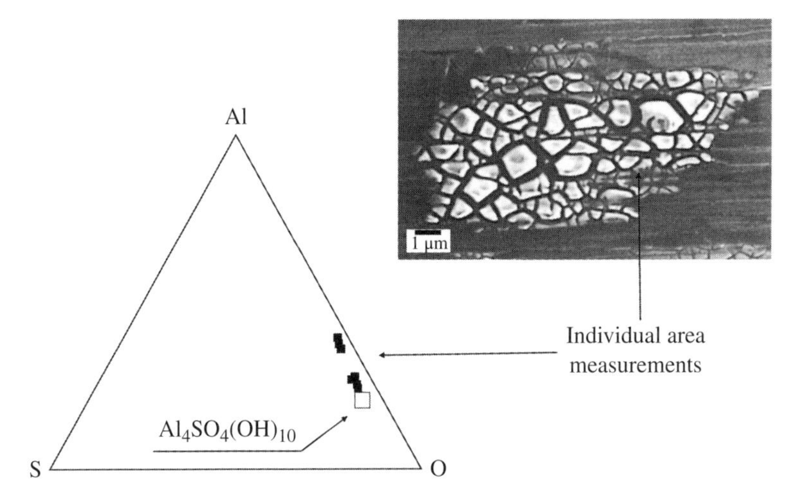

FIGURE 9.14 Morphology and elemental composition of corrosion products gradually formed on aluminum sheet up to 1 year of exposure at unsheltered and sheltered conditions in an urban environment. Similar findings have been found and reported at rural and marine sites.

is a metastable compound shown to exist for long-term periods at atmospheric conditions and to be a predominating corrosion product at both urban and marine conditions. Amorphous basic aluminum sulfate is gradually formed via reactions either at the solid–solution hydrate oxide interface or in the aqueous layer. In a field study conducted by Odnevall in Sweden, aluminum sheet was exposed at sheltered rural and urban conditions for 1 month, 3 months, and 1 year. The aluminum oxide/hydroxide layer gradually evolved in thickness, more rapidly at the humid rural site compared with the urban site. Whitish crackled sulfate-rich deposits gradually formed locally over the aluminum oxide/hydroxide surface (evident via SEM/EDS, XPS, and IRAS measurements after both 1 and 3 months) in both environments, as illustrated via SEM/EDS findings in Figure 9.14. After 1 year of exposure, they compositionally resembled amorphous $Al_4SO_4(OH)_{10}$ (findings supported via different surface-sensitive analytical tools including XPS, GI-XRD, and IRAS).

The chloride ion has the capacity to locally destroy the passive surface oxide. However, as the formed chloride-rich corrosion products are highly soluble, their evolution and detection at the corroded surface on aluminum at outdoor conditions are scarce. It has been proposed that $AlCl_3$ forms together with an intermediate compound $AlCl(OH)_2{\cdot}4H_2O$ (cadwaladerite), both soluble in weak acid, through a stepwise incorporation of chloride ions into $Al(OH)_3$, and that $AlCl_3$ can be retransformed to $Al(OH)_3$ in the presence of H_2O. No evidence exists though that these chloride-rich corrosion products form at outdoor conditions. However their existence are likely, and they may act as precursors for the formation of $Al(OH)_3$, predominantly formed in chloride-rich environments. The corrosion products that mostly form on aluminum in chloride-rich environments are hence $Al(OH)_3$ together with amorphous $Al_2(SO_4)_3{\cdot}nH_2O$ ($n=4$, 5) and poorly soluble aluminum hydroxysulfates ($Al_4SO_4(OH)_{10}$).

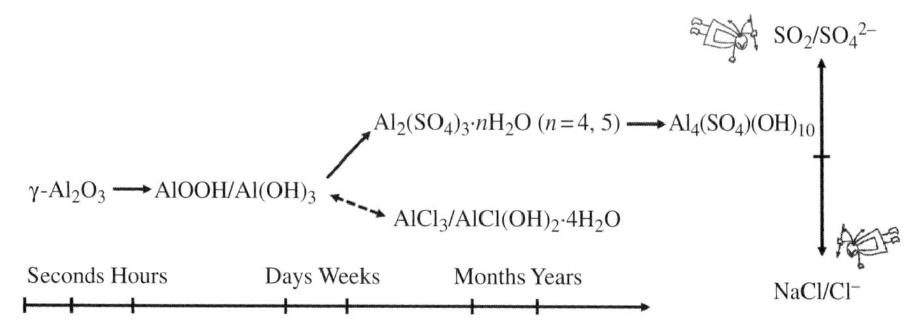

FIGURE 9.15 Schematic description of the evolution of corrosion products on aluminum in atmospheres of varying extent of SO_2/sulfate surface interaction and chloride deposition. The time scale indicates approximately the shortest time period after which a certain corrosion product has been observed.

In all, corrosion sequences of aluminum at atmospheric conditions can generally be summarized in four steps:

1. Interaction between water and the initial aluminum surface oxide forming a porous AlOOH layer and the additional formation of other aluminum oxides/hydroxides
2. Adsorption of sulfate or chloride ions on the oxidized surface causing oxide thinning or local pit formation and the concomitant dissolution of aluminum ions
3. Formation of poorly soluble amorphous aluminum sulfate hydrate in the presence of sulfate ions and its potential transformation into aluminum hydroxysulfate
4. Local formation of soluble aluminum chloride or hydroxychlorides and the potential conversion back to aluminum hydroxide

A general reaction sequence of corrosion products formed on aluminum at sheltered and unsheltered outdoor conditions is proposed in Figure 9.15 based on literature findings.

9.6 SUMMARY

The corrosion product sequences of copper, zinc, carbon steel, and aluminum form evidence of a gradual evolution of corrosion products. These sequential processes appear, although highly complex, in many ways similar for all metals. Triggered by cyclic environmental conditions, the sequential growth proceeds through a sequence of consecutive steps that each involve dissolution of species followed by ion pairing and subsequent reprecipitation.

The dissolution process is accelerated by protons or other ions, including sulfate and chloride. Once dissolved in the aqueous adlayer, the metal ions tend to generally coordinate with counterions according to their classification as hard or soft Lewis acids and bases (see Section 2.3.4). Dissolved Zn^{2+} and Cu^{2+} ions are classified as intermediate acids from which follows the ability to coordinate with a large variety of hard, intermediate, and soft bases including sulfate, nitrate, carbonate, hydroxyl, and sulfide ions. The composition of counterions in the aqueous adlayer is highly dependent on the atmospheric environments. Thus, the composition of zinc, copper, and aluminum corrosion products is strongly influenced by environmental conditions, as evidenced by the corrosion product sequences presented in Figures 9.1, 9.7, and 9.15.

If the ion pairs reach supersaturation, they will precipitate into a solid phase. Compounds observed within the patina described in the corrosion product sequences of copper and zinc have all precipitated according to the sequence of principal steps outlined in Sections 9.2 and 9.3. This begins with the nucleation of the first phase in each sequence (zinc hydroxide and cuprite, respectively) and continues by renewed dissolution of part or all of the starting phase, followed by ion pairing of metal ions with sulfate or carbonate ions, reprecipitation and rearrangement of new compounds in which the previous compounds act as seed crystals, renewed dissolution, ion pairing, etc. The result is a patina that gradually becomes less soluble with time and hence more corrosion resistant.

Eventually, after years or decades of exposure during which the corrosion rate has continuously decreased, the patina will not significantly change its composition. The long-term corrosion rate becomes very low and practically constant, and the corrosion products reach a constant thickness. At this final stage of corrosion product evolution, the runoff rate of metals from corrosion products within the patina equals the rate of renewal of corrosion products. Descriptions of metal runoff processes are given in Chapter 10.

Corrosion products formed on zinc and copper consist mostly of layered structures. Such structures are common in hydroxy salts of divalent cations such as Zn^{2+} and Cu^{2+}. The similarity in structure promotes the growth of one crystalline phase from a previous crystalline phase. It is anticipated that other divalent metals such as cadmium, lead, and nickel will be found to form similar corrosion product sequences. At least one sequence has been observed for nickel, which includes amorphous and crystalline nickel hydroxysulfates, possibly with a crystalline hydrated nickel sulfate ($NiSO_4 \cdot 6H_2O$) as precursor (cf. Figure 8.6).

Iron or carbon steels dissolve into the liquid layer as Fe^{2+} that is subsequently oxidized to Fe^{3+}, a hard Lewis acid. As a result, its ability to coordinate with anions in the liquid layer seems to be limited to oxygen- and hydrogen-containing species. Hence, the most common phases found in rust are a number of iron oxides and hydroxides. The rust reaction sequence presented in Section 9.4 starts with γ-FeOOH (lepidocrocite), proceeds via an amorphous phase, and ends with less soluble α-FeOOH (goethite). As opposed to most phases in the zinc and copper corrosion product sequence, the crystalline γ- and α-phases bear no structural resemblance, and the α-phase cannot directly precipitate onto the γ-phase. The only way the two phases can structurally adapt to each other seems to be through an amorphous phase. Partly because of the crystal mismatch, iron corrosion layers tend not to be very protective.

Considerable changes occur in the rust layer during dry–wet–dry cycles with respect to the oxidation state of Fe, the electron conductivity in the γ-FeOOH lattice, and the transport of O_2 through pores of the rust layer to sites for the cathodic reaction. Such changes in corrosion product properties are likely to occur during cyclic exposure conditions of other metals as well.

In chloride-rich environments the β-FeOOH (akaganeite) formation is favored over α-FeOOH (goethite). These corrosion products are believed to evolve from initially formed amorphous hydrated Fe(II)–Fe(III) oxyhydroxides present at the interface between the rust layer and the steel substrate. β-FeOOH is stabilized by the presence of chlorides within channels of the FeOOH structure.

Corrosion of aluminum involves the hydration of the initial aluminum surface oxide and the subsequent formation of a porous outer layer of AlOOH and other oxides/hydroxides. In the presence of sulfate and chlorides, the oxide undergoes thinning or local destruction. This results in the formation of poorly soluble amorphous $Al_2(SO_4)_3 \cdot nH_2O$ that may be transformed into $Al_4SO_4(OH)_{10}$ and/or in the formation of soluble $AlCl_3$ that can be retransformed to $Al(OH)_3$ due to structural similarities.

FURTHER READING

Evolution of Corrosion Products on Zinc

E. Almeida, M. Morcillo, and B. Rosales, Atmospheric corrosion of zinc Part 1: rural and urban atmospheres, *Corrosion Engineering, Science and Technology, 35*(4), 284–288, 2000.

E. Almeida, M. Morcillo, and B. Rosales, Atmospheric corrosion of zinc Part 2: marine atmospheres, *Corrosion Engineering, Science and Technology, 35*(4), 289–296, 2000.

Z.Y. Chen, D. Persson, and C. Leygraf, Initial NaCl-particle induced atmospheric corrosion of zinc—effect of CO_2 and SO_2, *Corrosion Science, 50*(1), 111–123, 2008.

D. de la Fuente, J.G. Castaño, and M. Morcillo, Long-term atmospheric corrosion of zinc, *Corrosion Science, 49*(3), 1420–1436, 2007.

D. Lindström and I. Odnevall Wallinder, Long-term use of galvanized steel in external applications. Aspects of patina formation, zinc runoff, barrier properties of surface treatments and coatings and environmental fate, *Environmental Monitoring and Assessment, 173*, 139–153, 2011.

I. Odnevall and C. Leygraf, Reaction sequences in atmospheric corrosion of zinc, in *Atmospheric Corrosion*, ASTM STP 1239, eds. W.W. Kirk and Herbert H. Lawson, ASTM, Philadelphia, pp. 215–229, 1995.

J.J. Santana, B.M. Fernández-Pérez, J. Morales, H.C. Vasconcelos, R.M. Souto, and S. González, Characterization of the corrosion products formed on zinc in archipelagic subtropical environments, *International Journal of Electrochemical Science, 7*, 12730–12741, 2012.

Evolution of Corrosion Products on Copper

K. FitzGerald, J. Nairn, G. Skennerton, and A. Atrens, Atmospheric corrosion of copper and the colour, structure and composition of natural patinas on copper, *Corrosion Science, 48*(9), 2480–2509, 2006.

A. Krätschmer, I. Odnevall Wallinder, and C. Leygraf, The evolution of outdoor copper patina, *Corrosion Science, 44*(3), 425–450, 2002.

R.E. Lobnig, R.P. Frankenthal, D.J. Siconolfi, J.D. Sinclair, and M. Stratmann, Mechanism of atmospheric corrosion of copper in the presence of submicron ammonium sulfate particles at 300 and 373 K, *Journal of Electrochemical Society, 141*(11), 2935–2941, 1994.

I. Odnevall and C. Leygraf, Atmospheric corrosion of copper in a rural atmosphere, *Journal of the Electrochemical Society, 142*(11), 3682–3689, 1995.

I. Odnevall Wallinder, T. Korpinen, R. Sundberg, and C. Leygraf, Atmospheric corrosion of naturally and pre-patinated copper roofs in Singapore and Stockholm—runoff rates and corrosion product formation, in *Outdoor and Indoor Atmospheric Corrosion*, ASTM STP 1421, ed. H.E. Townsend, ASTM, West Conshohocken, pp. 230–244, 2002.

J.R. Vilche, F.E. Varela, E.N. Codaro, B.M. Rosales, G. Moriena, and A. Fernández, A survey of Argentinean atmospheric corrosion: II—copper samples, *Corrosion Science, 39*(4), 655–679, 1997.

Reaction Sequences on Iron or Carbon Steel

K. Asami and M. Kikuchi, In-depth distribution of rusts on a plain carbon steel and weathering steels exposed to coastal-industrial atmosphere for 17 years, *Corrosion Science, 45,* 2671–2688, 2003.

Y. Ma, Y. Li, and F. Wang, Corrosion of low carbon steel in atmospheric environments of different chloride content, *Corrosion Science, 51,* 997–1006, 2009.

M. Morcillo, D. de la Fuente, I. Díaz, and Y.H. Cano, Atmospheric corrosion of mild steel (review), *Revista de Metalurgica, 47*(5), 426–444, 2011.

M. Stratmann, The atmospheric corrosion of iron: a discussion of the physico-chemical fundamentals of this omnipresent corrosion process, *Berichte der Bunsengesellschaft/Physical Chemistry Chemical Physics, 94,* 626–639, 1990.

M. Yamashita, H. Miyuki, Y. Matsuda, H. Nagano, and T. Misawa, The long term growth of the protective rust layer formed on weathering steel by atmospheric corrosion during a quarter of a century, *Corrosion Science, 36,* 283–299, 1994.

Evolution of Corrosion Products on Aluminum

D. de la Fuente, E. Otero-Huerta, and M. Morcillo, Studies of long-term weathering of aluminum in the atmosphere, *Corrosion Science, 49,* 3134–3148, 2007.

J.J. Friel, Atmospheric corrosion products on Al, Zn, and AlZn metallic coatings, *Corrosion, 42*(7), 422–426, 1986.

M. Natesan, G. Venkatachari, and N. Palaniswamy, Kinetics of atmospheric corrosion of mild steel, zinc, galvanized iron and aluminium at 10 exposure stations in India, *Corrosion Science, 48,* 3584–3608, 2006.

I. Odnevall, Characterization of corrosion products formed on rain sheltered aluzink and aluminum in a rural and an urban atmosphere, *13th International Corrosion Congress,* Cape Town, South Africa, 1999.

S. Oesch and M. Faller, Environmental effects on materials: the effect of the air pollutants SO_2, NO_2, NO and O_3 on the corrosion of copper, zinc and aluminium. A short literature survey and results of laboratory exposures, *Corrosion Science, 39,* 1505–1530, 1997.

S. Sun, Q. Zheng, D. Li, and J. Wen, Long-term atmospheric corrosion behavior of aluminum alloys 2024 and 7075 in urban, coastal and industrial environments, *Corrosion Science, 51,* 719–727, 2009.

R. Vera, D. Delgado, and B.M. Rosales, Effect of atmospheric pollutants on the corrosion of high power electrical conductors: Part 1. Aluminium and AA6201 alloy, *Corrosion Science, 48,* 2882–2900, 2006.

10

ENVIRONMENTAL DISPERSION OF METALS FROM CORRODED OUTDOOR CONSTRUCTIONS

10.1 INTRODUCTION

Atmospheric corrosion studies of metals and alloys in outdoor applications traditionally focus on how the climate, environment, and presence of pollutants influence their corrosion behavior and their service life and aesthetical appearance. The release of metals from these corroded surfaces into the environment requires also an opposite perspective, that is, how corrosion of metals influences the environment. The Brundland report *Our Common Future* from 1987 emphasized the necessity of political and economic counteractions to obtain a long-term sustainable society, for example, to reduce environmental issues and effects induced by the industrialized society, effects that may arise from the diffuse dispersion of metals from corroded surfaces. From this report followed a gradually increasing global environmental concern in many countries related to potential adverse effects induced by the environmental dispersion of metals from different sources in the society. Legislative actions were implemented that restricted and regulated the dispersion of metals from industrial point sources, and the precautionary principle was often adopted to limit their diffuse dispersion from various sources, when known. Building and outdoor constructions were early identified as potential sources of metal dispersion, also the transport sector and tap water systems as they are closely related to the global population growth. Regulatory actions are still ongoing and, for the European Community, even more restricted within the framework of the Registration, Evaluation, Authorization and Restriction of Chemicals (REACH) directive. REACH was implemented in 2007 to regulate the use of chemicals (including metals and alloys, in force since 2010) on the market and their safe use.

Atmospheric Corrosion, Second Edition. Christofer Leygraf, Inger Odnevall Wallinder, Johan Tidblad and Thomas Graedel.

Whereas long-term corrosion and patina formation processes and their underlying mechanisms on outdoor metal constructions are relatively extensively investigated, the extent of released metals from such corroded surfaces is significantly less explored and suffers from knowledge gaps of underlying mechanisms and key parameters that govern these processes. This novel research area within atmospheric corrosion has primarily been subject for investigations during the last decades for a limited number of metals and alloys and by a limited number of research teams with the authors as one of the most active groups, as reflected in this chapter.

In the following sections the state-of-the-art knowledge will be summarized from a combined fundamental and an applied mechanistic perspective from source to recipient. This is primarily based on research efforts undertaken and reported in the scientific literature since the late 1990s on the diffuse dispersion and environmental fate of metals used in outdoor building applications. These aspects will primarily be illustrated with long-term (up to 17 years) findings for bare copper and zinc exposed to atmospheric corrosion degradation conditions. Aspects of metal runoff from alloys used in architectural applications are discussed in Chapter 13. The influence of prevailing environmental conditions, exposure conditions, and surface characteristics on the extent of released metals will be discussed as well as their fate upon environmental interactions with solid surfaces present in the close vicinity of buildings. Further, the importance of the chemical state of the released metal (speciation) and its changes upon environmental interaction with different settings will be discussed, as speciation and the bioavailability of the released metal, that is, its possibility for uptake and potential toxic effects, are essential factors to consider in risk assessment and management.

10.2 METAL DISPERSION (RUNOFF): ATMOSPHERIC CORROSION

Metals and metal alloys used in outdoor constructions interact with the environment forming surface oxides and other corrosion products (patina), either instantaneously or gradually. As discussed in Chapter 9, daily repeated wet–dry humidity/temperature cycles and differences in prevailing environmental conditions, which result in recurring dissolution–reprecipitation processes, change with time the chemical composition, thickness, and characteristics of corrosion products within the patina. Most dissolved metals rearrange within the patina, whereas a small fraction can be released and transported from the surface to the environment by the action of rainwater (or dew droplets at damp conditions), either directly or indirectly via internal or external drainage and stormwater systems. This process is slow and has a minor effect on, for example, the thickness and protective properties of the adherent patina. The amount of dissolved and released metals from these surface oxides or corrosion products to the environment is denoted the *metal runoff* (or *metal release*) and is defined as the total amount of metal released per geometric area and time from a given surface ($g\,m^{-2}\,year^{-1}$). The term metal leakage, which implies a damaged material or the complete chemical extraction of a metal into solution, should be avoided in this context as it is not a relevant description of metal runoff processes at atmospheric conditions.

Dispersion of metals from the patina can to some extent also take place via wear processes induced by, for example, ice, hail, and sandstorms; however such contributions to the total amount of released metal have on a long-term perspective been found minor compared with electrochemical and chemical processes. There may be other exceptions, though not covered in this chapter, such as runoff induced as a result of large deposits of dust and conditions of high relative humidity in, for example, road tunnel systems.

The patina characteristics and composition determine the corrosion resistance of the material. Even though prevailing environmental and climatic conditions influence both the corrosion and the metal release processes, corrosion data (from mass loss measurements) or corrosion resistance findings cannot be used to predict the extent of metal runoff (total amount of metal released and dispersed from a corroded surface) at atmospheric conditions. Whereas corrosion (oxidation of the metal/alloy) primarily is electrochemically governed, metal release is the result of both electrochemical and chemical processes that predominantly take place at the interface between the surface oxide (corrosion products) of the metal/alloy and the environment.

Corrosion rates are often determined for one or possibly a few years of exposure following the ISO 9226 standard, which aims to assess the corrosivity of a given site and specific metal, or predicted using available dose–response functions (cf. Chapter 8). Since these rates predominantly reflect the initial formation of a protective corrosion patina and not the long-term rate of corrosion, the direct use of these measured or predicted short-term corrosion rates may underestimate the actual service life of the material (aspects handled by the ISO 9224:2012 standard).

The ratio between corrosion rates and runoff rates during the first year of exposure is illustrated in Figure 10.1 for copper sheet exposed at different urban and marine sites in Europe. The corrosion rate is highly time dependent with generally reduced rates with time as the corrosion patina gradually evolves and acts as a physical barrier for the substrate, whereas the annual runoff rate (integrated metal runoff from all rain episodes over a year) is significantly lower, relatively constant or slightly reduced with time. These observations, supported by literature findings for urban, rural, and marine sites, show significantly lower runoff rates (runoff/corrosion ratios <20%) during the first year of exposure compared with corresponding 1-year corrosion rates. From this follows that most of the corroded (oxidized) metal during the first year of exposure is retained within the patina as relatively surface adherent corrosion products of low solubility (not to be confused with easily soluble metal salts) and that only a fraction of its constituents are dissolved and released from the surface (primarily by the action of rainwater). On a long-term perspective the discrepancy between the average corrosion rate and the runoff rate becomes less pronounced as the former generally decreases with time and the runoff rate remains relatively constant or gradually decreases. Copper roofs on ancient buildings are still in service even after centuries of outdoor exposure. Cross-sectional studies of a church copper roof aged for approximately 350 years in the old town of Stockholm, Sweden, revealed an average thickness reduction of the bulk material with approximately 25%, that is, most of the metal sheet was still intact.

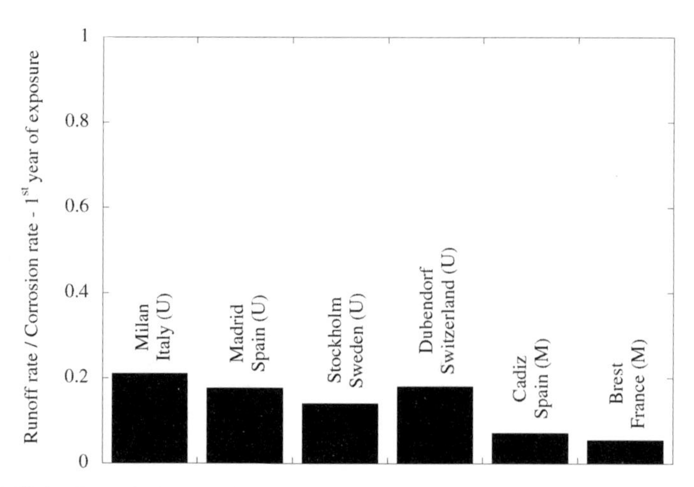

FIGURE 10.1 Ratio between observed runoff rates and corrosion rates for copper sheet during the first year of exposure at the urban sites of Milan, Italy; Madrid, Spain; Dübendorf, Switzerland; and Stockholm, Sweden, and the marine sites of Cadiz, Spain, and Brest, France (exposed at 45° from the horizontal, facing south).

Similar observations are reported for other parts of the world. Lower annual runoff rates compared with initial corrosion rates have also been reported for other metals and alloys, for example, zinc.

Direct comparisons between different exposure sites are not straightforward due to varying environmental characteristics, different surface inclinations and orientations, and different years of exposure. As rainfall is one of the main parameter that governs the released metal quantity, rough comparisons can be made by normalizing the mass of released metal per surface area to the annual rainfall quantity ($g\,m^{-2}\,mm^{-1}$) and to the surface inclination (further discussed in Section 10.4). The methodological approach on how to perform runoff rate measurements is now standardized, and how to report data is described in the ISO 17752:2012 standard.

Since the extent of metal release from metal alloys is significantly different as compared with their pure metal constituents, it is imperative to treat alloys as unique materials of disparate intrinsic properties and not as simple mixtures of the pure constituents. Differences in metal runoff from metal alloys compared with bare metals are elucidated in Chapter 13.

10.3 TIME-DEPENDENT ASPECTS AND IMPORTANCE OF RAIN AND ENVIRONMENTAL CONDITIONS

Continuous long-term (based on integrated rain episodes over a year and normalized on an annual basis) runoff measurements of copper and zinc released from bare sheet and their alloys in outdoor constructions clearly show relatively constant or slightly reduced runoff rates with time per given rainfall quantity as a result of the gradual evolution of corrosion products of poor solubility and improved barrier properties. Reduced air pollution levels can result in the same effects (Chapters 4 and 8). Improved or relatively constant barrier properties with time are illustrated in Figure 10.2 for bare copper sheet and zinc sheet per surface area and given rainfall unit exposed at two urban sites in Europe for periods up to 17 and 6 years, respectively. Similar trends with reduced runoff rates with time are also evident for aged natural patinas with two-layer structures (e.g., an inner layer of cuprous oxide and an outer layer of basic copper sulfates/chlorides, illustrated in Chapter 9), exemplified for a naturally green patinated surface preexposed for approximately 130 years at the same urban site. Initially slightly higher runoff rates compared with one-layer (cuprous oxide) patinas are explained by longer contact periods for water and corrosive species to interact within a spatially heterogeneous, porous, and thick patina and hence a larger amount of released copper in the first flush portion, further discussed later. Reduced total runoff rates with time are also observed for metal alloys such as brasses, bronzes, and zinc-based alloys used in outdoor constructions (cf. Chapter 13).

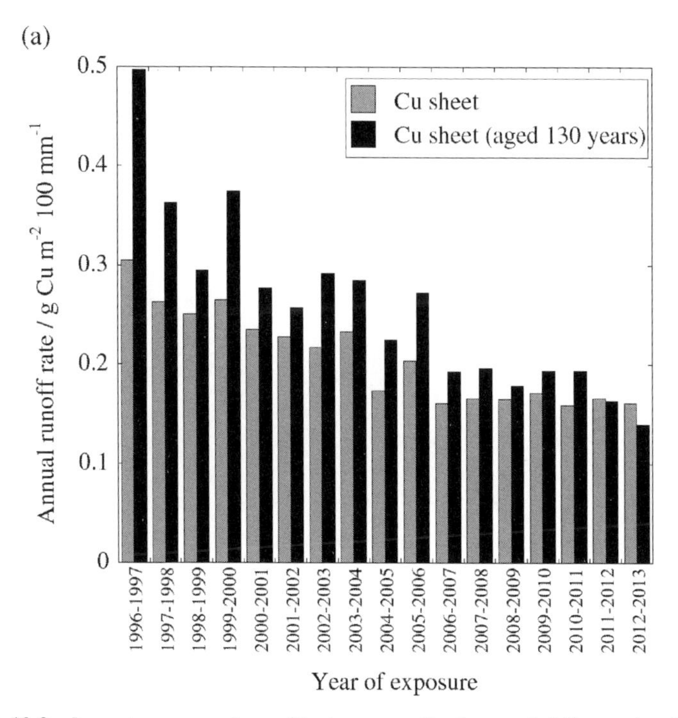

FIGURE 10.2 Long-term annual runoff rates normalized per rainfall quantity (integrated for all rain episodes of varying characteristics over a year) for bare and aged (130 years) copper sheet exposed at 45° inclination facing south for 17 years in Stockholm, Sweden (a); for bare copper sheet exposed for 6 years in Milan, Italy (b); and for bare zinc sheet exposed for 15 years in Stockholm, Sweden (c).

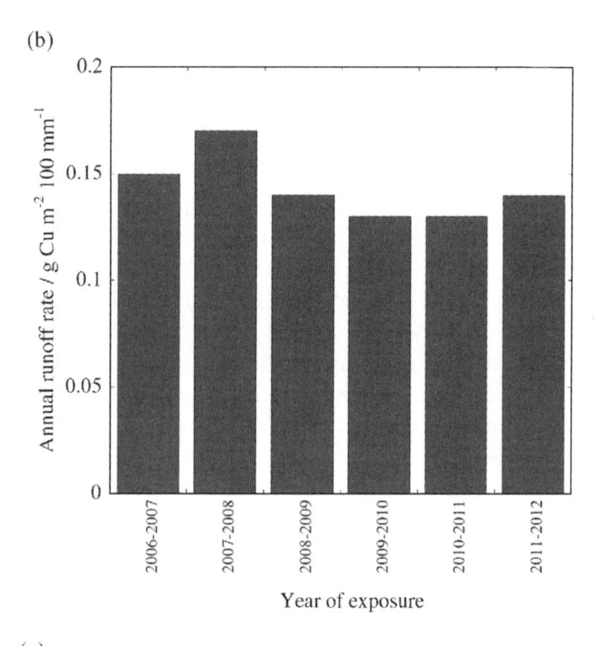

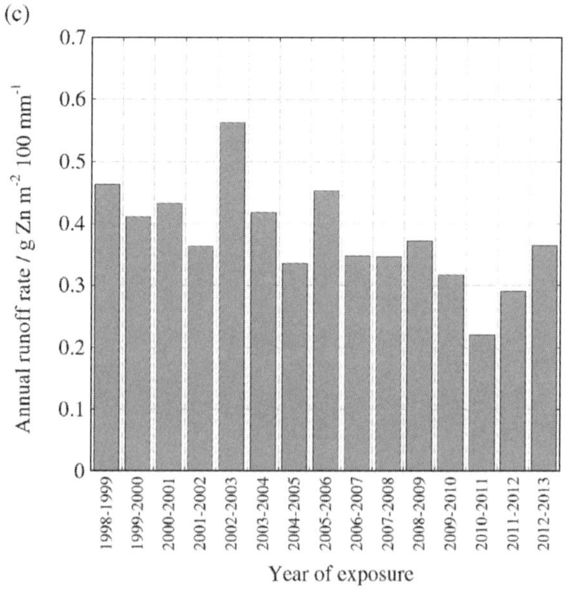

FIGURE 10.2 (Continued)

On a short-term perspective, that is, during single or a few, in time separated, rain events, the release of metals is highly time and rain volume dependent with initially relatively more metals released during the first flush portion of a rain event compared with lower, relatively constant levels during the remaining rain event, schematically illustrated in Figures 10.3a and c. The definition of the first flush portion varies in the

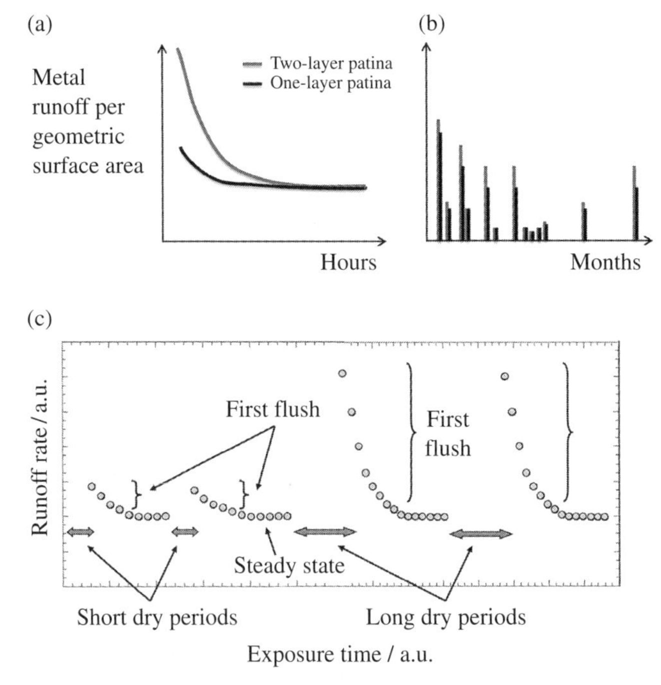

FIGURE 10.3 Schematic illustrations of differences in magnitude of the first flush portion of released metal from a patina of different characteristics during a given rain event (a) and during different rain episodes over a longer time period (months) (b) and effects of dry periods in between different rain events on the first flush portion (c).

scientific literature and refers to rates during the initial rainfall volume (1–2, 2–4, or 4–5 mm), during the first 3 h of a rain event, or to strongly reduced rates until reaching steady-state conditions. The released amount of metals per given rainfall volume increases with prolonged contact period between the impinging rainwater and the patina (lower rain intensity and/or small rain volume), an effect seen as a higher first flush contribution. The magnitude of the first flush portion is more pronounced during rain events preceded by long periods without rainfall, increased deposition of corrosive species, and humid conditions prior to the rain event. It is also more pronounced for increased acidity of the rain pH and for patinas of increased thickness, heterogeneity, and porosity able to retain water and corrosive species for long time periods. The first flush effect is less pronounced, or minor, during repeated successive rain events separated by very short dry periods. The initial volumes of the rain event and of the snow melt water in contact with the patina are more acidic compared with succeeding rain events since both rain and snow are efficient scavengers of atmospheric pollutants and particles. From this follows a pronounced first flush effect with initially high release rates of dissolved metals. As previously discussed, first flush effects are generally more pronounced for aged two-layer patinas compared with one-layer patinas due to thicker and more porous patinas, schematically illustrated for copper in Figure 10.3.

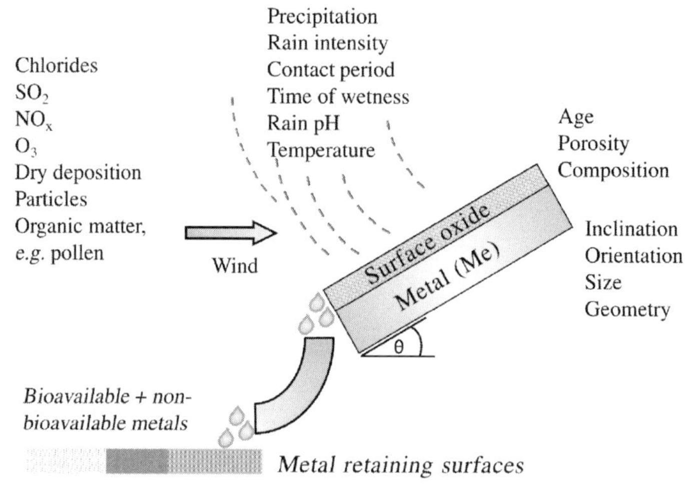

FIGURE 10.4 Schematic illustration of main exposure, surface, and environmental parameters that influence the runoff (and corrosion) process from metals and alloys.

Long-term runoff rates (based on integrated rain episodes during longer time periods, e.g., over a year) are though relatively constant, despite time- and volume-dependent metal release processes during each rain event. They may even be slightly reduced with time, as long as prevailing environmental conditions do not change dramatically, Figures 10.2a and b. Even if the runoff rate is highly varying between short periods and during a given rain event, the distribution of episodes of high and low runoff rates is reasonably equal on a long-term perspective (years).

Besides patina characteristics (composition/solubility, thickness, porosity), the amount of released metals during the first flush portion from a surface is influenced by prevailing environmental (e.g., length of dry periods between rain events, rain quantity, intensity, duration, pH, humidity, dry deposition, pollutants, etc.) and exposure conditions (e.g., building geometry, surface inclination, orientation, prevailing wind direction). This is further discussed in Chapter 13 and schematically illustrated in Figure 10.4. Gaseous pollutants such as SO_2, NO_2, O_3, and atmospheric aerosols containing, for example, NaCl or $(NH_4)_2SO_4$ determine the rate of initial corrosion reactions and the formation of corrosion products of different solubility and barrier properties within the patina and influence as a consequence also the runoff process. However, in contrast to corrosion the release of copper from copper sheet and zinc from zinc sheet seems not to be enhanced by the presence of chlorides. This is reflected by very similar annual copper runoff rates per given rainfall quantities during long-term field conditions (up to 7 years) at urban and marine sites: 0.0009 and $0.0012\,g\,m^{-2}\,mm^{-1}$ for the marine sites of Cadiz, Spain, and Brest, France, respectively, compared with 0.0020 and $0.0013\,g\,m^{-2}\,mm^{-1}$ for the urban sites of Stockholm, Sweden, and Milan, Italy, respectively. Similar observations have been reported for the marine site of Newport, United States, with rates of $0.0009\,g\,Cu\,m^{-2}\,mm^{-1}$.

The release of copper and zinc was instead depressed due to long wet periods, which resulted in rapid crystallization of corrosion products, limited first flush contribution of released metals, and fewer dissolution–reprecipitation events. This example illustrates that even though similar parameters govern both the metal runoff process and prevailing atmospheric corrosion reactions, their influence is dissimilar. Environmental and exposure parameters of importance for the metal runoff process (and corrosion process) and of importance for the environmental fate of dispersed metals are presented in Figure 10.4. Effects of these parameters will be discussed in the following sections.

10.4 INFLUENCE OF CONSTRUCTION GEOMETRY ON THE METAL RUNOFF AND RUNOFF RATE PREDICTIONS

The action of rainwater, snow melting water, and/or dew is essential to enable dissolved metals from the patina to be transported from the surface and potentially become environmentally dispersed. Without these processes, dissolved metals form and precipitate as corrosion products within the patina during dry periods. The capacity of the aqueous media to transport dissolved metals from the surface is for a given surface strongly connected to the amount, intensity, and duration of the impinging rain event, or the extent of dew formation, effects that are strongly linked to the surface area and inclination, prevailing wind directions, and degree of surface sheltering (from construction details or nearby buildings). The effect of sheltering has been proposed to reduce the total exposed surface area of roofing and facades with as much as 25–30%. No or negligible metal runoff will take place when the total impinging rainfall is of insufficient volume or when the surface inclination hinders its removal from the surface.

Rainfall quantities are for a given site expressed in the meteorological entity, millimeter per time unit (e.g., hours or years), and reflect the total rain volume (L) impinging on a 1 m^2 large horizontal surface area. The amount of rain that actually impinges surfaces of outdoor buildings and constructions, often of very complex geometries, varies hence significantly but can roughly be calculated based on trigonometry. However, effects of prevailing wind directions and sheltering must also be taken into account. Such effects may not be straightforward since they largely influence the actual amount of impinging rainwater and consequently the extent of metal runoff. Normalization to the actual impinging rainfall quantity is hence crucial. Examples where these aspects largely influence the released amount of metals are for facades that can be exposed to large rainfall quantities despite vertical surfaces, if positioned toward the prevalent wind direction, that is, wind-driven rain, whereas another surface orientated in a direction opposite to the wind direction may be exposed to very limited rainfall quantities. The effect of surface inclination on the released amount of copper normalized to the impinging rainfall volume is illustrated in Figure 10.5a for copper sheet exposed at different inclinations and orientations during 4 years in the urban city of Milan, Italy. The figure also illustrates how differences in surface inclination influence the released amount of copper for a given surface orientation (illustrated for surfaces facing south), Figures 10.5b and c.

(a)

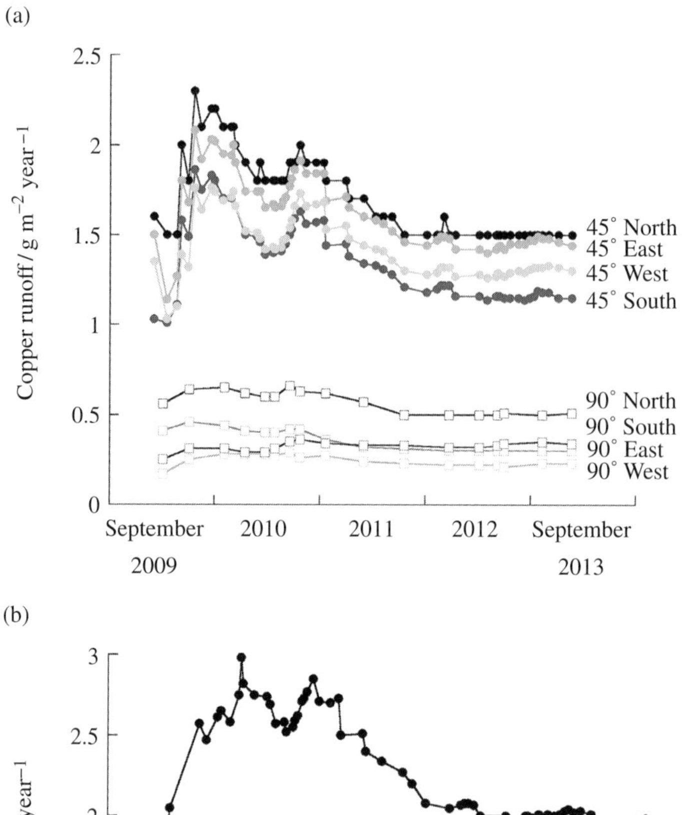

(b)

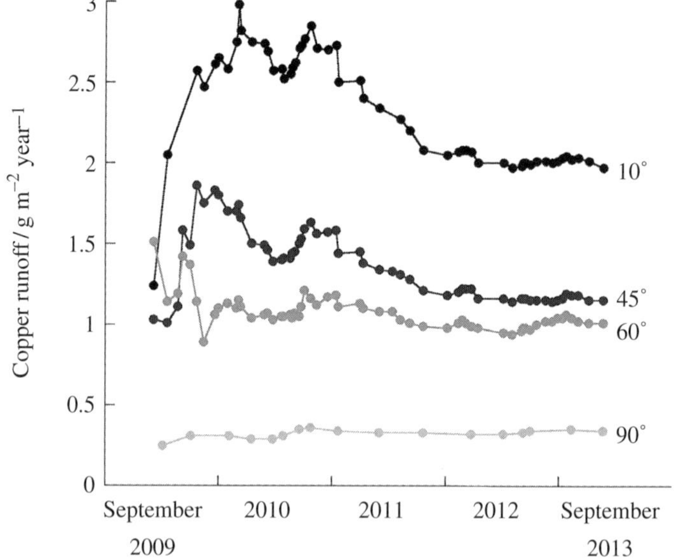

FIGURE 10.5 Effects of orientation on the copper runoff per time unit for surfaces of copper sheet inclined 45° and 90° from the horizontal during 4 years of unsheltered exposure in Milan, Italy (a), and effects of inclination for surfaces facing south (b). (Printed with permission from Elsevier; Y.S. Hedberg, S. Goidanich, G. Herting, and I. Odnevall Wallinder, Surface-rain interactions: differences in copper runoff for copper sheet of different inclination, orientation, and atmospheric exposure conditions, *Environmental Pollution, 196*, 363–370, 2015.)

Surfaces of different orientation experience varying extent of dry deposition, duration of wet–dry cycles, prevailing wind direction, etc., conditions that influence both corrosion and metal release processes, pronounced during the first flush portion.

The results clearly illustrate the importance of both inclination and orientation on the amount of metal runoff for a given surface area, parameters that govern the contact time and amount of water in contact with the exposed surface, as well as deposition of corrosive species. It is hence essential that observed runoff rates also are related to the surface inclination. Direct comparison of released amounts of metals without such a normalization is highly erroneous, for example, when comparing findings of different sites. Standardized runoff rate measurements stipulate investigations of surfaces inclined 45° from the horizontal facing south (at the northern hemisphere) and the use of trigonometry to consider surfaces of different inclination. Long-term field investigations propose that predictions of copper runoff from facades (vertical surfaces) should be considered as surfaces inclined 60–80° due to wind-driven effects. Some models have been elaborated and discussed in the literature to predict runoff rates of copper and zinc based on the pH of the rainwater, the annual rainfall quantity, the surface area and contact time and duration of impinging rainwater, surface inclination, and atmospheric SO_2 concentration. For example, the model that is currently used to predict copper runoff for urban and rural conditions in tempered zones is based on physically explainable parameters (Eq. 10.1):

$$R = \left(0.37\mathrm{SO_2}^{0.5} + 0.96\,\mathrm{rain}\,10^{-0.62\mathrm{pH}}\right)\left(\cos(\theta)/\cos(45°)\right) \tag{10.1}$$

where R is the annual copper runoff rate (g m^{-2} year^{-1}), SO_2 the atmospheric concentration (μg m^{-3}), rain the annual rainfall quantity (mm year^{-1}), pH the rain pH, and θ the surface inclination angle from the horizontal (°). (Table 4.2 displays conversion factors when going from ppbv to μg m^{-3} for different gases.)

Available models are so far limited to certain environments/conditions. Some refinements, especially in terms of inclination and orientation, and effects of chlorides still need to be taken into consideration. It is proposed that the predictive copper runoff model could be applicable also for marine sites although it tends to overestimate copper runoff rates compared with measured rates at marine sites (<156%) due to faster corrosion and formation of a more protective patina. At sites of very high chloride deposition rates, the model underestimates the copper runoff compared with measured rates only during the first year of exposure (<48%). As these predictive models reflect the immediate release situation and not any chemical changes upon environmental entry or the environmental fate of released metals, they are not directly applicable to predict, or assess any environmental risks. A model that includes aspects of environmental fate and changes in speciation of copper has been developed for watersheds. However this model is only based on a single field investigation. Parameters of importance for the environmental fate of released metals from outdoor constructions, and for environmental risk assessment and management, will shortly be discussed in Section 10.5.

10.5 ENVIRONMENTAL FATE AND SPECIATION: IMPORTANCE FOR RISK ASSESSMENT AND MANAGEMENT

Released metals (e.g., zinc, copper, iron, and chromium) from oxidized/corroded outdoor constructions can be directly dispersed into the environment or indirectly distributed via internal or external dewatering and drainage and stormwater systems. However, metal-containing runoff water will become diluted with stormwater, and chemical transformation and retention processes of released metals take place that alter their total concentration and chemical speciation (form). These processes are decisive for their bioaccessibility (the pool of released and potentially bioavailable (see Section 10.1) metal species). Released concentrations of metals are at the immediate release situation (e.g., at the rooftop) significantly higher compared with concentrations measured after interactions with organic matter and different dewatering surfaces such as concrete in pavement and stormwater pipes, limestone, soil, etc., surfaces that act as efficient sinks in the near vicinity of outdoor constructions or buildings. Rapid reduction of total amounts of released metals to background levels due to irreversible retention processes and transformation into nonavailable metal complexes and poorly soluble metal compounds have found evidence in both laboratory and field investigations, schematically illustrated in Figure 10.6.

Significant irreversible retention of copper, zinc, nickel, and chromium has been shown for surfaces such as concrete (in pavement, stormwater piping) and limestone that increase the pH of the runoff water. Other materials, solid surfaces, and soils of varying characteristics (pH, cation exchange capacity, etc.) have also been reported

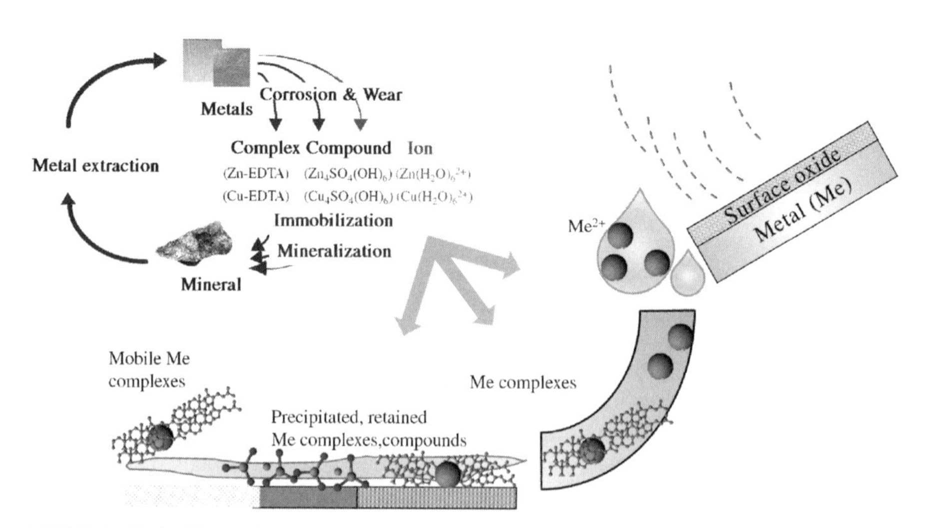

FIGURE 10.6 Illustration of the life cycle of and transition of extracted metals and illustration of changes in metal speciation, retention, and mobility of metals released from outdoor buildings in contact with metal retaining surfaces in the near vicinity of a building. (*See insert for color representation of the figure.*)

to act as efficient sinks for released metals when exposed to concentrations relevant for roof runoff. Countermeasures may be taken if roof runoff is directed into a small volume of receiving water such as a pond.

As clearly stated in the Environmental Health Criteria (EHC) Programme from 1973 and the International Programme on Chemical Safety (IPCS) from 1976, which from 1989 also include metals, measures of total metal concentrations of essential metals such as copper and zinc cannot predict either their bioavailability or toxicity. Aspects such as metal speciation, natural occurrence (background levels), and prevailing environmental conditions must be taken into account.

Metals such as zinc, copper, iron, and chromium that are released to different extent from outdoor constructions are both essential and potentially toxic toward, for example, water organisms, effects that depend on their concentration, speciation (oxidation state, chemical form), and target organism. The chemical form (speciation) of released metals before and after surface and environmental interactions is therefore of utmost importance when assessing potential environmental risks. Studies at urban and rural sites in Stockholm, Sweden, at an urban site in Storrs, Connecticut, United States, and at a marine site in Brest, France, show that 14–100% (depending on the presence of organic matter such as pollen, pine needles, and leafs present at the rooftop) of the total amount of released copper at the immediate release situation (e.g., at the rooftop) is present in its most bioavailable forms (as free Cu^{2+} ions and weak complexes). This released fraction in runoff water comes into contact with different solid surfaces, for example, downspouts and drainage systems, or, for example, soils and pavements in the close vicinity of a building. Such surfaces are able to retain and change the chemical speciation of released metals to nonbioavailable species. Similar effects have been shown for zinc.

The importance of natural interactions, dilution effects, and chemical speciation of released metals in runoff water from outdoor constructions needs to be considered in any environmental risk assessment and management. Risk assessments that employ flow analyses to address metal dispersion need to consider these aspects and consider whether anthropogenic and natural sources actually have shown and proven any adverse effects or not.

FURTHER READING

Several papers have been published related to metal dispersion during the last decades of which a few are listed as follows.

Introduction

UN, Report of the World Commission on Environment and Development: our common future. Transmitted to the General Assembly as an Annex to document A/42/427—Development and International Co-operation: Environment, 1987.

Metal Dispersion, General

S.D. Cramer, S.A. Matthes, J.B.S. Covino, S.J. Bullard, and G.R. Holcomb, Environmental factors affecting the atmospheric corrosion of copper, in *Outdoor Atmospheric Corrosion*, ed. H.E. Townsend, ASTM, West Conshohocken, pp. 245–264, 2002.

W. He, I. Odnevall Wallinder, and C. Leygraf, A laboratory study of copper and zinc runoff during first flush and steady-state conditions, *Corrosion Science, 43*, 127–146, 2001.

Y.S. Hedberg, J.F. Hedberg, G. Herting, S. Goidanich, and I. Odnevall Wallinder, Critical review: copper runoff from outdoor copper surfaces at atmospheric conditions, *Environmental Science & Technology, 48*(3), 1372–1381, 2014.

ISO, ISO 17752, Corrosion of Metals and Alloys—Procedures to Determine and Estimate Runoff Rates of Metals from Materials as a Result of Atmospheric Corrosion, 2012.

S. Jouen, B. Hannoyer, A. Barbier, J. Kasperek, and M. Jean, A comparison of runoff rates between Cu, Ni, Sn and Zn in the first steps of exposition in a French industrial atmosphere, *Materials Chemistry and Physics, 85*(1), 73–80, 2004.

A.U. Leuenberger-Minger, M. Faller, and P. Richner, Runoff of copper and zinc caused by atmospheric corrosion, *Materials and Corrosion, 53*, 157–164, 2002.

L. Veleva, E. Meraz, and M. Acosta, Zinc corrosion runoff process induced by humid tropical climate, *Materials and Corrosion, 58*(5), 348–352, 2007.

Metal Runoff—Kinetic Aspects, Environmental Factors

K. Athanasiadis, H. Horn, and B. Helmreich, A field study on the first flush effect of copper roof runoff, *Corrosion Science, 52*(1), 21–29, 2010.

G. Bielmyer, W.R. Arnold, J. Tomasso, J. Isely, and S. Klaine, Effects of roof and rainwater characteristics on copper concentrations in roof runoff, *Environmental Monitoring and Assessment, 184*(5), 2797–2804, 2012.

M. Faller and D. Reiss, Runoff behaviour of metallic materials used for roofs and facades—a 5-years field exposure study in Switzerland, *Materials and Corrosion, 56*(4), 244–249, 2005.

D. Lindström and I. Odnevall Wallinder, Long-term use of galvanized steel in external applications. Aspects of patina formation, zinc runoff, barrier properties of surface treatments, and coatings and environmental fate, *Environmental Monitoring and Assessment, 173*, 139–153, 2011.

J. Sandberg, I. Odnevall Wallinder, C. Leygraf, and N. Le Bozéc, Corrosion-induced zinc runoff from construction materials in a marine environment, *Journal of the Electrochemical Society, 154*(2), C120–C131, 2007.

Construction Geometry, Predictions

Y.S. Hedberg, S. Goidanich, G. Herting, and I. Odnevall Wallinder, Surface-rain interactions: differences in copper runoff for copper sheet of different inclination, orientation, and atmospheric exposure conditions, *Environmental Pollution, 196*, 363–370, 2015.

I. Odnevall Wallinder, B. Bahar, C. Leygraf, and J. Tidblad, Modelling and mapping of copper runoff for Europe, *Journal of Environmental Monitoring, 9*(1), 66–73, 2007.

I. Odnevall Wallinder, P. Verbiest, W. He, and C. Leygraf, Effects of exposure direction and inclination on the runoff rates of zinc and copper roofs, *Corrosion Science, 42*, 1471–1487, 2000.

Environmental Fate

S. Bertling, I. Odnevall Wallinder, D.B. Kleja, and C. Leygraf, Long-term corrosion-induced copper runoff from natural and artificial patina and its environmental impact, *Environmental Toxicology and Chemistry, 25*(3), 891–898, 2006.

S. Bertling, I. Odnevall Wallinder, C. Leygraf, and D.B. Kleja, Occurrence and fate of corrosion-induced zinc in runoff water from external structures, *Science of the Total Environment, 367*(2–3), 908–923, 2006.

B. Boulanger and N.P. Nikolaidis, Mobility and aquatic toxicity of copper in an urban watershed, *Journal of the American Water Resources Association, 39*(2), 325–336, 2003.

B. Boulanger, and N.P. Nikolaidis, Modeling framework for managing copper runoff in urban watersheds, *Journal of the American Water Resources Association, 39*(2), 337–345, 2003.

L. Landner and R. Reuther, *Metals in Society and in the Environment: A Critical Review of Current Knowledge on Fluxes, Speciation, Bioavailability and Risk for Adverse Effects of Copper, Chromium, Nickel and Zinc*. Kluwer Academic Publishers, Dordrecht, 2004.

I. Odnevall Wallinder, Y. Hedberg, and P. Dromberg, Storm water runoff measurements of copper from a naturally patinated roof and from a parking space. Aspects on environmental fate and chemical speciation, *Water Research, 43*(20), 5031–5038, 2009.

J. Pettersen and E.G. Hertwich, Critical review: life-cycle inventory procedures for long-term release of metals, *Environmental Science & Technology, 42*, 4639–4647, 2008.

11

APPLIED ATMOSPHERIC CORROSION: ELECTRONIC DEVICES

11.1 INTRODUCTION

Electronic equipment and devices are an increasing part of our everyday life, and the electronics industry has been the largest industry, worldwide, for the last two decades. With computers and other electronic equipment being spread to locations in emerging markets around the world with high pollution levels, a dramatic increase has been observed on corrosion-related failures. The electronic devices can be degraded by a variety of electric, magnetic, mechanical, and chemical effects, often through simultaneous action of two or more of these effects. Among the most important causes of chemical degradation is atmospheric corrosion. Corrosion of critical parts or components of electronic devices can produce a wide range of consequences, from intermittent electrical faults to complete functional breakdown.

Corrosion of electronic components or parts is often more critical than corrosion of other objects, such as materials in the road environment (Chapter 12) or architectural materials (Chapter 13), because even small amounts of this degradation affect the electrical properties of the device (e.g., conductance, resistance), and thus its performance. Device performance is quite dependent on the requirements of the circuitry in which the device exists and corrosion-induced failures cannot, in general, be defined in a straightforward way. On connector or contact materials, minor amounts of corrosion products—hardly visible to the eye or even through a microscope—may cause failures, partly because of the trend toward smaller signal voltages. Small residues of humidity or chloride contaminants in an encapsulated integrated circuit can result in serious corrosion of the conductive pattern and in ionic migration between conductor paths. The problem is aggravated by dissimilar metal couples that create conditions for bimetallic corrosion, small metallization features, electric bias, and large electrical potential gradients across metallization

Atmospheric Corrosion, Second Edition. Christofer Leygraf, Inger Odnevall Wallinder, Johan Tidblad and Thomas Graedel.

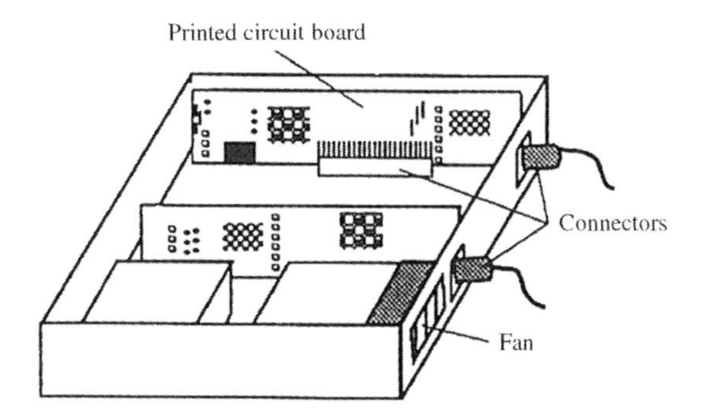

FIGURE 11.1 A typical configuration for electronic equipment.

features, and the use of materials that have been selected without corrosion resistance is the primary selection criterion.

Examples of environments with reported corrosion-induced failures of electronic devices include pulp, paper, and chemical industries, agricultural and coastal environments, telecommunication facilities, and during transportation (automobile, sea, aircraft). Corrosion-related problems in such environments are crucial factors in the reliability of electronics because of the extremely high susceptibility of the circuitry to the presence of gaseous or particulate pollution. Failures of electronics are nowadays connected with airborne pollutants at concentrations that are comparable with those related to health issues, or even lower. As one example, the highest recommended concentration of sulfur dioxide (SO_2) for electronics in the most benign corrosivity category has been set at 10 ppbv by the International Society of Automation (ISA) (see further in Section 11.5), while the European Environment Agency has defined 125 $\mu g\, m^{-3}$ (almost 50 ppbv) as an upper limit average value for a 24 h period. In general it is difficult, however, to compare highest recommended values for safety of electronics and humans since they apply over different time periods and can act in a variety of synergistic ways with other pollutants. Nevertheless, exposure in relatively safe environments from a health perspective can result in significant debasement of electronic components or equipment. Even environments normally characterized as benign have been known to cause problems.

Typical electronic equipment consists of a variety of components such as printed circuit boards, integrated circuits, discrete devices (switches, resistors, diodes, solder joints), connectors, a surrounding box, and fans, Figure 11.1. The components inside the shielding box are usually exposed to the same corrosive species as outside the box, although the concentrations can be different.

Atmospheric corrosion mechanisms in electronic components and equipment are similar to those of atmospheric corrosion in other situations. Considering that a wide variety of corrosion-induced failures can occur, the discussions in this chapter are restricted to some of the most vital parts of electronic components, including contacts, connectors, and integrated circuits. In addition, the chapter briefly discusses accelerated tests of electronics, environmental classification, and the most relevant corrosion protection methods.

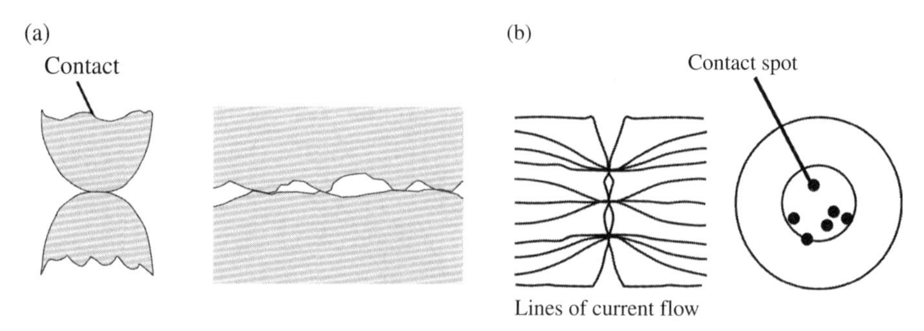

FIGURE 11.2 Current flow through contact spots.

11.2 CORROSION-INDUCED FAILURES OF CONTACTS AND CONNECTORS

Most metallic parts of a printed circuit board, such as the conductive pattern, are normally protected from environmental influences by a tin–lead plating, an organic coating termed a "solder resist," or a passivation treatment. An additional conformal coating may be applied after the components have been mounted. Other metallic parts of a printed circuit board, such as contacts and connectors, are exposed to the environment without any protection and are therefore highly susceptible to atmospheric corrosion effects. They commonly consist of a substrate of copper, brass, copper beryllium, or tin bronze alloy coated with a thin layer of plated gold or tin and often with a plated interlayer of nickel.

During contact between two surfaces, true electrical contact only occurs at small spots where the elevated irregularities of the two surfaces meet, Figure 11.2. The area of mechanical contact is mainly determined by the contact force and the hardness of the materials. When the current passes through the small contact spots, it is constricted to these spots, which causes a voltage drop. The resulting resistance is called constriction resistance. During corrosion of the contact surface, the result in some cases is a uniform film of corrosion products, which cause an additional voltage drop. More commonly, the corrosion attack results in distinct corrosion products that reduce the area and/or number of contact spots, Figure 11.3. A reduction in contact area increases the risk of intermittent contact failures. Should surface films of corrosion products or organic contaminants be present, the current through the contact experiences an additional film resistance, and the area through which the current passes may be even smaller than the area of mechanical contact. When determining the total contact resistance, the measured value includes the constriction resistance, the film resistance, and the bulk resistance of the contact substrate.

Pore corrosion is a special type of bimetallic corrosion often observed on gold-coated electric contacts. The noble metal top coating not infrequently has pores and other defects reaching to the undercoating or even to the substrate. This may create a galvanic couple in which the substrate corrodes and corrosion products fill the pores, migrate along the pore walls, and finally reach the contact surface, resulting in

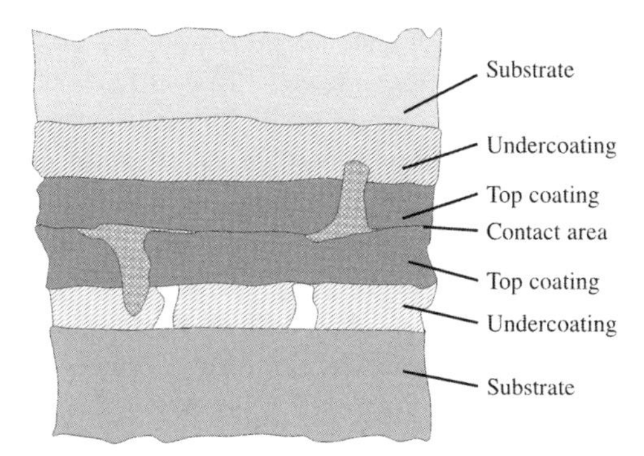

FIGURE 11.3 Corrosion products and influence on contact area.

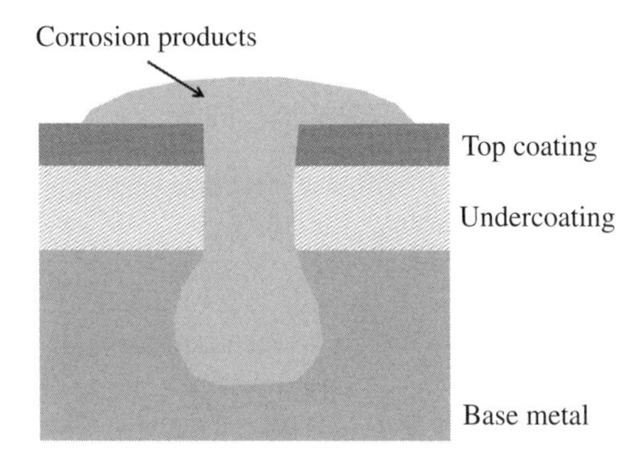

FIGURE 11.4 A schematic view of pore corrosion.

increased contact resistance, Figure 11.4. Because of the pores in the noble metal top coating, gold-coated contacts in particular have fairly short service life in severe environments.

Corrosion products, formed at cut edges of the contact surface, sometimes creep over the noble metal top coating and form an insulating film on its surface. Silver corrosion products creep mainly on gold-coated surfaces, whereas copper corrosion products creep on both silver- and gold-coated surfaces. A driving force for creep corrosion may be the difference in potential between the cut edges, where the anodic reaction (metal dissolution) occurs, and the noble metal coating, where the cathodic reaction (oxygen reduction) occurs. As a result, an electric field is created in the electrolyte between anodic and cathodic sites, in which cations, such as metal ions, migrate along the noble metal coating toward the cathodic sites and anions toward the anodic sites. The metal ions react with anions and form reaction products that precipitate on the noble metal coating. Nickel, used as an undercoating, often reduces the creep of copper corrosion products. In some environments, however, corrosion and the creep of corrosion products may be higher on nickel than on copper. This is the case in atmospheres containing SO_2 and chlorine (Cl_2) as principal gaseous pollutants.

Fretting corrosion involves simultaneous corrosion and oscillatory slip between two vibrating surfaces in contact. It can occur if two surfaces of normally stationary contacts are subjected to low-amplitude oscillatory motion relative to each other. This relative motion can be caused by repetitive electrical or magnetic variations or by mechanical or thermal movements. Fretting corrosion effects are triggered by wear that removes passivating layers, thereby increasing the oxidation rate, and wear production of fine metal particles that rapidly oxidize. This results in the formation of insulating films that will separate the surfaces electrically, thus increasing the contact resistance. Contact materials such as tin, which rapidly form oxides, are most susceptible to fretting corrosion.

Different failure mechanisms of contacts or connectors can be expected depending on the materials selected and on the actual environment. If the surface is gold plated, pore corrosion and creep of corrosion products are probable corrosion-related failure mechanisms. If the surface is tin plated, fretting corrosion is the dominant failure mechanism.

The different types of corrosion effects of gold-plated contact surfaces tend to be stimulated by different pollutants. Pore corrosion on gold-plated surfaces, resulting from corrosion of a nickel undercoating or of a copper or copper alloy substrate, is promoted by high levels of relative humidity, SO_2, Cl_2, and H_2S. The creep of corrosion products is mainly stimulated by the presence of H_2S. Polluted indoor industrial or agricultural environments have a detrimental effect on gold-plated surfaces. Pulp and paper industry environments often include Cl_2 and H_2S and are among the most severe environments for gold-plated surfaces.

On palladium- and platinum-plated surfaces, organic contaminants from the air can result in vibration-induced formation of insulating films. In this case the surface film is an organic polymer rather than a film of oxide or other corrosion products. To ameliorate this effect, palladium is sometimes plated with a thin gold flash. The gold reduces the probability of frictional polymerization, acts as a solid lubricant, and provides some corrosion protection.

11.3 CORROSION-INDUCED FAILURES OF INTEGRATED CIRCUITS

A wide variety of metallic materials are used in integrated circuits. The structure that forms the conductive pattern connects chips, resistors, and other components. It consists of narrowly spaced conductors, traditionally of aluminum or aluminum alloys but increasingly of copper. The conductor material may be coated with gold, tin, palladium, or silver.

The conductor paths have extremely small dimensions, with widths down to 50 nm in advanced devices, and form a complex geometric pattern that is highly sensitive to the presence of humidity and pollutants. To provide both chemical and mechanical protection, most parts of the conductor paths are coated with an insulating layer of silicon nitride and/or an organic polymer film. The chip is encapsulated in a rigid, water-impermeable plastic or placed in a hermetic package to provide further

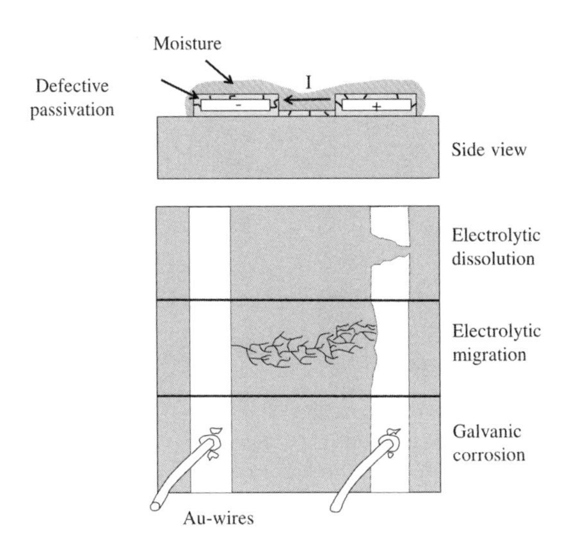

FIGURE 11.5 Corrosion failure mechanisms of metallic conductors on integrated circuits. (Adapted with permission from Taylor & Francis; G. Frankel, Corrosion of microelectronic and magnetic data-storage devices, in *Corrosion Mechanisms in Theory and Practice*, 3rd edition, ed. P. Marcus, CRC Press/Taylor & Francis Group, Boca Raton, p. 834, 2012.)

protection. The performance of the chip may, however, be influenced by defects in the insulating layer such as microcracks and pinholes, through which humidity and pollutants can penetrate. Pollutants may also originate from the encapsulation material or from residues of processing chemicals within the chip package itself. Chips operating at continuously high power levels corrode less frequently than those in unpowered conditions. The reason is that the temperature at the chip surface is higher and therefore the relative humidity under powered conditions is lower than in the ambient environment.

As in other examples of applied atmospheric corrosion, various corrosion processes may occur when the metallic conductor paths are exposed to humidity and pollutants. These include electrolytic dissolution, pitting corrosion, ionic migration, and galvanic corrosion. Electrolytic dissolution and pitting corrosion take place, for example, on aluminum conductors in the presence of electric potential differences between the separated conductor paths. When moisture is present, the conductor with the higher electric potential forms the anode of a corrosion cell and the metal conductor may be dissolved, Figure 11.5. When the contaminants contain chlorides, the dissolution of aluminum proceeds in a manner similar to pitting corrosion, and the device may fail because of a discontinuity in the aluminum conductor path. Reaction products such as $Al(OH)_3$ may easily precipitate because of the small volume of electrolyte. These products may influence further electrochemical reactions of the aluminum conductor. Conductors made of other metals, such as copper, may also dissolve anodically and fail because of pitting corrosion and electrolytic dissolution.

FIGURE 11.6 An example of electrochemical migration and dendrite formation on a chip capacitor. (a) SEM image of dendrite at high magnification. (b) Optical image showing the overall view of how the dendrite has grown from right (cathode) to left (anode). (Reproduced with permission from Elsevier; D. Minzari, M.S. Jellesen, P. Möller, and R. Ambat, On the electrochemical migration mechanism of tin in electronics, *Corrosion Science, 53,* 3366–3379, 2013.)

The dissolved metal ions migrate toward the opposing conductor. Even if the voltage difference between conductors is quite small, ionic migration (often referred to as electrochemical migration or electrolytic migration) may occur because the small distance between the paths creates high electric field strengths. When the metal ions reach the opposite, cathodically polarized, conductor, the metal ions can deposit, often in a dendritic pattern. This dendritic pattern is much more voluminous than the corresponding volume of dissolved metal and may result in a short circuit between the adjacent conductor paths, Figure 11.5. As the dimensions of integrated circuits continue to decrease, electrolytic dissolution and ionic migration have the potential to become even more important failure mechanisms in the future. Metals that are susceptible to electrochemical migration include tin, lead, copper, silver, and gold. Figure 11.6 displays an example of electrochemical migration and dendrite formation.

Electrical leakage on insufficiently protected parts of integrated circuits may occur not only by dissolved metal ions but also by humidity in combination with contaminated dust and other particles, many of which have hygroscopic properties. The result is an increase in surface conductance as the different particles dissolve in the thin surface electrolyte. Results presented in Figure 11.7 show that the surface conductance in the presence of indoor particles increases approximately exponentially as a function of relative humidity.

Corrosion-induced failures on integrated circuits also occur in the absence of any applied voltage if corrosive species are present in the environment. Another process, galvanic corrosion (bimetallic corrosion), takes place in the presence of humidity when two dissimilar metals are in electric contact, forming a galvanic couple. The driving force is their difference in electrochemical nobility. Galvanic corrosion on integrated circuits may exist, for example, at aluminum bond pads connected to gold wires, Figure 11.8.

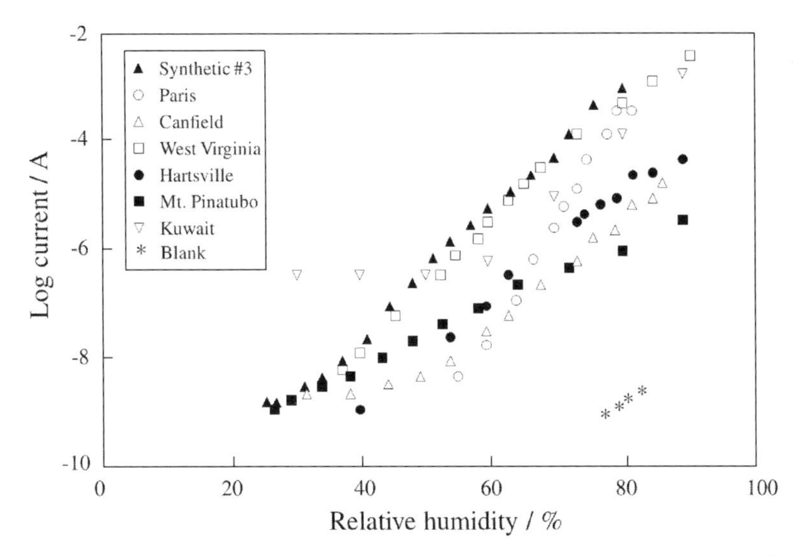

FIGURE 11.7 Leakage current induced on an integrated circuit test pattern at different relative humidities in the presence of ionic contaminants originating from particles collected at various indoor sites. The increase in leakage current with relative humidity is usually exponential because of the hygroscopic properties of the particles. The exception is the Kuwait site, where most particles consist of soot, which is not hygroscopic. (Reproduced with permission from The Electrochemical Society; R.B. Comizzoli, R.P. Frankenthal, R.E. Lobnig, G.A. Peins, L.A. Psota-Kelty, D.J. Siconolfi, and J.D. Sinclair, Corrosion of electronic materials and devices by submicron atmospheric particles, *The Electrochemical Society Interface, 2*(3), 26–34, 1993.)

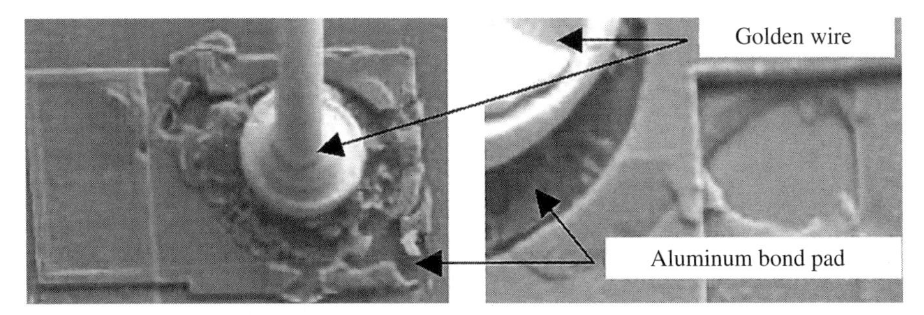

FIGURE 11.8 An example of galvanic corrosion on an integrated circuit: a gold wire connected to aluminum bond pad. (Reproduced with permission from Maney Publishing; O.V. Elisseeva, A. Bruhn, J. Cerezo, A. Mavinkurve, R.T.H. Rongen, G.M. O'Halloran, R. Ambat, H. Terryn, and J.M. Mol, Novel electrochemical approach to study mechanism of Al-Au wire-bond pad interconnections, *Corrosion Engineering, Science and Technology, 48*, 409–417, 2013.)

11.4 ACCELERATED TESTS OF ELECTRONICS

Various aspects of accelerated corrosion tests have been treated in some detail in Chapter 6. Table 11.1 presents important parameter values of selected and frequently used accelerated corrosion tests of contacts and connectors. More detailed descriptions of these and other tests are given in the standard ASTM B845-97(2013). All tests are the result of extensive investigations of electronic equipment or components aiming at possibly correlating the corrosion products and mechanisms of relevant materials found in the accelerated tests and in application conditions, respectively.

The Battelle (Class II) is one of several tests developed by the Battelle Columbus Laboratories and accelerates exposure conditions found in business offices or control rooms located in industrial areas without continuous environmental control. The IBM G1(T) test was developed by IBM for qualifying connectors and electrical components in IBM business office applications. The IEC (68-2-60) test was developed by the International Electrotechnical Commission and based on investigations conducted by Siemens Corporation in Germany in public buildings belonging to the German Federal Railways and Post Office Department. An ASTM designation number identifies a unique version of an ASTM standard.

The severity level of each accelerated test is assessed through the corrosion effects of copper coupons used as reference sample (techniques for evaluating corrosion effects of copper and other metals; see Appendix A). This allows each user of the test to determine whether the corrosivity of the atmosphere that has been generated is within the desired range. A principal difference between the tests listed in Table 11.1 is the exclusion (Battelle) or inclusion (IBM and IEC) of SO_2. Examinations of corrosion products formed on copper exposed according to these tests show that the main crystalline substances in all tests consist primarily of cuprite (Cu_2O) and atacamite ($Cu_2Cl(OH)_3$) with minor amounts of other phases including posnjakite ($Cu_4SO_4(OH)_6{\cdot}H_2O$). The difference in corrosion product

TABLE 11.1 A Selection of Accelerated Corrosion Tests of Electronic Devices Based on Mixtures of Gaseous Corrosive Species

Test	SO_2 (ppbv)	NO_2 (ppbv)	Cl_2 (ppbv)	H_2S (ppbv)	Relative Humidity (%)	Temperature (°C)
Battelle		200	10	10	70	30
Class II		±25	+0/−2	+0/−4	±2	±2
IBM G1(T)	350	610	3	40	70	30
	±5%	±5%	±15%	±5%	±2	±0.5
IEC (68-2-60)	200	200	10	10	75	25
	±20	±20	±5	±5	±3	±1

composition with and without SO_2 is in the relative proportions of the corrosion products: SO_2 seems to stimulate the formation of more sulfate-containing corrosion products. The difference is most clearly demonstrated on nickel, which is a strong sulfate former.

All tests include Cl_2 and H_2S. Analyses of a broad range of industrial indoor environments have shown, however, that HCl rather than Cl_2 is the most important gaseous chlorine-containing pollutant in almost all circumstances. The analyses have furthermore shown that HCl and H_2S concentrations vary over three orders of magnitude and are uncorrelated, suggesting that their sources are unrelated. Hence, sites with high levels of both corrosive substances are unlikely. This implies that in field exposures the reaction of H_2S seldom competes with that of HCl (or Cl_2), whereas in accelerated laboratory tests, with high delivery rates of both corrosive substances to the metal surface, there is a strong competition between the reactions. As a consequence, while field-exposed silver forms both silver sulfide (Ag_2S) and silver chloride (AgCl), presumably at different times during which corrosion conditions change, silver exposed in accelerated tests primarily forms Ag_2S only.

Accelerated tests for integrated circuits, finally, are based on conditions of electric bias and with the electronic device exposed to elevated relative humidity and temperature. A frequently used test operates for 1000 h at 85% relative humidity and 85°C and with the conducting lines under bias at normal operating conditions. Increased device reliability has resulted in the development of tests with more severe conditions, including 85% relative humidity and 130°C under bias or 100% relative humidity and 100°C under bias.

So far no universal corrosion test method has been developed that is capable of fully simulating corrosion processes or of predicting the overall performance of an electronic component in all kinds of environments. An important criterion in the search for better accelerated tests of contacts, connectors, and integrated circuits is that the failure mechanisms that occur in the tests should be the same as under normal operating conditions (Section 6.5).

11.5 CLASSIFICATION OF ENVIRONMENTS WITH RESPECT TO CORROSIVITY

In order to predict possible corrosion-induced failures of electronic components or devices, different classification systems have been developed with the goal of predicting the corrosivity of a given indoor environment. One of the most widely established standard procedures for indoor environment corrosivity classification has been adopted by the ISA. Although liquids and solids are known to have great influence on indoor corrosivity, the influence of gases has more often been considered. The contaminants in the ISA standard include liquids, solids, and gases, with the liquids subdivided into vapors and aerosols and the solids into particles of different sizes. The standard follows two different approaches. The first

TABLE 11.2 Corrosivity Classification According to the International Society of Automation (ISA)

Corrosivity Level	Thickness of Corroded Copper Layer (nm)	Description of Severity Level
G1 mild	<30	An environment sufficiently well controlled such that corrosion is not a factor in determining equipment reliability
G2 moderate	<100	An environment in which the effects of corrosion are measurable and may be a factor in determining equipment reliability
G3 harsh	<200	An environment in which there is a high probability that corrosive attack will occur. These harsh levels should prompt further evaluation, resulting in environmental controls or specially designed and packaged equipment
GX severe	>200	An environment in which only specially designed and packaged equipment would be expected to survive. Specifications for equipment in this class are a matter of negotiation between user and supplier

The left column gives the severity level, the middle column the maximum thickness of the corroded copper layer, and the right column is an explanation of each level. All thickness values refer to 1 month of exposure.

is based on the exposure of copper samples and subsequent quantitative analysis of corrosion attack, and the second on measurement of the concentration of pollutants and of climatic parameters. Copper is used to define the environment in the first approach, because it is a common metal in electronics and because its corrosion behavior is influenced by many different gaseous corrosive species rather than only by a few (see Table 4.1).

According to the standard, each indoor environment belongs to one of four corrosivity classes, labeled G1, G2, G3, and GX, Table 11.2. The copper corrosion related to these classes is determined through exposure of freshly abraded copper coupons for at least one month. To obtain a correct value for low levels of corrosivity, exposure for at least three and preferably 12 months is recommended. The amount of corrosion products formed is determined by cathodic reduction (see Appendix A).

For each corrosivity class the ISA standard also provides maximum levels of certain gaseous corrosive species (Table 11.3). The data is based on a compilation of many field observations. The classes of indoor environments that are defined in this way are believed to result in the corresponding copper corrosion according to Table 11.2, provided the relative humidity is less than 50%. Following the ISA standard, the corrosivity class is expected to increase by one level if the average relative humidity increases by 10%.

TABLE 11.3 Maximum Concentrations of Important Gaseous Corrosive Species in the Four Corrosivity Levels According to ISA Corrosivity Classification

	Gaseous Corrosive Specie and Maximum Concentration in ppbv						
Corrosivity Level	H_2S	SO_2	Cl_2	NO_X	HF	NH_3	O_3
G1	<3	<10	<1	<50	<1	<500	<2
G2	<10	<100	<2	<125	<2	<10,000	<25
G3	<50	<300	<10	<1250	<10	<25,000	<100
GX	≥50	≥300	≥10	≥1250	≥10	≥25,000	≥100

If the gases in a given environment correspond to different corrosivity levels, then the overall level is determined by the gas designating the most severe level.

The ISA standard can be regarded as only a crude guide for corrosivity prediction. The maximum concentrations depend on many unmeasured parameters, such as the presence of other corrosive species in gaseous, liquid, or solid form, the airflow rate, and variations in temperature and relative humidity. Accordingly, experience has shown that a corrosivity classification based on exposure and analysis of copper samples or of a combination of metals results in more reliable results than a classification based on measurements of gaseous corrosive species. Nevertheless, the ISA standard can be a useful guide if treated with caution. As one example of corrosivity classification, Figure 11.9 displays how the corrosivity levels are distributed across a number of indoor sites that have been studied in detail.

11.6 METHODS OF PROTECTION

All electronic components, devices, or equipment can be more or less effectively protected against corrosion-induced failures. The protective measures can be grouped into:

- Selection of materials and design of components
- Protective measures during manufacture, transport, and storage
- Protective measures during installation and service

11.6.1 Selection of Materials and Design of Components

An analysis of the expected ambient indoor environmental conditions is an essential part of the precautionary measures. Foreseeable changes of the environment during the anticipated life cycle of the equipment must also be judged and taken into account. A recommended practice is to determine the corrosivity class of the environment before and after installation of electronic equipment to make sure that the environment matches the corrosivity level for which the equipment is designed. The measurement period should include the most corrosive time of the year, usually

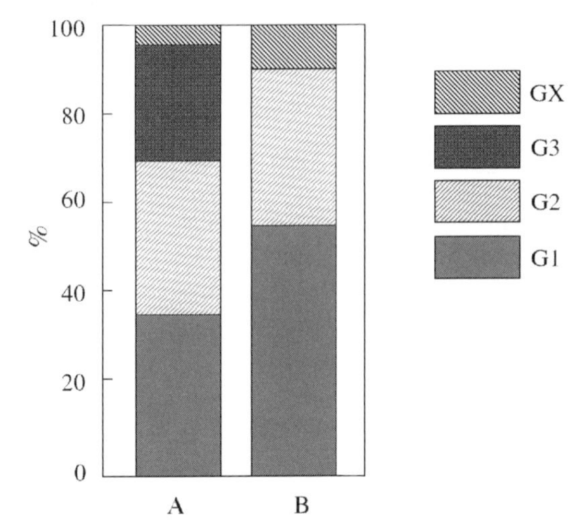

FIGURE 11.9 Examples of distributions of severity levels (G1 to GX) among different environments. The distributions are based on 45 sites within a Nordic research project (A) and 20 air-purified instrument rooms in pulp and paper industries (B). The corrosivity classification was based on exposure of copper samples. (Adapted with permission from J. Henriksen, R. Hienonen, T. Imrell, C. Leygraf, and L. Sjögren, *Corrosion of electronics*: a handbook based on experiences from a Nordic research project. Bulletin No 102, Swedish Corrosion Institute, Stockholm, Sweden, 1991.)

the most humid months. If the environment is found to be of severity level G2, G3, or GX, it should be further analyzed in order to determine the reason for the high corrosivity and to explore ways to ameliorate the situation.

In environments of corrosivity level G2, the reliability and service life of electronics are strongly influenced by the design and by the choice of material and manufacturing processes. In more corrosive environments, corresponding to G3 or GX, protection by means of a barrier such as a protective housing or an organic coating becomes necessary.

The corrosion susceptibility of assembled units can be markedly influenced by the choice of certain design parameters. Equipment housings and component encapsulations can reduce the levels of pollutants and humidity by lowering air convection and diffusion and thus the risk of corrosion. Fans used for cooling may simultaneously speed up the corrosion rate, because increased air velocity increases the deposition rate of pollutants and dust on components. An adverse effect of encapsulations may be the outgassing of organic corrosive vapors. Condensed moisture inside encapsulations and housings may also cause serious corrosion. Increasing the internal temperature through self-heating reduces the relative humidity and, usually, the corrosivity. Materials that outgas corrosive substances (such as certain cardboard spacers containing small amounts of H_2S) should be avoided in sealed or poorly ventilated enclosures or packages.

11.6.2 Manufacture, Transport, and Storage

During the manufacture and assembly of electronic equipment, corrosion problems are easily added. Surfaces may be contaminated with fingerprints or with metallic or organic contaminants. Reactive species from local pollutant sources, inside or outside the plant, may attack metal surfaces. Residues from soldering (corrosive flux) or the cleaning process (cleaning chemicals or salts) are occasionally left on components and boards. Hence, careful cleaning of components and assembled units is essential.

During transport, storage, and installation, environmental conditions may be significantly different from those in other phases of the service life of electronics. For electronics used in mild environments, transport and storage are often the most severe phases of the service life. Precautions necessary for minimizing corrosion effects include various means of preventing the exposure of electronics to humidity or pollutants such as hermetically sealed transport packages, desiccants, or vapor-phase inhibitors.

A transport package typically consists of a mechanical protection component, a barrier, and shock absorbing materials. The barrier—commonly a polymeric film but sometimes one containing reactive flakes of metal—serves primarily to reduce the rate of exchange of air, water vapor, and, if the metal flakes are present, corrosive gases. The films sometimes permit transport of water vapor, however, and diffusion of water into the package may cause corrosion problems unless a desiccant is used. It is essential that all materials are dry during the packing process and that the atmosphere of the packing room has a low water content.

Vapor-phase inhibitors are sometimes used in order to reduce corrosion during transport, storage, and service. The inhibitors vaporize gradually because of their low vapor pressure and are subsequently deposited on the electronic surface to form a vapor barrier. They are usually effective only in closed compartments for limited periods of time and have found application mainly in connection with the transport or storage of electronics. There are possible harmful effects, however, such as the deterioration of certain plastic materials, a decrease in the insulating properties of printed boards, and the formation of thin insulating layers on contact surfaces.

11.6.3 Installation and Service

A complete indoor air treatment system involves a special design of the room. The most important parts of the system are temperature and humidity control, air filtration, and room sealing. Sometimes all these parameters must be combined to reach an acceptable corrosivity level. To reduce penetration of corrosive gases and water vapor, the room will need to be sealed, new doors and/or air locks installed, and cable entries, floor, wall, and ceiling joints sealed.

High relative humidity can be reduced by heating or drying the air through sorption drying using moisture-absorbing materials or by other methods. If highly polluted outside air is to be drawn into the room, an air purification system may need to be installed. The most common filters for such a system include active carbon, impregnated active carbon, or aluminum oxide impregnated with potassium permanganate. The reaction time between filter and pollutant is critical for absorption efficiency. The most effective method is to use a deep-bed carbon filter and to pressurize the room with purified air.

FURTHER READING

Corrosion-Induced Failures

R. Ambat, S.G. Jensen, and P. Möller, Corrosion reliability of electronic systems, *ECS Transactions, 6*, 17–28, 2008.

R.B. Comizzoli, R.P. Frankenthal, R.E. Lobnig, G.A. Peins, L.A. Psota-Kelty, D.J. Siconolfi, and J.D. Sinclair, Corrosion of electronic materials and devices by submicron atmospheric particles, *The Electrochemical Society Interface, 2*(3), 26–34, Fall 1993.

G. Frankel, Corrosion of microelectronic and magnetic data-storage devices, in *Corrosion Mechanisms in Theory and Practice*, 3rd edition, ed. P. Marcus, CRC Press/Taylor & Francis Group, Boca Raton, pp. 825–861, 2012.

J. Henriksen, R. Hienonen, T. Imrell, C. Leygraf, and L. Sjögren, Corrosion of electronics: a handbook based on experiences from a Nordic research project. Bulletin No 102, Swedish Corrosion Institute, Stockholm, Sweden, 1991.

Accelerated Tests

ASTM, ASTM B845-97(2013), Standard Guide for Mixed Flowing Gas (MFG) Tests for Electrical Contacts. Standard, American Society for Testing and Materials, 2013.

Classification of Environments

ANSI/ISA, ANSI/ISA-S.71.04-2013, Standard: Environmental Conditions for Process Measurements and Control Systems: Airborne Contaminants. The International Society of Automation, 2013.

Methods of Protection

J.B. Jacobsen, J.P. Krog, A. Hjarbaek Holm, L. Rimestad, and A. Riis, Climate-protective packaging, *IEEE Industrial Electronics Magazine*, pp. 51–59, September 2014.

12

APPLIED ATMOSPHERIC CORROSION: AUTOMOTIVE CORROSION AND CORROSION IN THE ROAD ENVIRONMENT

12.1 INTRODUCTION

Corrosion of cars is perhaps the most well-known aspect of atmospheric corrosion to the general public, and significant costs are associated with corrosion of vehicles, for the society as a whole, for the manufacturers and their suppliers, and for the individual users. The road environment is also a significant part of our infrastructure including a variety of metal constructions like safety rails, lamp posts, and many other constructions.

The corrosivity of the road environment is varying significantly going through open landscapes, over bridges, into tunnels, and not the least inside cities. Each of these is inherently different in terms of type of environment, covering open exposures in all types of environments and sheltered exposures in dirty tunnels. However, even if tunnels are protected from natural rain, it does not mean that they are unaffected by running water as regular cleaning and unexpected leakage need to be accounted for. In urban environments with tall buildings, street canyon formation affects the transportation of pollutants resulting in corrosion rates highly depending on the local environment. The importance of the microclimate is a significant characteristic of corrosion in the road environment, both with regard to corrosion at different positions on a vehicle and to corrosion at different places in the vicinity of the road.

Even if the road environment shares many aspects of outdoor and indoor environments in general, there is one factor that makes it unique, at least in part of our world—application of deicing salts. Safety of winter roadways is the primary concern, and many different types of requirements are put on a freeze point depressor

Atmospheric Corrosion, Second Edition. Christofer Leygraf, Inger Odnevall Wallinder, Johan Tidblad and Thomas Graedel.

TABLE 12.1 Typical Corrosion Rates of Carbon Steel and Zinc after 1 Year of Exposure in Different Environments

Environment	Steel (μm)	Zinc (μm)
Off-road stationary exposure (open urban)	10–20	0.5–1
Off-road stationary exposure (marine)	50[a]	1–2
Near-road stationary exposure (open urban)	35	5
Near-road stationary exposure (shelter and tunnel)	50	5–8
On-vehicle exposure (general)	10	1–2
On-vehicle exposure (areas with deicing salts)	50	5–8[b]

From Sederholm and Almqvist (2011), Kreislova et al. (2011), and Jonsson (2012) (see Further Reading).
[a] The corrosion in marine environments varies substantially (see text).
[b] Occasional values of extreme corrosion >20 μm may be observed.

including cost-effectiveness and nontoxicity. Therefore, the solution has been and continues to be to spread salt, most commonly NaCl, on our roads, one of the most efficient corrosion stimulators we know. Thus, the primary aim of this chapter will not be to reiterate the general aspects of atmospheric corrosion, including effects of gaseous and particulate pollutants, but to focus on the effect of deicing salts in stationary environments and on different components of a running vehicle. Furthermore, and as for the remaining applied parts of this book, the primary aim is not to give a full overview of all important aspects to consider regarding automotive corrosion and corrosion in the road environment but rather to illustrate how some of the basic principles described in the initial chapters can be relevant for important applications of atmospheric corrosion.

Testing is very important in automotive applications, and the use of accelerated corrosion tests is so frequent and has resulted in such a variety of test methods that it is an industry in itself, with testing chambers from simple cabinets to sophisticated equipment affordable to few. Nevertheless, there are many common denominators to the methods, and they can be understood in terms of general characteristics of atmospheric corrosion as described in this book.

12.2 TYPICAL CORROSION RATES IN THE ROAD ENVIRONMENT

To put the corrosion rates measured in road environments into context, it is worthwhile to compare rates measured in the road environment with other values obtained at other outdoor conditions. Table 12.1 shows such typical values for different conditions. As should be evident at this stage of reading, each of these entries depends on a variety of environmental parameters, including the occurrence of gaseous, particulate pollutants and climatic conditions. As will be seen in the continuation of this chapter, the corrosion rates in the road environment vary considerably depending on a variety of parameters. Nevertheless, Table 12.1 gives a first indication of results and also a comparison of the different sensitivities of carbon steel and zinc to different environments.

The values for carbon steel are about equally high in off-road stationary marine conditions, near-road stationary exposure, and on-vehicle exposures in areas with deicing salts (about 50 µm) but depend largely on the distance to the coast (marine exposures), the salting frequency (road environments), and local wind conditions. For zinc, there is a much larger relative variation in corrosion rates depending on the environment, ranging from less than 1 µm in less polluted off-road environments to extreme values on vehicles in areas with deicing salts. Parameters of importance for this variation in corrosion include salting frequency, especially in the beginning of the exposure, degree of sheltering, effect of different cations in salts, deposition of road mud, and the temperature–relative humidity complex. These parameters will be discussed in some detail in the next sections.

12.3 PARAMETERS AFFECTING CORROSION IN ROAD ENVIRONMENTS

12.3.1 Effect of Salting Frequency

Normally when temperature increases it is expected that corrosion also increases. For example, when comparing corrosion rates for carbon steel exposed at a marine station at the Swedish West Coast (Bohus-Malmön), it can be around 30 µm year^{-1} for a cold year and twice as much for a warm year (60 µm year^{-1}). For the same years, however, the corrosion in road environments may be reversed so that corrosion rates will reach lower values during a warmer year, with lower frequency of deicing. Figure 12.1 shows the effect of the number of salting events on

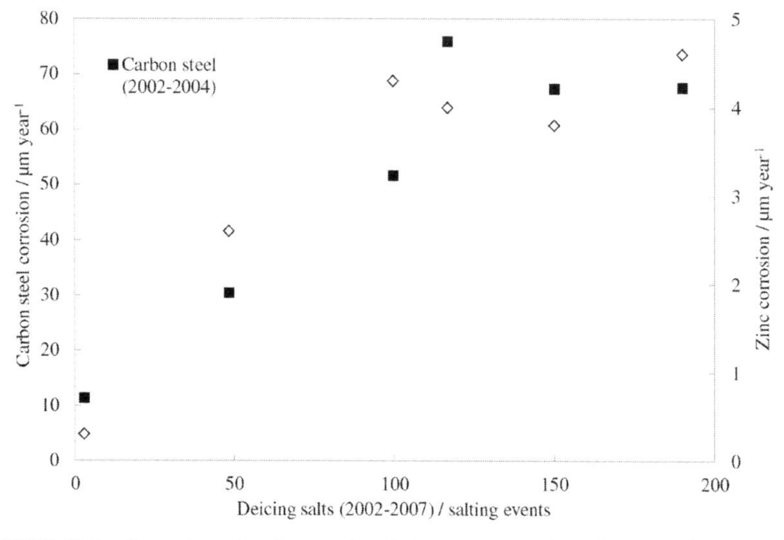

FIGURE 12.1 Corrosion of carbon steel and zinc versus number of events with spreading of deicing salts based on measurements in two city road environments and four bridges in Sweden. (B. Sederholm and J. Almqvist, unpublished data.)

corrosion rates of carbon steel and zinc exposed at different positions in roads and bridges in Sweden.

It is evident from Figure 12.1 that the number of deicing events is a crucial parameter. It seems like the corrosion reaches a plateau after about 100 events per winter at 70 $\mu m\ year^{-1}$ for carbon steel and 4 $\mu m\ year^{-1}$ for zinc. It should be noted that the number of salting events is a very difficult parameter to obtain from authorities and is probably very uncertain and that if possible a simultaneous measurement of chloride deposition is to be preferred.

12.3.2 Effect of Sheltering

As was stated for outdoor exposures in general (Chapter 8), empirical evidence suggests that the washing effect is more important for heavily polluted sites. This is especially true for sites with significant chloride deposition where sheltered samples are expected to have a higher corrosion rate than unsheltered samples. The road environment is obviously aggressive since in addition to the chloride deposition, road mud is heavily deposited on sheltered samples contributing to increased time of wetness. This is illustrated in Figure 12.2 for three cities in the Czech Republic, where the corrosion rate in sheltered position is almost twice as high as in open position for both carbon steel and zinc.

12.3.3 Deicing Salts and Effect of Different Cations

A comparison of corrosion rates in marine and near-road environments shows clear differences for both carbon steel and zinc (Table 12.2). Deicing salts is most commonly NaCl while sea salt contains predominantly NaCl but also $MgCl_2$, $MgSO_4$, K_2SO_4, and $CaCO_3$. One possible explanation for the different behavior can therefore be the influence of cations. Furthermore, both $CaCl_2$ and $MgCl_2$ are also frequently used for deicing as alternative to NaCl. The complex effect of Na^+, Ca^{2+}, and Mg^{2+} in combination with Cl^- is shown in Figure 12.3. For carbon steel, the mass gain is the highest in presence of $MgCl_2$, followed by $CaCl_2$ and NaCl, while for zinc the situation is reversed with the highest mass gains for NaCl. The interpretation of this difference is related to pH and the different corrosion products that form on carbon steel and zinc. Mg^{2+} and Ca^{2+} form hydroxy complexes resulting in an expected lower pH value for aqueous adlayers containing $MgCl_2$ and $CaCl_2$ compared to NaCl. This is also in accordance with the relative ranking of the corrosivity of these cations for carbon steel where a decrease of pH is detrimental. The decrease in corrosion rate for zinc at lower pH is instead related to the increased formation of simonkolleite ($Zn_5(OH)_8Cl_2{\cdot}H_2O$), which is stable at lower pH values as opposed to hydrozincite ($Zn_5(OH)_6(CO_3)_2$), which is stable at higher pH values. Under these conditions simonkolleite is assumed to possess a higher protective ability compared to hydrozincite, and therefore zinc exhibits a lower corrosion attack despite the lower pH levels.

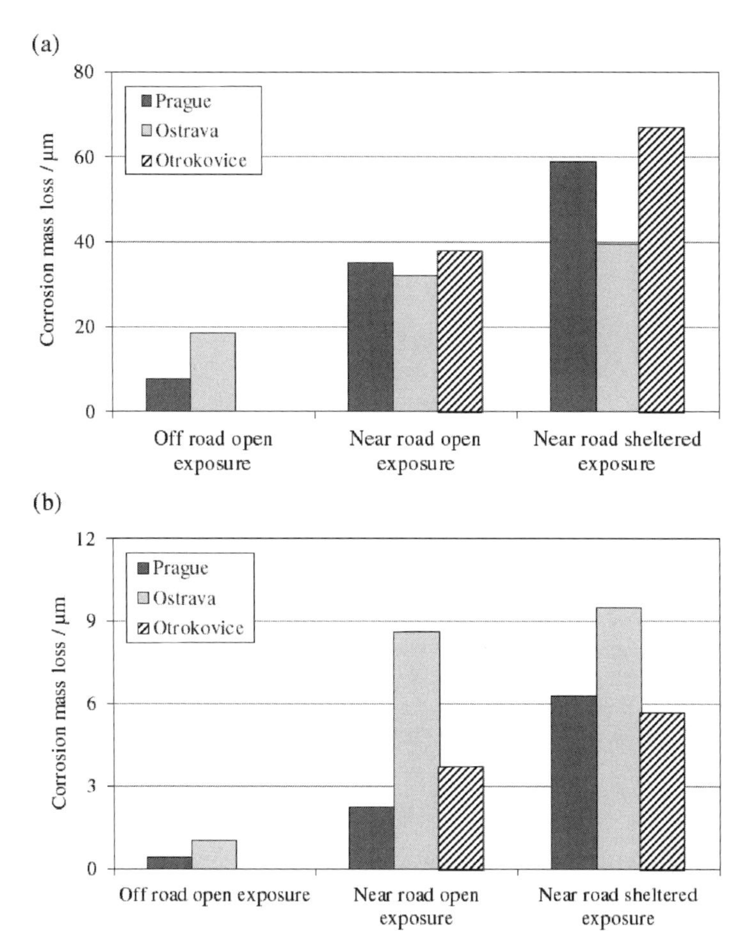

FIGURE 12.2 Corrosion of carbon steel (a) and zinc (b) after 1 year of open and sheltered exposure at different off- and near-road positions. (Adapted with permission from K. Kreislova, H. Geiplova, T. Link, and K. Hovorkova, The effect of road environment on corrosion of the infrastructure constructions, in *Proceedings of the EUROCORR 2011*, Stockholm, Sweden, September 4–8, 2011.)

TABLE 12.2 Weight Loss of Carbon Steel and Zinc with Deposited $CaCl_2$, $MgCl_2$, and NaCl at a Chloride Concentration of 1400 mg m^{-2} after 28 Days at 20°C and 80% RH; Weight Loss in g m^{-2}

		Weight Loss (g m^{-2})	
Substrate/Deposits	Calcium Chloride	Magnesium Chloride	Sodium Chloride
Zinc	5.1	1.4	12.9
Carbon steel	52.8	69.2	30.2

Adapted from T. Prosek, D. Thierry, C. Taxén, and J. Maixner, Effect of cations on corrosion of zinc and carbon steel covered with chloride deposits under atmospheric conditions, *Corrosion Science*, *49*, 2676–2693, 2007.

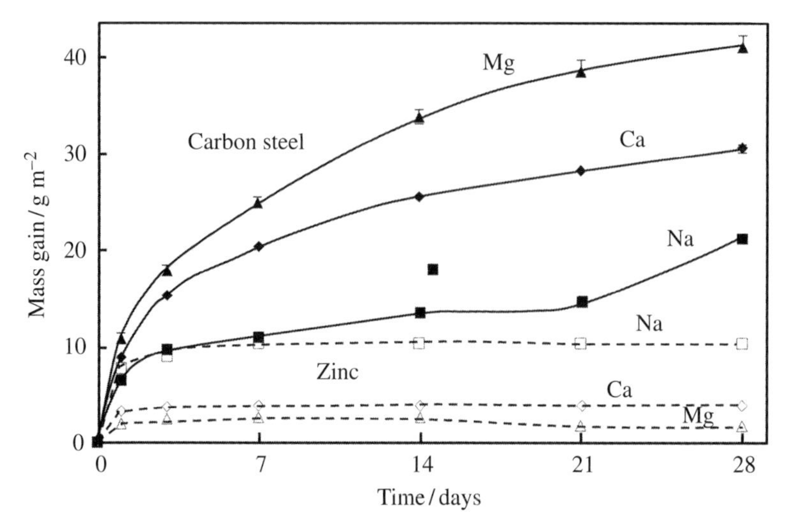

FIGURE 12.3 Wet mass gain of carbon steel (full curves) and zinc (dashed curves) with deposited $CaCl_2$, $MgCl_2$, and NaCl at the chloride concentration of 1400 mg m^{-2}. (Reproduced with permission from Elsevier; T. Prosek, D. Thierry, C. Taxén, and J. Maixner, Effect of cations on corrosion of zinc and carbon steel covered with chloride deposits under atmospheric conditions, *Corrosion Science, 49,* 2676–2693, 2007.)

12.3.4 The Temperature and Relative Humidity Complex

The influence of temperature and relative humidity has been described previously, both in connection with stationary outdoor, Chapter 7, and indoor, Chapter 8, exposures. It is important to realize, however, that there are differences in temperature and relative humidity variations for an on-vehicle situation, as opposed to stationary exposures, as is shown in Figure 12.4. Furthermore, the variations on a driving vehicle can vary substantially depending on the location on the vehicle, in particular the distance to the engine.

Looking first at the absolute values in Figure 12.4, the temperature on the vehicle can reach higher values, corresponding of course to periods when the vehicle is run for a longer time. To increase the temperature is an important acceleration factor for corrosion, as will be discussed later in connection with accelerated testing. Another important difference between the temperature and relative humidity variations is the rate of change for short-term variations (hours). For normal variations during most of the time, there is not much difference. However, when looking at extremes (99 percentiles), the differences are more prominent. For temperature, these extremes are about (both positive and negative changes) $\pm 1\,°C\,h^{-1}$ for stationary and $\pm 3°C\,h^{-1}$ for mobile exposures. For relative humidity the corresponding values are $\pm 3\%\,h^{-1}$ and $\pm 6\%\,h^{-1}$, respectively. The sudden changes associated with the mobile exposure can be exemplified by the situation when a

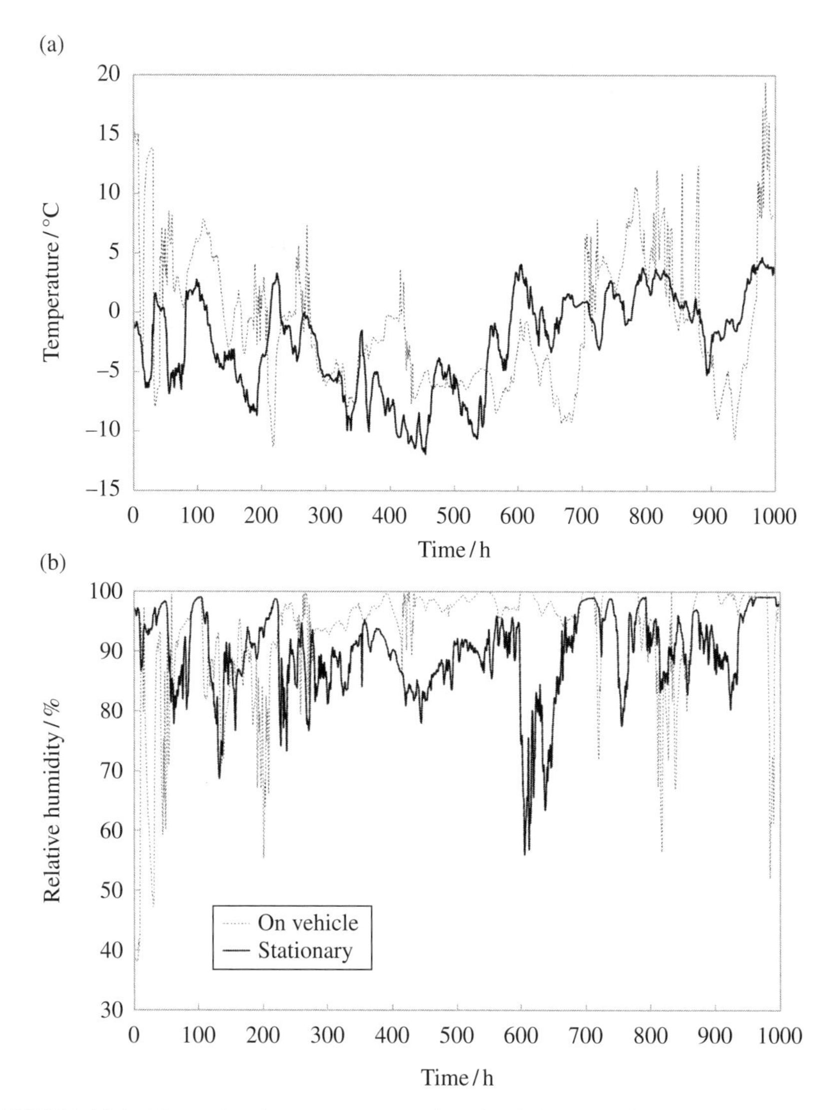

FIGURE 12.4 Example of temperature (top) and relative humidity (bottom) variations (winter 2010/2011) for a driving truck (dashed curves) and a marine stationary site (solid curves) in Sweden. (B. Rendahl, unpublished data.)

cold vehicle enters a warm garage. Dew will then form on the metal surface, which most likely is dirty and contaminated with pollutants. The resulting aqueous adlayer will then as a consequence be saturated with pollutants and be very corrosive with ample time to react with the metal in the garage at a higher temperature compared to the outside temperature.

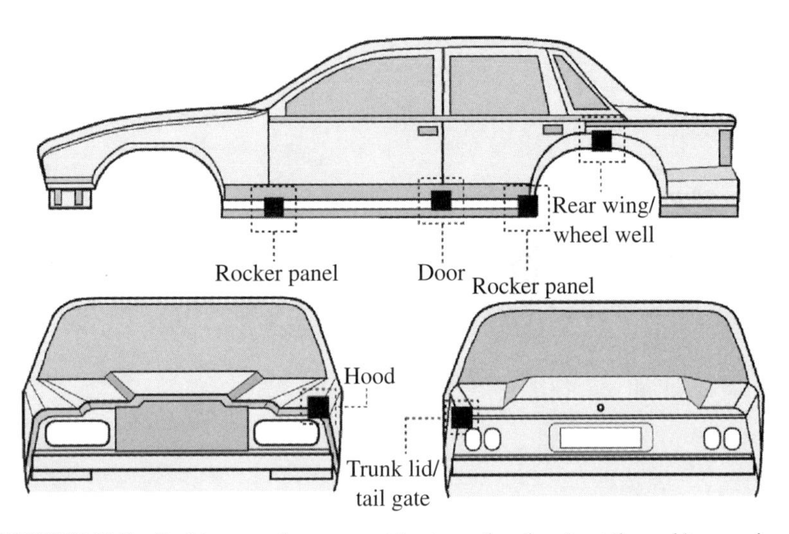

FIGURE 12.5 Positions on the car sensitive to perforation (rust through) corrosion.

12.4 CORROSION OF VEHICLES

The rapid and frequent variations in relative humidity and temperature together with the deposition of deicing salts and road mud make it a challenge to design a car that has a cost-efficient protection system against corrosion. Positions on the car with sensitive crevice surfaces include spot-welded joints (rocker panel and rear wing) and hem flanges (door, hood, and trunk lid), Figure 12.5. Of specific importance for automotive applications are also corrosion and other properties of joined materials and joining technologies. (For more details on these issues, see "Further Reading.")

A typical hem flange with critical crevices is shown in Figure 12.6. A series of investigations have been performed since the 1990s on more than 30 car models from the most prominent car manufacturers by sawing out critical parts from cars and inspecting them for corrosion. Figure 12.7 shows an example of all results including parts sawn out from the door.

Looking at Figure 12.7, the situation has improved a lot since the 1990s where almost half of the cars produced in 1989 after 3 years of service had considerable or superficial corrosion as opposed to less than 20% of cars produced in 2005. The reason for this is increased understanding of the most important factors determining a good resistance of the car body and their practical implementation. This includes, but is not limited to, good design, metallic surface coating, properly applied electro-coated primer with good adhesion and penetration, and, after that, sealer ensuring protection from moisture and mud. Treatment with anticorrosion agents in cavities can also increase the performance and will protect surfaces not fully covered with adhesives.

Similar as to what was described in Chapter 8 where outdoor corrosivity could be determined by exposure of standard specimens on racks, it is also possible to expose

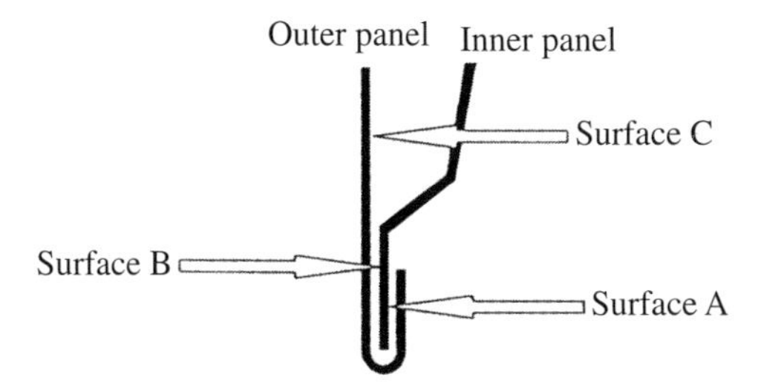

FIGURE 12.6 Schematic hem flange on door, hood, or trunk lid and selected crevice surfaces (A–C).

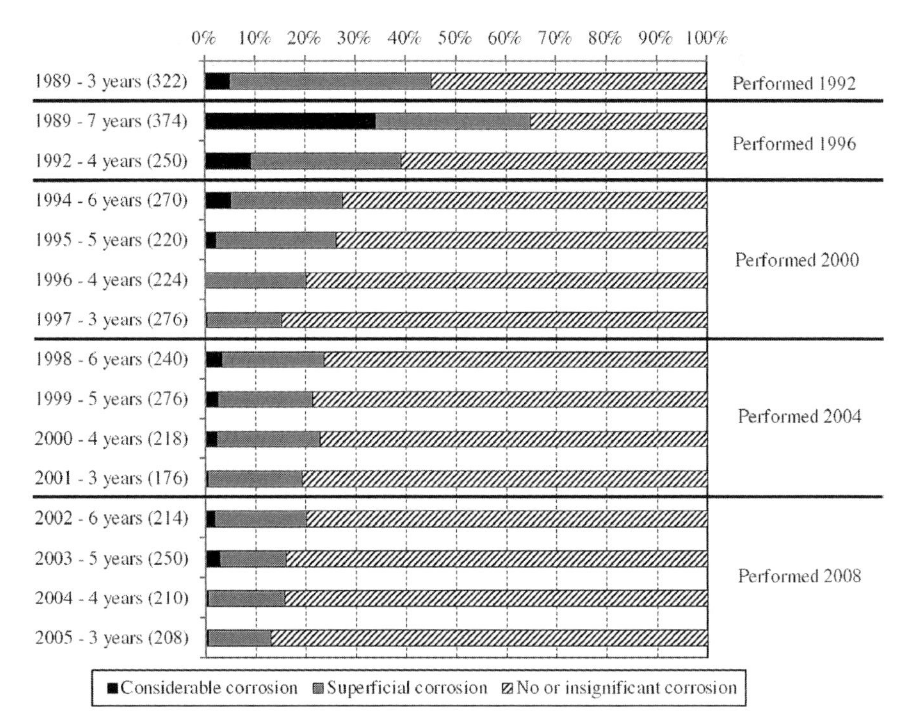

FIGURE 12.7 Results from inspections performed in different years (1992–2008) of the door hem flange on cars showing percentage of observations with considerable, superficial, and no/insignificant corrosion for cars of different year models (1989–2005), run for 3–7 years. The number of inspected cars varies, depending on year model and years of service, from 176 to 322. (B. Rendahl, unpublished data.)

standard specimens and specimens for particular applications on vehicles, thereby representing mobile exposures in the road environment. The principle is the same as described in Chapter 8, but naturally higher demand is placed on the robustness of the sample racks and attachment to the vehicle. Figure 12.8 shows corrosion of zinc

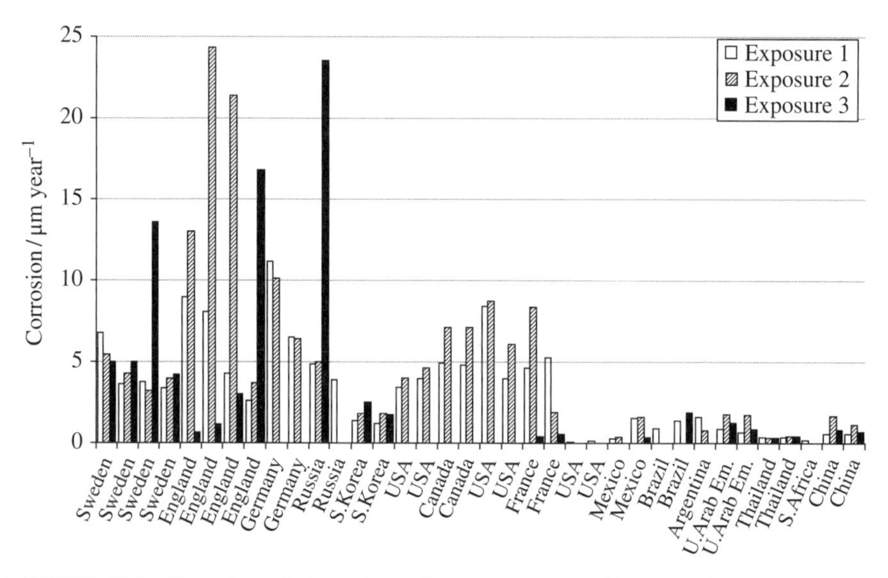

FIGURE 12.8 Corrosion of zinc after being exposed in mobile conditions by attaching racks on vehicles running different traffic routes in Europe, America, Asia, and Africa. (B. Rendahl, unpublished data.)

exposed in mobile conditions worldwide. Corrosion rates are highest in the colder climates, where the use of road salts is more frequent. It is also evident that even for the same route large variations from year to year can occur, most likely depending on the specific details of road salt application for that particular year. These corrosion rates are especially difficult to predict since they not only depend on the climate conditions but also on the local decisions on when to apply road salts, which can vary depending on changes in legislation and budgetary constraints.

12.5 ACCELERATED CORROSION TESTING FOR AUTOMOTIVE APPLICATIONS

Today it is easy to forget that when the salt spray test was introduced 100 years ago, it was a significant improvement compared to other tests performed at that time, which mainly consisted of chemical attack in different solutions not at all representative for atmospheric corrosion conditions. To expose samples in a chamber with an atomized spray of water contaminated with salt in solution was an innovative approach that in principle still is used today but with several important modifications. The neutral salt spray (NSS) has resulted in Worldwide (ISO 9227), American (ASTM B117), and German (DIN 50021) standards and is the most popular corrosion test in general, and this includes also automotive applications. It is known since a long time that the salt spray test fails to reproduce almost all test conditions encountered in the real world and the reaction to this has been particularly strong in the

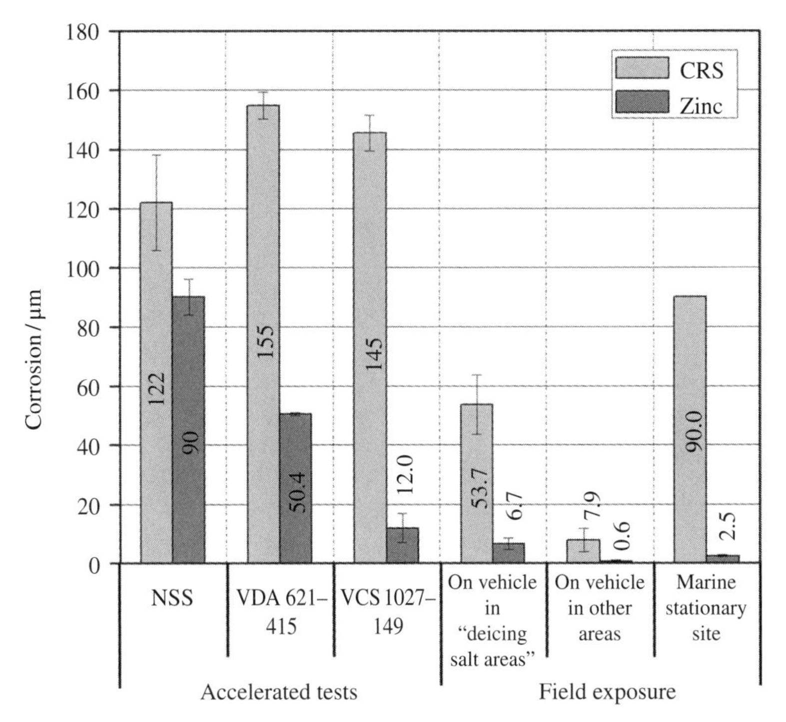

FIGURE 12.9 Corrosion rates of carbon steel (CRS) and zinc after exposure in different accelerated tests and field conditions. (Adapted with permission from John Wiley & Sons; N. Le Bozec, N. Blandin, and D. Thierry, Accelerated corrosion tests in the automotive industry. A comparison of the performance towards cosmetic corrosion, *Materials and Corrosion*, *59*, 889–894, 2008.)

automotive sector, resulting in a variety of cyclic accelerated tests where almost all important car manufacturers have developed their own test cycle but based on variations of the same theme: intermittent salt spray combined with temperature and relative humidity fluctuations.

An intermediate solution to accelerated testing, which is frequently used, is the so-called scab test where a natural outdoor environment is accelerated by spraying of a salt solution on the samples two times a week (ISO 11474). While this method has the benefit of having natural fluctuations of relative humidity and temperature, it does not utilize the possibility of accelerating corrosion by increasing the temperature. Furthermore, scab testing cannot be reproduced, only repeated, since the results will depend on the location of the actual site.

A selection of these accelerated tests and typical testing results are shown in Figure 12.9. The figure allows for comparison of relative levels of zinc and carbon steel corrosion for individual tests. However, the absolute values of zinc and carbon steel corrosion varies between the tests also due to different exposure times. First, look at the results of the natural salt spray compared to the results obtained in different outdoor conditions. As was illustrated in Table 12.1, corrosion rates of carbon steel in natural environments are typically much higher than corrosion rates of zinc. In the natural salt spray, however, corrosion attack of zinc is similar to that of carbon steel. Two major factors are responsible for this. First, the constant wet conditions in the salt spray test as opposed to the cyclic variations occurring in the field do not allow zinc to dry out and form naturally occurring corrosion products with respect to morphology and protective ability. Even for steel, the constant humidity conditions made the NSS unsuitable for addressing adhesion problems on mild steel and therefore the "Verband der Automobilindustrie (VDA)" launched in 1982 an alternative with cyclic humidity conditions, VDA 621-415. However, and similar to the NSS, the salt load was still too high, 5 wt% NaCl, to be able to predict the corrosion performance, Figure 12.9, which is the second reason why NSS fails.

Subsequent studies show that it is possible to accelerate corrosion without changing the corrosion mechanisms significantly only by increasing the temperature to not more than about 50°C in the warm phase. This is the case for the Volvo VCS 1027.149 (launched in 2002; Figure 12.9), Renault D17 2028, GM 9540P, and Daimler KWT standards as examples, which all are cyclic tests with sufficient drying periods, salt concentrations in the range of 1 wt%, and temperature acceleration to about 50°C. When looking into the details of these standards, they seem very different, but as is pointed out, they all have several common characteristics. This is maybe one of the reasons for the NSS to still be used today so frequently, despite the fact that everyone familiar with basic corrosion principles agrees with its drawbacks. There is simply no alternative that is commonly agreed throughout the automotive sector.

One such recent attempt to achieve a *de facto* standard is the VDA 233-102, commonly referred to as New VDA or NVDA, shown in Figure 12.10.

The VDA 233-102 test consists of three daily cycles (A, B, and C), each with their own characteristics. Cycle A (repeated three times per week) contains a salt spray of 3 h (NaCl 1 wt%) followed by lowering the relative humidity to 50% to ensure sufficient drying. Cycle B (repeated three times per week) is the "normal" cycling of temperature and relative humidity with the intention to mimic outdoor conditions but with acceleration conditions accomplished through an increase in temperature. The period of "ambient conditions" (25°C) is there to allow for opening of the chamber and inspection of samples during operation without disturbing the test significantly. Cycle C (repeated one time per week) includes a freezing phase at −15°C. Worth noting is also the "knee" in the relative humidity curve on all cycles at 17–19 h. One of the shortcomings of modern tests is that they are not particularly suitable for reproducing the extent of filiform corrosion on coated aluminum. A moderate relative humidity level is one of the identified parameters for accelerating this mode, which was the intention by introducing this particular event in the cycle. Despite this, the NVDA test also fails to reproduce filiform corrosion on aluminum, which can be significant in field mobile testing conditions. Filiform corrosion manifests as threadlike mm-wide filaments under paints on different metals. It occurs preferentially in humid conditions and the filaments propagate where the "head" is actively corroding while the "tail" containing the corrosion products is inactive. The test may seem overly complicated considering that it performs similar to other much simpler tests used today, such as Volvo VCS 1027.149,

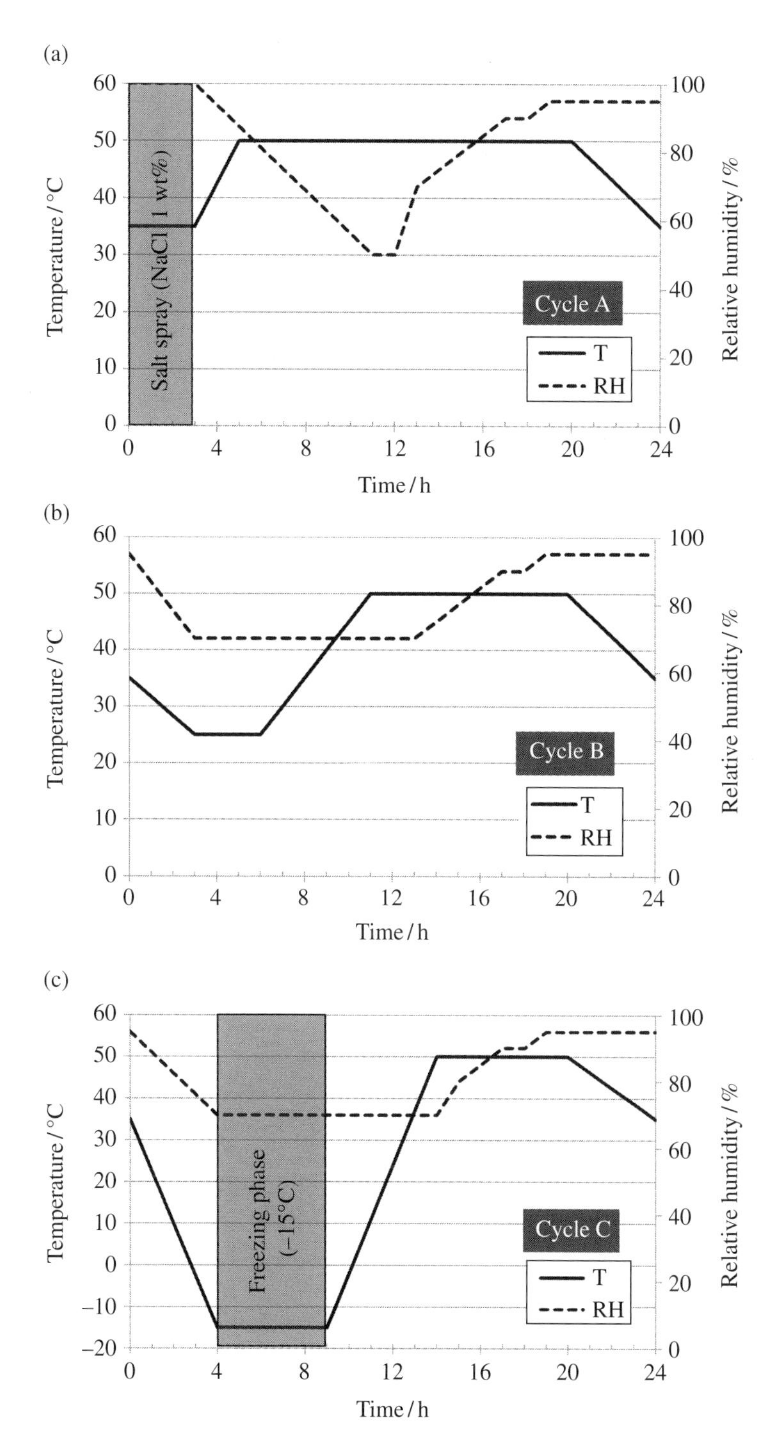

FIGURE 12.10 Graphical representation of VDA 233-102 sub-cycles A, B, and C. These are typically repeated in the sequence BACABBA (weekly cycle) and repeated for 6 weeks. See further in VDA 233-102. Cyclic corrosion testing of materials and components in automotive construction. German Association of the Automotive Industry (VDA), Berlin.

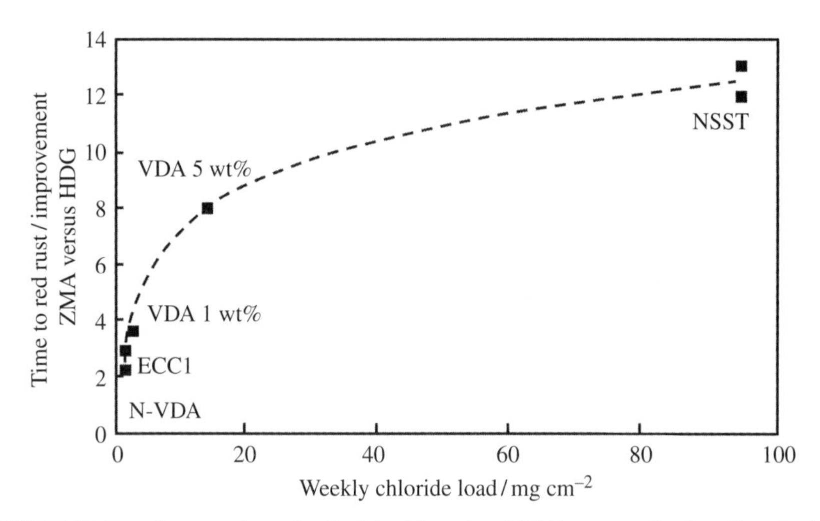

FIGURE 12.11 Time to red rust for Zn–Mg–Al coating (ZMA) compared to hot-dipped galvanized steel (HDG) in some accelerated tests with various chloride deposition: NVDA, Renault ECC1, VDA 621-415 (5 and 1 wt% chloride concentration), and natural salt spray (NSST). The improvement of ZMA over HDG is expressed as the ratio in time to observe red rust on ZMA compared to HDG. (Reproduced with permission from WILEY-VCH Verlag GmbH & Co. KGaA, Weinheim; N. Le Bozec, D. Thierry, A. Peltola, L. Luxem, G. Luckeneder, G. Merchiaro, and M. Rohwerder, Corrosion performance of Zn–Mg–Al coated steel in accelerated corrosion tests used in the automotive industry and field exposures, *Materials and Corrosion, 64,* 969–978, 2013.)

Figure 12.9, for materials like carbon steel and zinc. However, the point of the test is not only to assist in development of new products or pretreatments and select materials but also to allow for product control and be able to spot unexpected faults, which is the reason for the approach to include "all possible climates of the world." As indicated, the drawback of the method is that it is complicated and, especially due to the freezing phase, requires expensive and complicated equipment.

Corrosion testing is particularly important when developing new materials, and the corrosion performance of Zn–Mg–Al (ZMA) coatings has received attention since they have improved corrosion resistance compared to hot-dipped galvanized steel (HDG), but only under some conditions, as is illustrated in Figure 12.11.

The figure shows that the improved performance is closely related to the salt load. At high salt loads, corresponding to NSS, the performance is superior, but as the salt load decreases, the difference between ZMA and HDG becomes less obvious. This will be taken as one final example that the use of natural salt spray for testing natural field conditions is not to be recommended. Whether or not the NVDA test will be able to replace NSS or if it is necessary to find some middle ground requiring less sophisticated equipment remains to be seen.

One important development that shows promise in assisting development of modern corrosion testing methods is the use of automated corrosion sensors. These can measure corrosion rates, for instance, based on the change in resistance of a thin metal track and allow for more or less continuous measurement of corrosion.

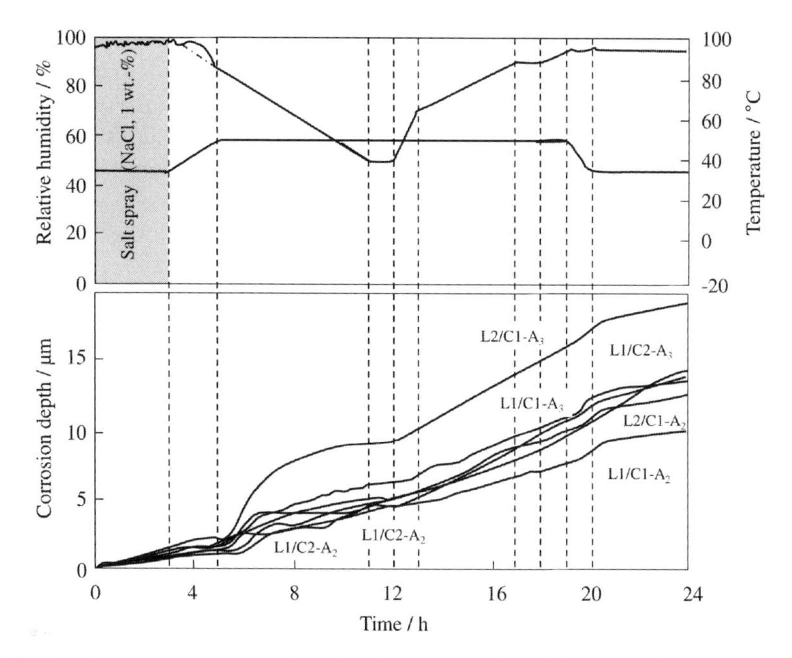

FIGURE 12.12 Temperature and relative humidity (top) and zinc corrosion depth (bottom) for sub-cycle A of the NVDA test. The dashed line for relative humidity shows the desired value (see Figure 12.10). L1 and L2 indicate two different sensors and C1–C3 three subsequent cycles. Logger L2 was removed during the second cycle. (Reproduced with permission from WILEY-VCH Verlag GmbH & Co. KGaA, Weinheim; T. Prosek, N. Le Bozec, and D. Thierry, Application of automated corrosion sensors for monitoring the rate of corrosion during accelerated corrosion tests, *Materials and Corrosion, 65,* 448–456, 2014.)

This can then be compared to temperature and relative fluctuations allowing systematic detailed information about corrosion rates in the different phases of the cycles. This is illustrated in Figure 12.12 showing measurements for one of the NVDA sub-cycles. Records from the initial cycle C1 are slightly higher than C2/C3. This corresponds to the initially higher corrosion on a fresh surface as opposed to corrosion after corrosion products have started to form. The result shows clearly that not only the actual conditions are important for the corrosion rate but also the history (previous cycles).

One final issue worth mentioning concerns the reproducibility of accelerated tests. Looking at Figure 12.12 there is one event where the desired and actual relative humidity differ (after around 4 h). In the high relative humidity range (constant conditions), requirements are typically specified as 95 ± 3%. It is unclear, however, what is actually meant by this. Does this imply that humidity at all times should be within these limits or that the relative humidity should be constant with an allowed systematic error within 95 ± 3%? Recalling the general influence of relative humidity on corrosion and in particular the importance of relative humidity in this interval, it is of course a very large difference between two tests carried out at 92 or 98%. Taking this into account, it is not surprising that when comparing results of accelerated tests from different laboratories, so-called round robin tests, the difference between results can be relatively high. Therefore, when trying to improve existing accelerated tests or to develop new tests, the robustness and clarity in information of the test should also be a primary concern, in addition to the ability of the test to reproduce corrosion modes observed in the field.

FURTHER READING

Corrosion of Infrastructure in the Road Environment and Important Parameters

K. Kreislova, H. Geiplova, T. Link, and K. Hovorkova, The effect of road environment on corrosion of the infrastructure constructions, in *Proceedings of the EUROCORR 2011*, Stockholm, Sweden, September 4–8, 2011.

T. Prosek, D. Thierry, C. Taxén, and J. Maixner, Effect of cations on corrosion of zinc and carbon steel covered with chloride deposits under atmospheric conditions, *Corrosion Science, 49*, 2676–2693, 2007.

B. Sederholm and J. Almqvist, Corrosion in the road environment. Classification and corrosion rates after long term exposure. Report KIMAB-2008-121, Swerea KIMAB, Stockholm, Sweden, 2008.

B. Sederholm and J. Almqvist, Corrosion and corrosion protection of metallic materials in road tunnels. Two years corrosion testing, Report KIMAB-2011-133, Swerea KIMAB, Stockholm, Sweden, 2011.

Corrosion of Vehicles

S. Fujita and D. Mizuno, Corrosion and corrosion test methods of zinc coated steel sheets on automobiles, *Corrosion Science, 49*, 211–219, 2007.

S. Jonsson, Corrosion of zinc in the automotive environment, Report KIMAB-2012-105, Swerea KIMAB, Stockholm, Sweden, 2012.

N. Le Bozec, A. LeGac, and D. Thierry, Corrosion performance and mechanical properties of joined automotive materials, *Materials and Corrosion, 63*, 408–415, 2012.

B. Rendahl, Investigation of perforation corrosion status of different car models, 2002–2005, Report KIMAB-2009-110, Swerea KIMAB, Stockholm, Sweden, 2009.

F. Zhu, B. Rendahl, and D. Thierry, Perforation corrosion of automotive materials. Comparison between laboratory and field exposures, *British Corrosion Journal, 35*, 195, 2000.

Accelerated Corrosion Testing for Automotive Applications

N. Le Bozec, N. Blandin, and D. Thierry, Accelerated corrosion tests in the automotive industry. A comparison of the performance towards cosmetic corrosion, *Materials and Corrosion, 59*, 889–894, 2008.

N. Le Bozec and D. Thierry, Influence of climatic factors in cyclic accelerated corrosion test towards the development of a reliable and repeatable accelerated corrosion test for the automotive industry, *Materials and Corrosion, 61*, 845–851, 2010.

N. Le Bozec, D. Thierry, A. Peltola, L. Luxem, G. Luckeneder, G. Merchiaro, and M. Rohwerder, Corrosion performance of Zn–Mg–Al coated steel in accelerated corrosion tests used in the automotive industry and field exposures, *Materials and Corrosion, 64*, 969–978, 2013.

T. Prosek, N. Le Bozec, and D. Thierry, Application of automated corrosion sensors for monitoring the rate of corrosion during accelerated corrosion tests, *Materials and Corrosion, 65*, 448–456, 2014.

M. Ström, A century with salt spray testing. Time for a final phasing-out by a replacement based on newly developed more capable test regimes, in *Proceedings of the EUROCORR 2014*, Pisa, Italy, September 8–12, 2014.

13

APPLIED ATMOSPHERIC CORROSION: ALLOYS IN ARCHITECTURE

13.1 INTRODUCTION

Metal alloys have been used by mankind for more than 5000 years. The first known alloy, arsenic copper, was produced around 3200 B.C. It was a harder and more useful material than pure copper. Later on copper was alloyed with tin forming bronze, and from the possibility to produce alloys at higher temperatures and with nonmetallic constituents such as silica and carbon followed the development of iron-based alloys. In the technological society of today are metals in engineering applications predominantly used as alloys and only to some extent in their pure state (Chapter 9). Metallic alloys are present in a vast quantity and appear in both highly advanced and everyday applications. Reasons for using metal alloys instead of pure metals are often related to their improved mechanical or chemical properties, superior corrosion resistance for a given application, or to gain a specific surface appearance as aesthetic aspects are of large importance, for example, to architects and the general public.

The overall aim of alloying is to design a material with improved intrinsic characteristics compared with its individual alloy components, for instance, to produce a more durable, more corrosion-resistant, less reactive, and easier material to machine and shape. Specific alloy properties such as corrosion resistance cannot, however, be predicted from corresponding knowledge of its constituent metals. Even very low amounts of alloying elements will change the corrosion properties, for example, the addition of gallium ($\leq$1 wt%) to silver reduces its indoor corrosion rate and the addition of chromium (>12 wt%) to iron changes the microstructure and surface oxide composition, which drastically reduces the corrosion rate by orders of magnitude. The type of alloying element and the intrinsic alloy properties and microstructure

Atmospheric Corrosion, Second Edition. Christofer Leygraf, Inger Odnevall Wallinder, Johan Tidblad and Thomas Graedel.

influence the corrosion/oxidation process and formation of surface oxides and other corrosion products in a given environment and the overall corrosion performance.

Metal alloys have different intrinsic physicochemical properties compared with their individual alloy components. The extent of corrosion/oxidation and degree of metal release are strongly dependent on the surface oxide/patina composition, which in turn is related to the characteristics and microstructure of the alloy and prevailing environmental and pollutant conditions. Many metals and alloys spontaneously form thin and compact passive surface oxides with self-healing properties, passive oxides, which act as efficient barriers for corrosion/oxidation. If the surface oxide becomes locally destroyed, for example, in the presence of chlorides or by scratching, the passive properties may locally be diminished or destroyed. Metals can be released from both highly corrosion-resistant alloys such as high-grade stainless steels and aluminum alloys (low corrosion rates) and low corrosion-resistant alloys such as magnesium alloys or low alloy steels at specific environmental conditions. However, the release is generally significantly lower compared with their pure metal constituents and not proportional to the nominal bulk alloy composition. Thick and porous surface oxides and corrosion products with nonpassive properties are typically formed on engineering alloys such as low alloy steels, copper, and zinc-based materials. These oxides evolve in thickness and composition with time upon environmental interaction. Even though these alloys corrode/oxidize at a significantly higher rate compared with passive alloys, the patina acts as a barrier from a corrosion perspective. The same effect may not necessarily be valid from a metal runoff perspective.

The following sections will briefly elucidate the short- and long-term atmospheric corrosion performance of selected alloys of relevance for outdoor architectural applications and constructions. Atmospheric corrosion mechanisms (corrosion and metal runoff) for the pure metals copper, zinc, iron, and aluminum are summarized in Chapters 9 and 10. The main focus of this chapter is placed on the formation of corrosion products and their influence on the amount of diffuse dispersion of metals from such alloy surfaces when exposed at unsheltered conditions and to some extent to application aspects. Alloys to be discussed include copper-based alloys, zinc–aluminum alloys, weathering steels, and stainless steels intended for and used in outdoor applications such as roofing and cladding/facades of both ancient and modern buildings in the society. The corrosion performances of hot-dipped galvanized steel and aluminum sheet, also commonly used architectural materials, are addressed from a corrosion product perspective in Chapter 9.

13.2 VARYING EXPOSURE CONDITIONS

Atmospheric corrosion of architectural and structural materials, including alloys, has been the subject of a vast number of investigations starting as early as in the 1920s. Numerous national and international long-term field exposure programs have been conducted to assess the role of various corrosive species on the corrosion rate with time on these materials. Such exposures are commonly performed according to standardized exposure conditions with test coupons exposed at an inclination of 45°

from the horizontal, facing south at the northern hemisphere (facing north at the southern hemisphere) at sheltered and/or unsheltered conditions (see Section 8.2).

Important, often overlooked, aspects on the corrosion performance of outdoor applications are their usually complex geometries with large variations in, for example, surface inclination, orientation, variations in environmental conditions, extent of surface soiling, and deposition of corrosive species due to differences in, for example, prevailing wind speed and wind directions, surface orientation, and degree of sheltering. Atmospheric corrosion reactions (and metal runoff) vary significantly over surfaces of the same building or structure, information that will not be provided via standardized testing. The corrosion rate is further largely dependent on the specific microclimate and chemical environment of a given surface. Vertical or groundward-facing surfaces exhibit lower surface temperature variations. This situation, together with less surface porosity and lower deposition rates of corrosive species, seems to disfavor the formation of a protective patina.

An extensive field investigation on variations in local environmental conditions (dry deposition rate of SO_2 and time of wetness) was performed at 44 different locations of a large building of complex shape and geometry in Montreal, Canada. The results showed that the SO_2 deposition rate varied by a factor of 3 and the time of wetness by a factor of 100 over the building. These variations were predominantly explained by differences in prevailing wind speed and wind direction as well as degree of sheltering. Observed differences resulted in varying corrosion rates for both copper (factor of 3) and steel (factor of 5) over the same building. The measured corrosion rates for both metals revealed the highest corrosion rates when determined at standardized conditions. Similar findings have been reported in other studies. The results are in harmony with the intention that standardized tests should be considered as a worst-case scenario when assessing corrosion rates and that most surfaces of a building of complex geometry receive some kind of sheltering effect.

Further information on important exposure and outdoor conditions that influence the corrosion process is provided in Chapter 8.

13.3 COPPER-BASED ALLOYS

Copper-based alloys including brasses (Cu–Zn) and bronzes (Cu–Sn) are common engineering materials in modern architecture. They are predominantly used for facade applications due to their weight, ductility, malleability, corrosion resistance, long-term performance, and visual appearance (important for architects during design or renovation of modern or ancient cultural buildings). Both bronzes and brasses naturally change their surface appearance with time and prevailing environmental conditions. Similar to copper sheet (see Chapter 9), a patina consisting of different corrosion products is gradually formed that reflects the alloy constituents and the chemistry of the environment.

13.3.1 Corrosion Mechanisms

The most common brass alloys used for outdoor architectural applications consist of 15–20 wt% Zn, although cases exist when using up to approximately 30 wt% Zn. These alloys all have a one-phase face cubic microstructure (α-phase, up to 35 wt% Zn) and are less susceptible for dezincification processes compared with alloys of higher zinc content and more complicated microstructure. Bronze alloys used for facade applications contain typically between 4 and 6 wt% Sn. Examples given in the following will primarily refer to these binary alloys. More complex alloys may behave differently.

Corrosion rates of brass are generally lower compared with copper sheet in low chloride-containing environments, Figure 13.1a. Less evident differences in corrosion rates between brass and copper sheet are observed at chloride-rich marine conditions, both at sites of high (sites 1 and 2) and relatively low (sites 3 and 4) deposition rates of chlorides, Figures 13.1b and c. Bronze shows generally significantly lower corrosion rates compared with copper sheet at low chloride polluted urban and rural conditions, Figure 13.1a, whereas the opposite situation prevails at marine conditions with high deposition rates of chlorides, Figures 13.1b and c but diminishes with increased distance from the seashore. Similar effects have been observed at urban and rural sites of relatively high chloride levels within the UNECE field exposure program (up to 8 years) with higher or similar corrosion rates for cast bronze (7 wt% Sn, 6 wt% Pb, 4 wt% Zn) compared with copper sheet, whereas lower rates were observed at sites of low chloride pollution.

Similar to copper sheet, described in Chapters 9 and 10, brass and bronze form an inner layer of Cu_2O (cuprite) and an outer layer of $Cu_4SO_4(OH)_6 \cdot H_2O$ (posnjakite) and $Cu_4SO_4(OH)_6$ (brochantite). Zinc-rich corrosion products, most probably amorphous $Zn_5(CO_3)_2(OH)_6$ (hydrozincite) and ZnO (zincite), form at the surface and are enriched within the patina of brass, whereas amorphous tin oxides exist within the patina of bronze. The zinc-rich corrosion products hinder to a different extent the corrosion process. The presence of small amounts of tin oxides reduces the corrosion rate of bronze to a different extent in low chloride-containing environments but lacks this beneficial effect at chloride-dominated conditions.

Uniform patinas generally evolve on architectural brass and bronze at urban and low polluted conditions, whereas nonuniform and flaking patinas initially tend to develop on bronze (and on copper sheet) in chloride-rich environments, effects significantly less pronounced on brass. Flaking of corrosion products from the patina explains the very high corrosion rates of bronze observed at the chloride-rich sites closest to the seashore, sites 1 and 2, Figure 13.2b. The flaking process is correlated to the formation of CuCl (nantokite), readily formed on bronze in chloride-rich environments, and its transformation into more voluminous $Cu_2Cl(OH)_3$ (atacamite/paratacamite), a process that induces physical stresses within, and volume expansion of the patina. Similar flaking, but less extensive, also takes place on copper sheet and to a relatively small extent on brass. The formation of $Zn_5(CO_3)_2(OH)_6$ within the patina on brass hinders the formation rate of CuCl and thereby the occurrence of flaking. A porous and discontinuous, loosely adherent patina on bronze composed of

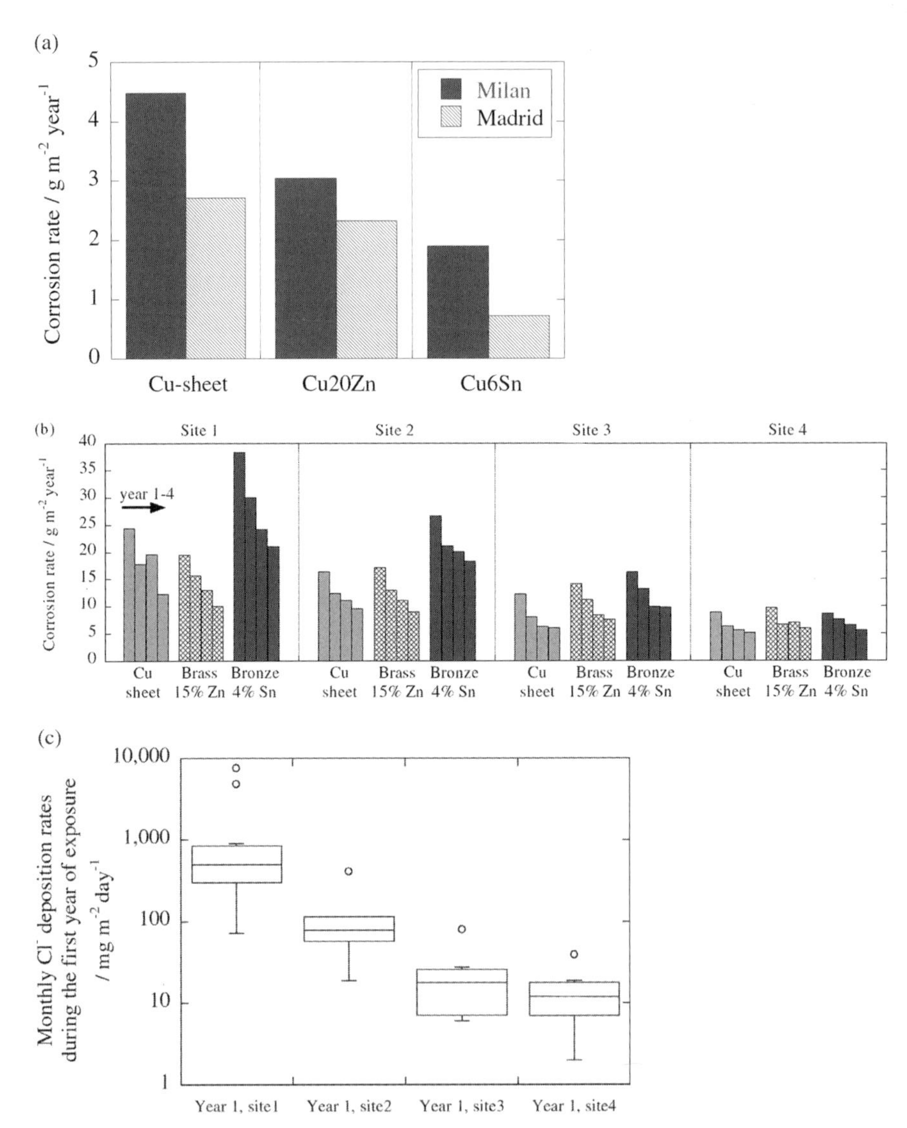

FIGURE 13.1 Annual average corrosion rates of brass (20 wt% Zn), bronze (6 wt% Sn), and copper sheet after 3-year exposure at unsheltered conditions at urban sites characterized by low deposition rates of chlorides (Milan, Italy, and Madrid, Spain) (a); corrosion rates of brass (15 wt% Zn), bronze (4 wt% Sn), and copper sheet exposed up to 4 years at unsheltered conditions at marine sites of reduced distance and deposition rates of chlorides from the seashore (Brest, France) (b); and differences in deposition rates of chlorides at the 4 marine sites (c). (Partly adapted with permission from Elsevier; Odnevall Wallinder, X. Zhang, S. Goidanich, N. Le Bozec, G. Herting, and C. Leygraf, Corrosion and runoff rates of Cu and three Cu-alloys in marine environments with increasing chloride deposition rate, *Science of the Total Environment, 472,* 681–694, 2014.)

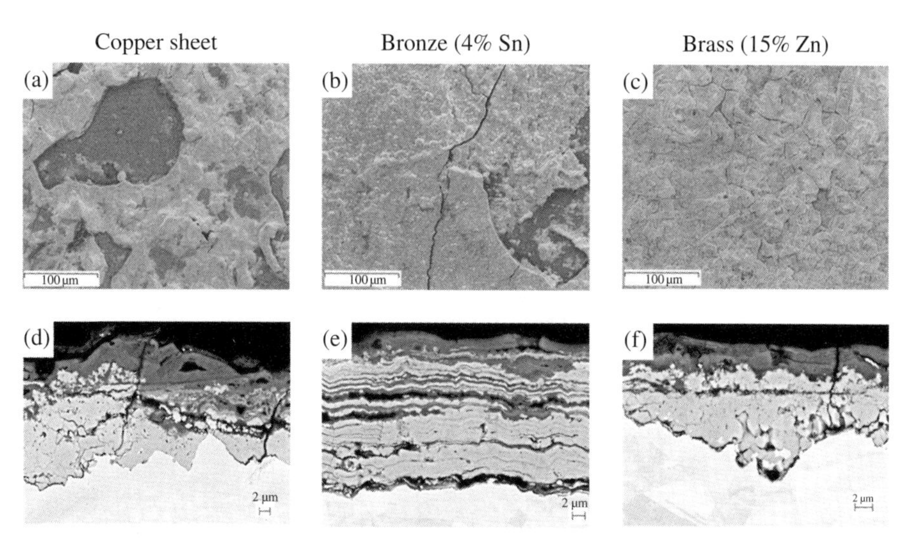

FIGURE 13.2 Light optical microscopy images of surface patina flaking (a–c) and scanning electron microscopy images (d–f) of cross sections of the patina formed on copper sheet (a and d), bronze (b and e), and brass (c and f) after 5 years of unsheltered exposure conditions. (I. Odnevall Wallinder, G. Herting, and S. Goidanich, unpublished data.)

noncontinuous streaks of copper- ($CuCl \rightarrow Cu_2Cl(OH)_3$) and tin-rich (probably SnO_2) corrosion products explains the higher extent of flaking and high corrosion rate compared with copper sheet in chloride-rich environments, Figure 13.2. The outermost patina layer becomes more adherent and compact with time, effects that improve the barrier properties of the patina and that reduce the corrosion rates with time also for the materials of severe flaking, that is, bronze and copper sheet.

The proportion of released alloy constituents depends on the corrosion process, the solubility and characteristics of formed corrosion products, and on their surface coverage. Most of the oxidized metals in the corrosion products retain within the patina. Since stable tin oxides of low solubility are located within the patina of bronze, copper-rich corrosion products dominate the outermost patina and act as the interface toward the environment. The amount of released copper (runoff) is consequently similar (or only slightly lower) compared with copper sheet when exposed to sites of low chloride levels, Figure 13.3a. No measurable amounts of tin are released at these conditions. At marine conditions where severe flaking takes place, more copper is released from bronze compared with copper sheet, at least during the first years of exposure. As the patina becomes more stable with time, the runoff rates of copper converge to similar values, Figure 13.3b.

Even though alloying with tin does not reduce the amount of released copper from bronze (4 wt% Sn), studies of other bronze alloys (3 wt% Sn, 9 wt% Zn) show such an effect. This is attributed to the formation of $Zn_5(CO_3)_2(OH)_6$ within the patina. Similar observations have been made for brass with 15, 20, or 31% Zn (per weight). Despite initial and short-term variations in the released proportion of zinc and copper,

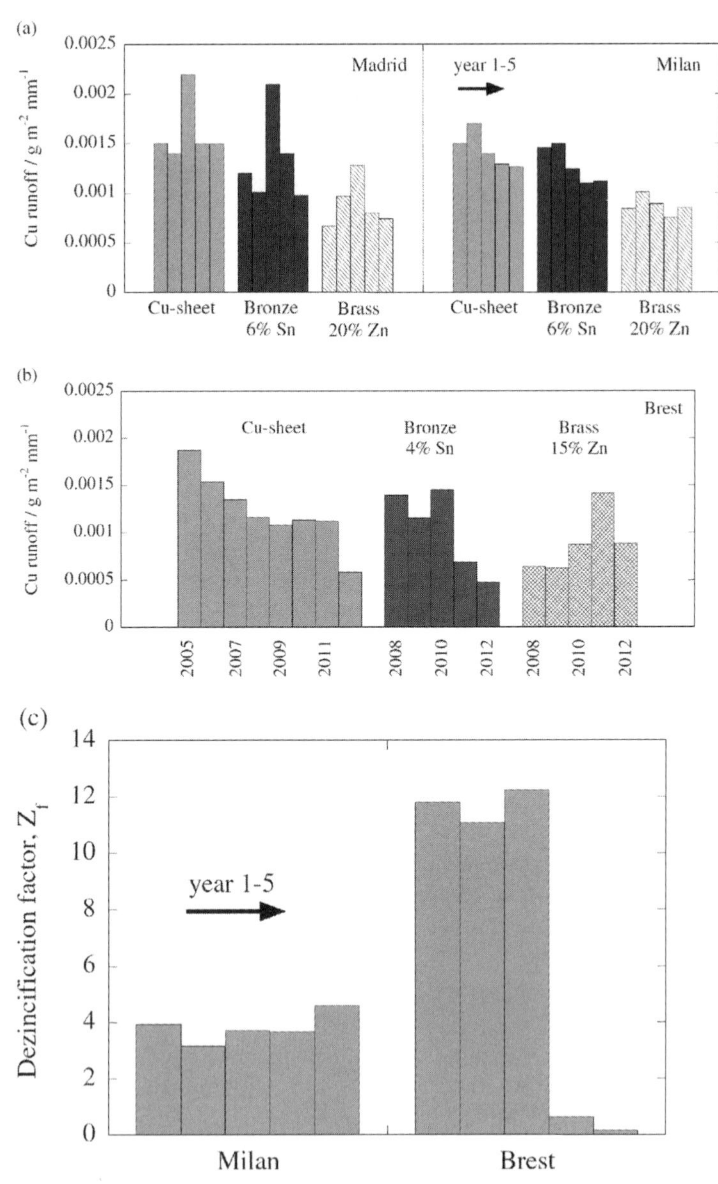

FIGURE 13.3 Annual copper runoff rates per given rainfall unit for copper sheet, bronze (4 or 6 wt% Sn), and brass (15 or 20 wt% Zn) exposed at unsheltered standard exposure conditions at urban (Milan, Italy, and Madrid, Spain) (a) and marine (Brest, France) (b) conditions up to 5 or 7 years and corresponding changes in the dezincification factor (Zn/Cu release ratio over Zn/Cu alloy ratio) for brass (15 wt% Zn) exposed in the urban (Milan, Italy) and the marine (Brest, France) environment for 5 years (c). (I. Odnevall Wallinder, G. Herting, and S. Goidanich, unpublished data.)

effects that are site specific (initially generally more zinc released compared with copper), it is evident that zinc on a long-term perspective (up to 5 years) is preferentially released compared to copper in both low and moderately polluted urban atmospheres. Similar trends are seen during the first few years of marine exposure, Figure 13.3c. The released proportion of copper and zinc depends not only on site and environmental conditions but also on alloy content as illustrated in Figure 13.4a and b for architectural brass with 15 or 31 wt% Zn. From the higher zinc alloy content follows a higher degree of dezincification, Figure 13.4c. With time it is anticipated that more copper-rich corrosion products dominate the surface and, hence, a changed proportion of released copper compared with zinc.

13.3.2 Application Aspects

The visual appearance is a complex function of many parameters, including patina morphology, composition, thickness, and deposits of soot and other particles. The discoloration of the same surface may vary considerably with time, type of environment, and exposure conditions. In general there is a change in visual appearance of copper and copper-based alloys from their original characteristic lustrous appearance to a more brownish, blackish, or greenish surface. The orientation of the surface is also of great importance for the visual appearance. A skyward-facing surface in general forms a green patina layer more easily than a vertical or downward surface. The main reason is that skyward surfaces are exposed to larger variations of surface temperature than vertical or downward-facing surfaces. From this follows that dry–wet–dry cycles with consecutive steps of dissolution, ion pairing, and reprecipitation become dominant processes in the formation of a green patina. Skyward surfaces are also more susceptible to corrosive pollutants from precipitation and from deposited particles, and the porosity of the patina increases generally with thickness. Thus, deposited corrosive pollutants and humidity are more easily trapped and retained within the patina, thereby further promoting the conditions for patina formation. The overall result is a relatively rapid formation of a greenish patina. A streaky pattern of green patina formation is common for complex geometries that allow channels for water flow over the surface.

Despite the fact that architectural brasses generally corrode at a more slow pace than copper sheet at most conditions and corrosion of bronze may be more rapid or of similar rate as copper sheet, depending on chloride deposition levels, differences in surface appearance (reflecting differences in patina thickness and composition) between the materials are only evident during the initial years of exposure (depending on site). The long-term (years, decade) appearance of the alloys is very similar to copper sheet.

As the visual appearance depends not only on patina and surface characteristics but also on prevailing weather and light conditions, it is perceived very differently. One way to illustrate such differences is to use spectrophotometry where the surface is illuminated at a given angle using a standard source of light (white light with equal amount of red, blue, green, and yellow light) and the reflected and/or scattered light is measured. The amount of absorbed light is a measure of the brightness of the

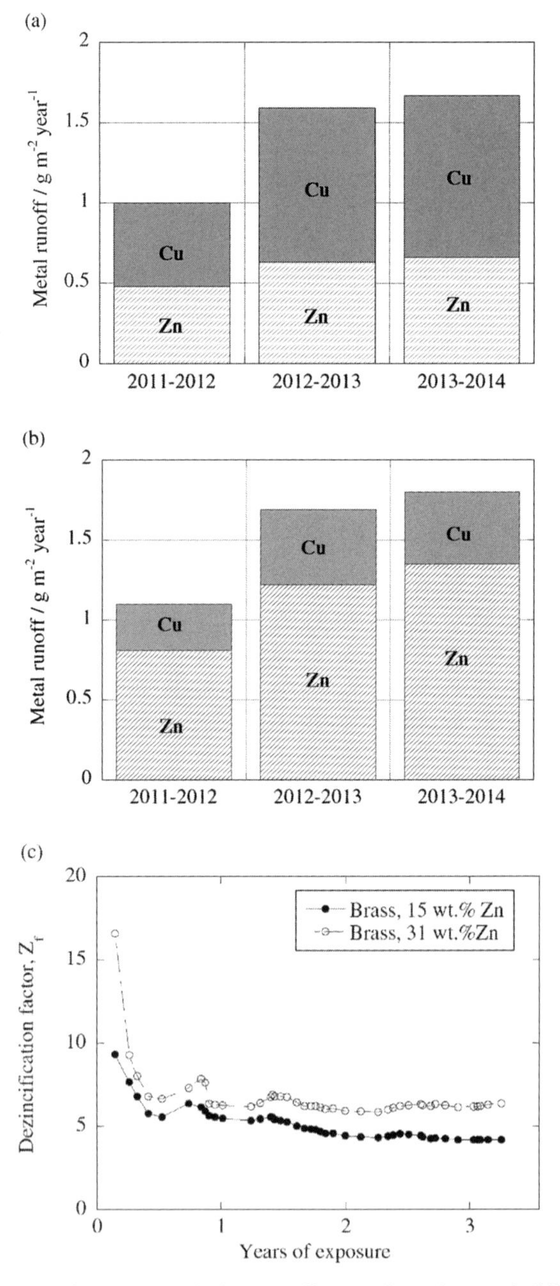

FIGURE 13.4 Annual copper and zinc runoff rates from brass of different zinc alloy contents exposed in parallel at unsheltered standard exposure conditions in an urban test site (Milan, Italy) for 3 years, 15 wt% Zn (a), 31 wt% Zn (b), and corresponding changes in the average dezincification factor with time (Zn/Cu release ratio over Zn/Cu alloy ratio) (c). (I. Odnevall Wallinder, G. Herting, and S. Goidanich, unpublished data.)

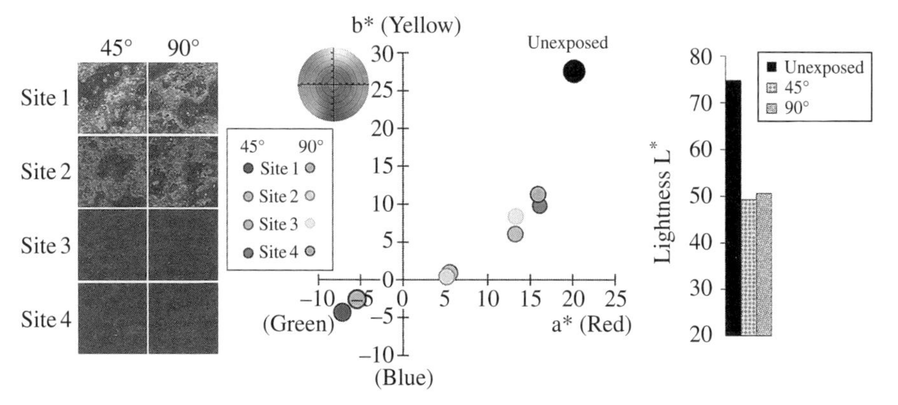

FIGURE 13.5 Spectrophotometric differences in visual appearance illustrated for copper sheet exposed at 45° and 90° from the horizontal at unsheltered conditions for 6 months at four different marine locations of increasing distance from the seashore (reduced deposition rates of chlorides from site 1 to 4; see also Figure 13.1c). Each color is represented by a point in the color space that is identified by three coordinates, a*, b*, and L*, with a* representing a color ranging from green to red, b* representing yellow to blue, and L* representing the brightness ranging from black to white. (I. Odnevall Wallinder, G. Herting, and S. Goidanich, unpublished data.) (*See insert for color representation of the figure.*)

surface, a parameter that depends on surface roughness and surface patina characteristics. Such measurements are illustrated in Figure 13.5 for copper sheet exposed for 6 months at marine sites of increased distance from the seashore (see Figure 13.1). This way to illustrate visual appearance may be used to predict the long-term appearance of copper and copper-based materials in different environments.

Long-term field exposure findings on the corrosion and runoff process presented earlier reveal very similar amounts of released copper on an annual perspective from bronze compared with copper sheet and approximately 30–40% lower annual release rates of copper from brass compared with copper sheet at rural and urban unsheltered conditions. These observations imply that the predictive runoff rate model, elaborated for copper sheet (Section 10.4), can be used for predictions of runoff rates of copper also from bronze and from brass in low chloride polluted environments, if taking these aspects into account. Recent findings for copper sheet imply that the predictive model could be used for long-term estimations of copper runoff rates from bronze at marine conditions. Any predictions of runoff rates of copper from brass will be highly overestimated but may possibly be used for conservative deliberations on copper dispersion from architectural materials.

Preoxidized copper sheet is commonly available with a different surface appearance for use in architectural applications such as facades and roofs and for replacement of certain sections of restored ancient buildings with naturally aged green patinated surfaces or to obtain a specific surface appearance already upon installation. These patinas are prefabricated via different chemical oxidation treatment steps with the aim to rapidly obtain patinas of similar composition and barrier properties as gained for natural patinas in different environments. Upon outdoor exposures follows a gradual transformation of these chemically obtained compounds into naturally formed corrosion products. Long-term field exposures at both marine and urban conditions have shown that this transformation takes place and that the patina with time behaves as a naturally formed patina both from a physical barrier and compositional and metal release perspectives. However, on a short-term perspective, the preoxidized surface may behave differently compared with a naturally corroded surface. Chemically preoxidized CuO (tenorite) has, for instance, been reported to be nonstable at marine conditions but with time observed to transform to more stable corrosion products.

Naturally oxidized surfaces on ancient buildings and historical surfaces in cities may suffer from permanent damage due to graffiti. The nature of the patina on copper and copper-based alloys makes any cleaning actions very difficult as any abrasion or cleaning activities with different solvents will cause irreversible surface damage and changes of the aesthetic appearance. Antigraffiti coatings are designed and used to hinder the adhesion of the graffiti paint and thereby reduce the surface damage, resist repeated cleaning procedures, at least for certain time periods, or be easily removed together with the graffiti upon cleaning. It is essential that these coatings do not damage, deteriorate, or change the appearance of the patinated surfaces. Most antigraffiti coatings are based on waxes, polysaccharides, paraffin, polyurethanes, silicon resins, or fluorinated polymers.

Recent research findings have shown that the patina composition is not influenced by the presence of wax-based antigraffiti coatings. The degree of soiling is more pronounced, especially for degraded coatings with local defects, areas where also nonuniform corrosion takes place. Except for antigraffiti properties, the coating acts as a barrier that on a short-term perspective improves the corrosion resistance and reduces the extent of copper runoff. The surface protection efficiency improves with coating thickness; however too thick layers may alter the aesthetic appearance.

Coating degradation is more rapid when applied on already patinated surfaces compared with fresh surfaces, an effect that is attributed to differences in surface roughness and adhesion properties.

13.4 ALUMINUM–ZINC ALLOYS

Zinc and zinc-based alloys are important materials in outdoor applications such as crash barriers, lamp poles, and fences (mainly as hot-dipped galvanized steel) and for architectural applications (mainly zinc sheet and zinc–aluminum coated steel) including roofs, sides, gutters, pipes, and facades. Galvanizing reduces the corrosion resistance by orders of magnitude compared with the bare steel surface. Its extent depends on prevailing exposure conditions and on the coating thickness. Coating of steel with aluminum–zinc alloys has been shown to reduce the corrosion rate of hot-dipped galvanized steel even more. One of the most commonly used alloys for facade and roofing applications except for zinc sheet (0.06–0.2 wt% Ti, 0.08–1 wt% Cu) is the 55 wt% Al, 43.4 wt% Zn and 1.6 wt% Si alloy coating (55% Al–Zn) on steel sheet. The coating is applied to steel via a continuous hot-dip process and given a typical thickness of 20–25 μm. Silicon is added to prevent the excessive growth and brittleness of the intermetallic layer formed during the hot-dip process and to improve the coating adherence to the steel substrate. Several field studies have been conducted in different environments worldwide to assess the long-term corrosion behavior of this Al–Zn alloy, but few have focused on the formation and evolution of corrosion products. The corrosion performance of this alloy will in the following be summarized.

13.4.1 Corrosion Mechanisms

The alloy microstructure of the 55 wt% Al–Zn alloy displays a network of aluminum-rich dendrites (80% of the coating volume) that contain dispersed particles of zinc and silicon and of zinc-rich interdendritic areas with both zinc-rich and aluminum-rich phases, Figures 13.6a and b. Scanning Kelvin probe measurements show a higher nobility of aluminum-rich dendrites compared with zinc-rich interdendritic regions. From differences in composition and nobility follow that the interdendritic areas will be more susceptible for corrosion and corrode at a different rate due to selective dissolution processes. Cross-sectional studies by Ramus Moreira et al. of 55% Al–Zn exposed for 5 years at urban conditions revealed preferential corrosion initiation of the aluminum-rich phase and not the zinc-rich phase within the interdendritic regions, whereas aluminum-rich dendritic regions remain relatively noncorroded.

Atmospheric exposure changes with time the overall outermost surface composition of the alloy. This oxidized surface layer on unexposed 55% Al–Zn is approximately 4–6 nm thick and predominantly composed of aluminum oxides. Detailed laboratory studies by Qiu et al. indicate an initial formation of amorphous Al_2O_3 and the parallel formation of basic zinc–aluminum hydroxycarbonates in the interdendritic areas at humidified CO_2-containing conditions. Additional deposition of chlorides results in the overall formation of aluminum-rich compounds, in particular AlOOH and $Al(OH)_3$, and ZnO, zinc–aluminum hydroxycarbonates, and/or hydroxychlorides in the interdendritic areas.

Short- (weeks) and long-term (up to 5 years) marine exposures clearly show changes in the overall surface coverage of aluminum- and zinc-rich corrosion products, Figure 13.6c and its relation to the amount of zinc and aluminum runoff, Figure 13.6d. Corrosion products of zinc evolve within the zinc-rich interdendritic areas during the first year; after which aluminum-rich corrosion products, promoted by deposited chlorides, gradually evolve over the surface (primarily $Al(OH)_3$ at aluminum-rich dendritic areas) and hinder further lateral growth of zinc-rich corrosion products. From a reduced surface coverage and formation of less soluble zinc–aluminum-rich corrosion products follow reduced released amounts of zinc with time, Figure 13.6c. Small amounts of aluminum are released from the surface throughout the exposure period. Similar processes take place at urban conditions. Preferential release of zinc compared with aluminum is expected at atmospheric conditions. However, the opposite situation can take place at anaerobic conditions in, for example, crevices.

FIGURE 13.6 Microstructure of a 55% Al–Zn coating on steel with aluminum-rich dendritic areas containing dispersed zinc particles and zinc-rich interdendritic areas containing both zinc-rich and aluminum-rich phases (a). Corrosion initiation and formation of corrosion products preferentially within interdendritic areas (dark features) exemplified after 1 year of unsheltered marine exposure (b), changes in the overall composition (two separate areas) of the outermost surface layer with time measured by means of X-ray photoelectron spectroscopy (c), and corresponding average runoff rates of zinc and aluminum with time (d). (Reproduced with permission from Elsevier; A. Ramus Moreira, Z. Panossian, P.L. Camargo, M. Ferreira Moreira, I.C. da Silva, and J.E. Ribeiro de Carvalho, Zn/55Al coating microstructure and corrosion mechanism, *Corrosion Science, 48,* 564–576, 2006 (a) and with permission from P. Qiu, C. Leygraf, and I. Odnevall Wallinder, Evolution of corrosion products and metal release from Galvalume coatings on steel during short and long-term atmospheric exposures, *Materials Chemistry and Physics, 133,* 419–428, 2012 (b–d).)

In both low polluted and chloride-rich marine environments, initial corrosion predominantly takes place in the zinc-rich areas, while the aluminum-rich dendritic areas initially remain relatively unaffected (nevertheless they corrode with prolonged time). The extent of corrosion on the aluminum-rich areas depends on the site corrosivity and presence of chlorides. The formation and evolution of corrosion products on the 55% Al–Zn coating follow the same general trends as previously described for bare zinc and aluminum, respectively (see Section 9.5). A general evolution sequence of corrosion product formation has been proposed by Zhang et al., Figure 13.7. Sequential formation of corrosion products is believed to proceed via chemical transformations as well as ion exchange mechanisms (see Chapter 9)

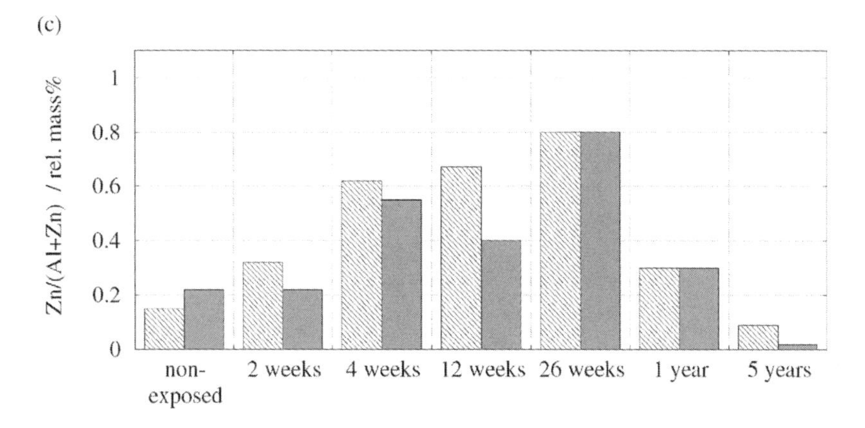

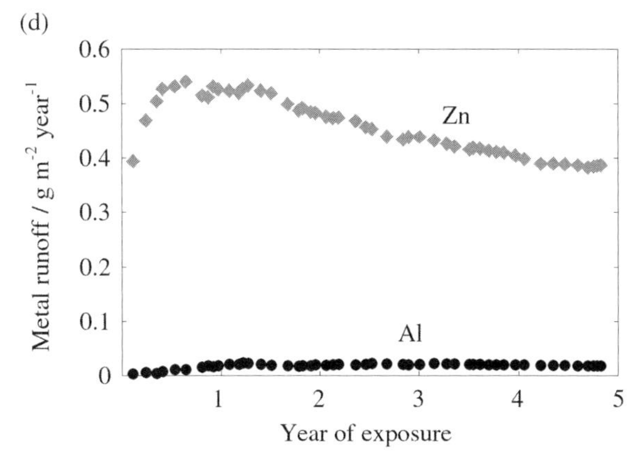

FIGURE 13.6 (Continued)

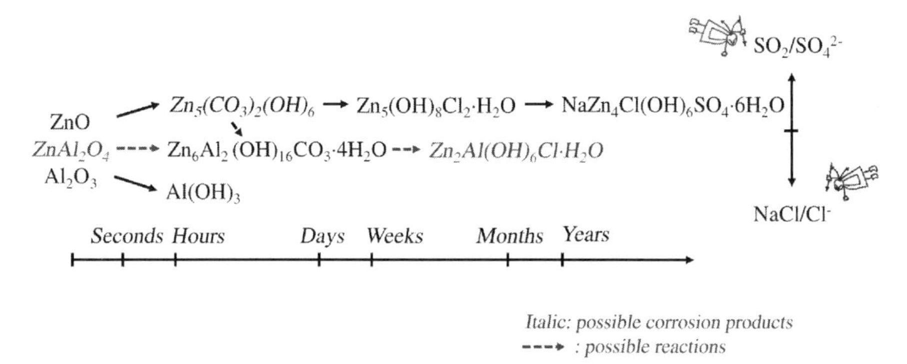

FIGURE 13.7 Evolution scheme of corrosion products on bare coatings of Al–Zn alloys on steel at atmospheric conditions.

due to the structural resemblance and layered structures of the different hydroxyl-based corrosion products. The same corrosion product sequence has been shown to operate also for other Al–Zn alloys containing different amounts of aluminum, such as 5% Al–Zn coatings on steel and possibly also Zn–Al–Mg and Zn–Al–Fe coatings exposed at similar conditions, although also other corrosion products may form and play an important role.

It can be concluded that the 55 wt% Al–Zn coating on steel from a corrosion perspective generally behaves as a zinc surface during initial stages of corrosion and more as an aluminum surface after longer time periods.

13.4.2 Application Aspects

The 55% Al–Zn coating is designed to combine the beneficial corrosion properties of aluminum, acting as a corrosion-resistant surface barrier, and zinc, providing cathodic protection (sacrificial protection) at sheared edges and scratches. Long-term (up to 36 years) corrosion testing in different rural, marine, severe marine, and industrial environments reveals a significantly improved corrosion resistance (2–6 times) compared with galvanized steel of similar coating thickness. From this follows an at least twice as long service life (estimated from the time to red rust staining) of 55% Al–Zn coated steel compared with bare galvanized steel at similar exposure conditions, Figure 13.8.

The amount of zinc runoff from Al–Zn alloy coatings is quantitatively different from what is released from bare zinc sheet at similar exposure conditions, Figure 13.9. These rates cannot be predicted based on alloy composition and runoff rates of zinc

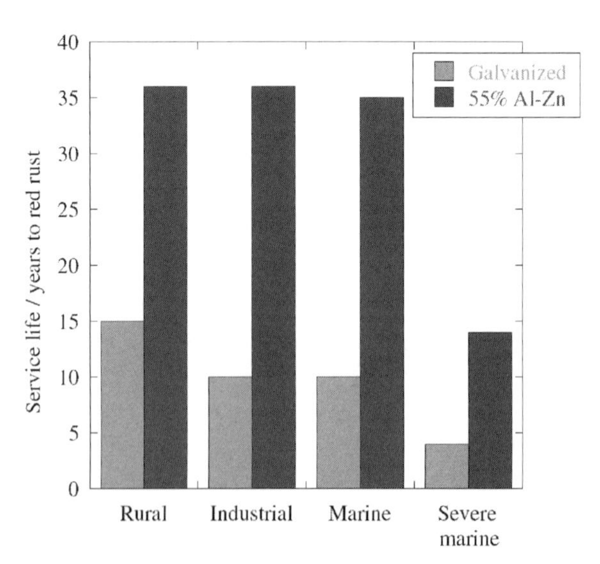

FIGURE 13.8 Differences in service life, expressed as time to red rust initiation, of galvanized steel and 55% Al–Zn coated steel (flat surfaces) of the same coating thickness exposed up to 36 years at different test sites in the United States and Canada. (Data from Bethlehem Steel.)

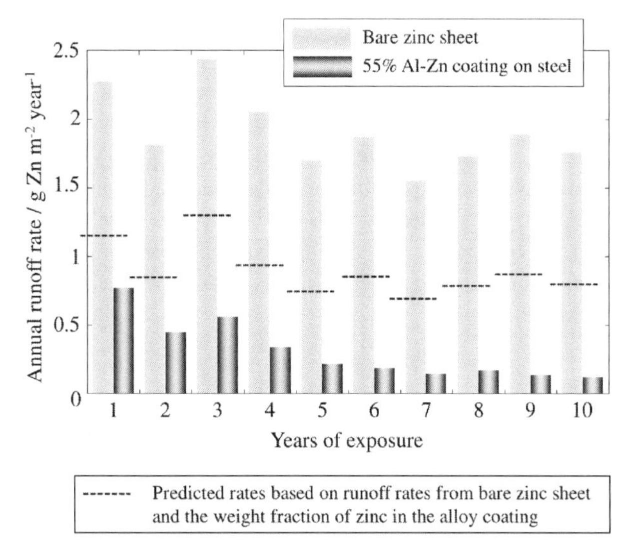

FIGURE 13.9 Annual runoff rates of zinc from 55% Al–Zn coated steel and bare zinc sheet exposed at unsheltered urban conditions in Stockholm, Sweden, up to 10 years. The dotted lines refer to corresponding calculated rates from the weight fraction of zinc in the alloy coating multiplied by corresponding annual zinc runoff rates from bare zinc sheet. (I. Odnevall Wallinder, unpublished data.)

from bare zinc sheet. Such predictions based on alloy composition are commonly made within the framework of environmental risk assessment of metals and are highly erroneous.

Similar to zinc sheet and galvanized steel, Al–Zn coated steel sheet is sensitive to wet storage staining, with rapid formation of visual corrosion products when, for instance, wrapped in coils or in crevices where water can condensate and remain for long time periods without drying. This corrosion process takes place when the airflow and access of atmospheric CO_2 are limited. It is mainly of cosmetic importance and will not influence the long-term corrosion material properties in any significant way. In contrast to the formation of "white rust" that predominantly consists of $Zn(OH)_2$ and to some extent $Zn_5(CO_3)_2(OH)_6$ on zinc sheet and galvanized steel at similar conditions, Al–Zn alloy coatings form dark grayish or blackish corrosion products predominantly composed of $Al(OH)_3$ (at aluminum-rich dendrite areas) and $Zn_6Al_2(OH)_{16}CO_3{\cdot}4H_2O$ (in interdendritic areas). The mixed zinc–aluminum hydroxycarbonate also form at aerated conditions, as described previously. The occurrence of black storage staining increases with temperature and with increased alkaline conditions. The formation of wet storage staining during transport and storage can be hindered to a different extent by applying corrosion-inhibiting oil systems, thin chemical surface treatments, or thick organic coating systems. However, its protective ability is gradually reduced as the coating degrades.

Al–Zn coatings on steel (and galvanized steel) are often applied with different paint systems with top coatings typically based on acrylics, different polyesters, and fluoropolymers to improve their service life or to obtain a certain appearance. The presence of coating systems improves the global corrosion resistance and the extent of released metals by the action of rainwater from the material, in particular at scratches and cut edges where the top coating may be damaged. Depending on the environment in terms of, for example, chloride deposition, UV radiation, temperature, relative humidity, etc., the coating will degrade and delaminate with time, a process that often takes place locally over the surface. Except for contributing to a poor surface appearance with possible flaking and delamination, these local weathering effects reduce the overall corrosion resistance of the material and increase the amount of released metals.

13.5 WEATHERING STEEL

Weathering steels have a low carbon content (<0.2 wt%) and are alloyed with up to 3–5 wt% minor elements such as Cu, Cr, Ni, P, Si, and Mn. These alloys were initially developed in the 1920s and have a generally improved corrosion resistance and possess self-healing properties compared with unalloyed steel in many environments and under certain weathering conditions. The material is predominantly used in outdoor heavy load-bearing constructions such as bridges and utility towers but also in architectural applications such as chimneys, facades, and roofing.

13.5.1 Corrosion Mechanisms

The corrosion mechanisms of weathering steels are very similar to rust formation on iron and carbon steel (Section 9.4 and Appendix F) with γ-FeOOH (lepidocrocite) predominantly formed at initial stages followed by the gradual growth of an amorphous ferric oxyhydroxide (after a few years) and later of α-FeOOH (goethite) (after decades). γ-FeOOH is primarily present in an outer layer and α-FeOOH in an inner layer. The long-term formation and transformation of corrosion products within the rust layer of weathering steel have schematically been summarized by Yamashita et al., Figure 13.10. The corrosion rate gradually decreases during these transformation processes. The difference between the rust layer formed on carbon steel and on weathering steel is that α-FeOOH on the latter forms a densely packed and uniform layer consisting of nanometer-sized particles that are closely attached to the underlying steel substrate, Figure 13.10. The formation of this layer is stimulated by repeated dry and wet cycles and the presence of alloying elements, in particular of Cr but also of Cu and P, which become enriched in α-FeOOH. It is generally considered that these elements promote the formation of amorphous FeOOH by suppressing the formation of a crystalline compound with time. Multianalytical studies of weathering steels exposed for 16 years in a rural environment revealed an outer layer with almost equal amounts of γ-FeOOH and α-FeOOH, a transformation believed to be controlled by local exposure conditions, and an inner layer of nanosized (5–30 nm) Cr-enriched α-FeOOH particles combined with small islands of maghemite and

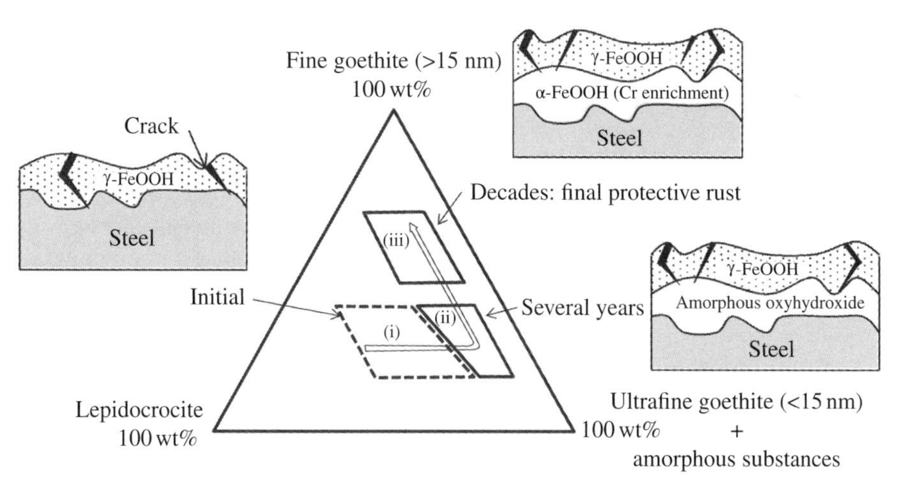

FIGURE 13.10 Schematic description of distribution and long-term changes between γ-FeOOH, amorphous ferric oxyhydroxide, and α-FeOOH during different stages of atmospheric corrosion of weathering steels. (Reproduced with permission from Elsevier; M. Morcillo, I. Díaz, B. Chico, H. Cano, and D. de la Fuente, Weathering steels: from empirical development to scientific design, *Corrosion Science, 83,* 6–31, 2014.)

magnetite. The distribution of corrosion products within the rust layer of carbon steel is more heterogeneous, which results in a mottled structure of lower corrosion resistance.

Long-term field exposures have shown that alloying with silicon increases the amount of nanometer-sized α-FeOOH particles, irrespective of environmental conditions, and that copper reduces the particle size and improves the corrosion resistance. Copper is the alloying element that has shown the most beneficial effect on the corrosion rate. Increasing the copper content from 0.01 to 0.4 wt% can reduce the corrosion rate by as much as 70%. As discussed earlier and in Section 9.4, three stages in rust layer formation can be distinguished: (i) from dry to wet conditions, (ii) wet conditions, and (iii) from wet to dry conditions. In each stage considerable changes in the rust layer can be observed with respect to the oxidation state of iron, the electron conductivity in the γ-FeOOH lattice, and the transport of oxygen through pores in the rust layer to sites for the cathodic reaction, Figure 9.13. Stratmann and coworkers have shown that copper alloyed in steel mainly influences stage III, the drying out of the aqueous layer. Copper may slow down the anodic dissolution process of iron or decrease the electronic conductivity of the rust layer in such a way that the flow rate of electrons reaching sites for the cathodic reaction is reduced. Weathering steels may also benefit from copper alloying in other ways. Cu(II) ions in the aqueous layer can oxidize Fe(II) to Fe(III), thereby promoting the formation of more protective ferric corrosion products. Copper can also form relatively insoluble copper hydroxysulfates such as $Cu_4SO_4(OH)_6$ or $Cu_3SO_4(OH)_4$ that can precipitate in pores of the rust layer, thereby improving its barrier properties. Direct evidence for such precipitation is difficult to provide due to their small quantities. In support of this explanation is the observation that the corrosion performance is improved in the presence of SO_2 or sulfate-containing aerosols combined with dry–wet–dry cycles.

Alloying with nickel (3–5 wt%) has been shown to reduce the formation of local chloride-rich porous areas, nests, which hinder the formation of the dense protective α-FeOOH layer. Similar to copper can nickel form small amounts of relatively protective nickel hydroxysulfates in pores of the rust layer that may improve its barrier properties.

13.5.2 Application Aspects

From their capacity to form protective corrosion products follows that weathering steels do not need to be coated or galvanized in a similar way as, for example, carbon steel to maintain a high long-term corrosion resistance as long as the surfaces are exposed to regular wet and dry cycles, periodic washing by rain, moderate concentrations of gaseous pollutants, and low deposition rates of chlorides, factors that, for example, depend on the surface orientation and wind direction as described in Section 13.2. These conditions promote, as described previously, the formation of dense and adherent corrosion layers. Surfaces exposed to conditions with long times of wetness, lack of sun and rain (repeated dry and wet periods), or chloride-rich conditions form more porous and loosely adherent thick rust layers with modest corrosion properties, similar to iron and carbon steel (see Section 9.4). This means that prevailing environmental conditions govern whether a protective layer will form or not and its formation rate.

A certain amount of deposited SO_2 or sulfate aerosols is beneficial for the formation of protective corrosion products, but large amounts result in an intense acidification of the aqueous layer, a condition that triggers dissolution and hinders precipitation of corrosion products. This is supported by findings from 8-year corrosion data on weathering steels exposed at urban and rural sites of different atmospheric SO_2 levels ($\leq 80\,mg\,SO_2\,m^{-2}\,day^{-1}$) within the UNECE research program, Figure 13.11. The results show that sites with deposition rates less than $20\,mg\,SO_2\,m^{-2}\,day^{-1}$ revealed steady-state corrosion rates less than $6\,\mu m\,year^{-1}$ whereas highly accelerated at higher levels. A corresponding limit of airborne deposition of chloride is $3\,mg\,Cl^-\,m^{-2}\,day^{-1}$ to ensure an annual corrosion rate of weathering steel less than 6 μm. However, this limit is based on findings from bridges in Japan.

A general criterion for the use of unpainted conventional weathering steels is to restrict their use to atmospheres of low corrosivity (class C2 and C3, cf. Chapter 8), thereby ensuring corrosion rates less than $6\,\mu m\,year^{-1}$.

Deposited chlorides and sulfates have often hygroscopic properties and retain water on the surface, often even at low relative humidity levels due to their combined points of deliquescence. Nonuniform deposition of these species may result in heterogeneous dissolution and precipitation of corrosion products.

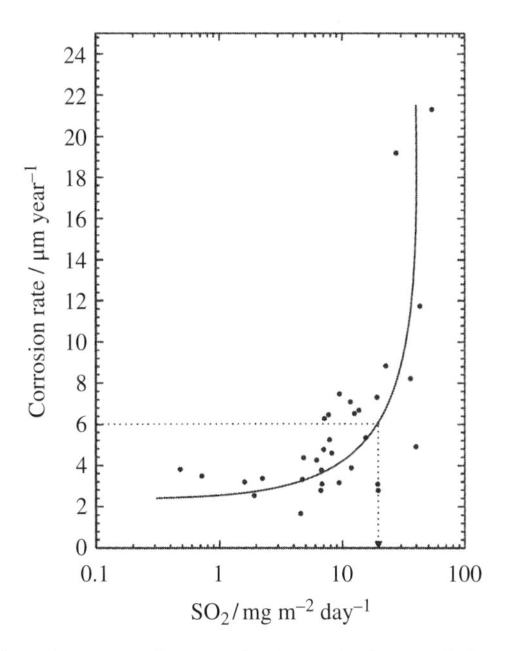

FIGURE 13.11 Corrosion rates of conventional weathering steel after 8 years of outdoor exposure at sites of different atmospheric SO_2 levels. (Reproduced with permission from Elsevier; M. Morcillo, B. Chico, I. Díaz, H. Cano, and D. de la Fuente, Atmospheric corrosion data of weathering steels: a review, *Corrosion Science, 77,* 6–24, 2013.)

During design and maintenance of weathering steel, special attention must be given to crevice areas that are able to retain water and corrosive species much longer than open surfaces and to horizontal surfaces where hygroscopic deposits accumulate and thereby prolong the times of surface wetness, conditions that result in a loosely adherent, porous, and nonprotective rust layer.

A drawback is that the gradual development of a protective rust layer on weathering steels takes years before steady-state conditions are reached (see Figure 13.10), a process that depends on prevailing environmental conditions. Protective rust layers form, as an example, more slowly on north-facing surfaces (at the northern hemisphere) that receive less hours of sunshine. Compilation of standardized long-term outdoor corrosion data for weathering steels shows that the time to obtain a stable rust layer (i.e., when the corrosion rate becomes relatively constant) decreases with increasing site corrosivity (see Chapter 8), being typically 6–8 years in low polluted environments and 4–6 years at more polluted, nonchloride-rich sites, Figure 13.12a. However, this time period does not provide any information regarding the protective capacity of the rust layer, which depends on test site corrosivity, corrosion product properties, exposure time, repeated dry–wet cycles, etc. The steady-state corrosion rates increase with increasing site corrosivity (chloride-rich sites excluded), Figure 13.12b.

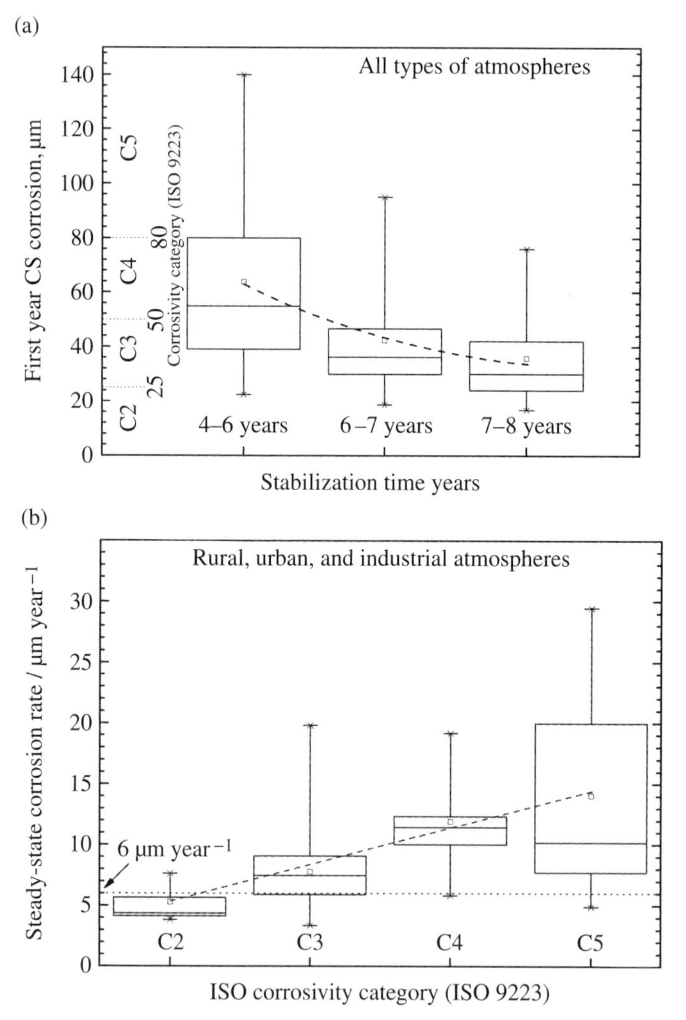

FIGURE 13.12 Rust stabilization time (a) and steady-state corrosion rates (b) of weathering steel at sites of different corrosivity categories (ISO 9223). (Reproduced with permission from Elsevier; M. Morcillo, B. Chico, I. Díaz, H. Cano, and D. de la Fuente, Atmospheric corrosion data of weathering steels: a review, *Corrosion Science, 77,* 6–24, 2013.)

13.6 STAINLESS STEEL

Since the 1920s stainless steel has been increasingly used for exterior architectural applications such as facades, wall claddings, roofing, handrailings, and decorations due to forming and aesthetic reasons and due to its long-term corrosion performance.

13.6.1 Corrosion Mechanisms

Stainless steel alloys form instantaneously a highly protective chromium-rich passive surface layer of amorphous structure with a thickness of only a few nanometers. Its corrosion resistance generally improves further during exposure at atmospheric conditions due to a gradual surface enrichment of chromium. Many investigations have focused on the nature of this outermost surface oxide (the passive layer). It is evident that its composition and structure, at least to some extent, are determined by prevailing exposure conditions. The bulk composition and microstructure are also important as the corrosion resistance generally increases with, for example, the chromium and molybdenum content of the steel. If corrosion takes place, it is generally localized in pits that initially form adjacent to inhomogeneities such as crevices close to inclusions or flaws in the passive layer. If these areas are not repassivated, the acidity and anion concentration within the pits become enhanced compared with the aqueous surface layer, due to hindered diffusion and convection, and the pits propagate. The corrosion attack can also be accelerated by the presence of free radicals. Corrosion continues if the rate of metal dissolution exceeds the rate of repassivation but ceases if the reverse is true.

Corrosion rates (determined as mass losses), pitting depths, or degree of surface coverage of rust stains have been determined worldwide at different environmental conditions ranging from low polluted rural sites to severe polluted marine conditions. Mass losses (representing longer than 5-year field data) range commonly between 0.2 and 0.5 $mg\,m^{-2}\,year^{-1}$ for AISI 304 (typically 18 wt% Cr, 8 wt% Ni) and 0.1 for AISI 316 (typically 18 wt% Cr, 10 wt% Ni, 2 wt% Mo), rates orders of magnitude lower compared with, for example, carbon steel and weathering steel exposed at identical conditions. The most corrosive environments for stainless steels are sites of high pollutant levels of SO_2, chlorides, and particles (in particular for sheltered surfaces), sites of low to moderate rainfall and moderate to high humidity, sites of high temperature and high humidity, and sites with frequent salt spray and low rainfall. The least corrosive conditions are sites of low pollutant levels, low amounts but high-intensity rainfall events, low temperature, and sites of high temperature but low humidity levels. Only under extreme conditions such as in highly acidic fogs or at severe marine conditions can some low alloy stainless steel grades degrade in a way similar to low alloy steels.

Site-specific variations in average annual corrosion rates (mass losses) have, as an example, been reported after 20 years of exposure at different sites in South Africa to range from 0.025 $\mu m\,year^{-1}$ (AISI 304, 316) at rural low polluted sites to 0.076 (AISI 304), to 0.025 (AISI 316) at moderate polluted marine sites, and to 0.41 (AISI 304) and 0.28 $\mu m\,year^{-1}$ (AISI 316) for severe moderate polluted marine sites.

Even though stainless steels usually possess high corrosion resistance due to the enrichment of chromium in the surface oxide, are metals from this interface dispersed to a different extent to the ambient environment (see Chapter 10). The amount of released metals (and the corrosion resistance) is strongly dependent on prevailing environmental conditions and on surface and bulk material characteristics such as microstructure, composition and nobility, differences in grain size, and presence of inclusions and other heterogeneities. These processes have been studied by Odnevall Wallinder et al. for planar surfaces of massive AISI 304 and 316 both at field and laboratory conditions. A 4-year unsheltered field exposure (inclined 45° from the horizontal, facing south) at an urban site (Stockholm, Sweden) clearly revealed a preferential release of Fe compared with Ni and Cr (see Chapter 10).

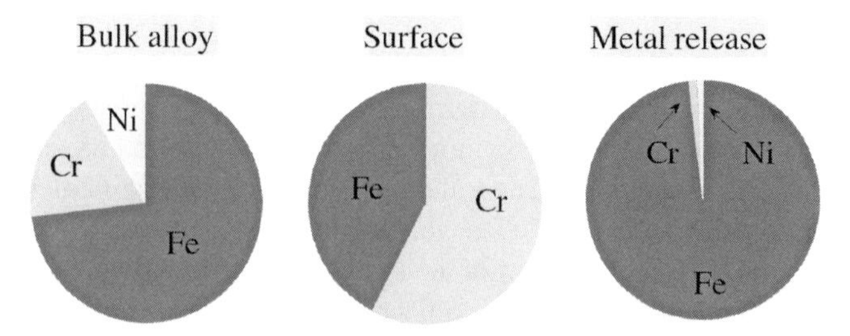

FIGURE 13.13 The extent of metal release of main alloying constituents (right) compared with the bulk (left) and the surface composition (middle) illustrated for stainless steel (AISI 316L) exposed at nonsheltered urban atmospheric conditions (Stockholm, Sweden) during 4 years.

Slightly more metals were, on an annual basis, released from AISI 316 (0.18–0.66 mg Cr m^{-2} $year^{-1}$; 0.30–0.83 mg Ni m^{-2} $year^{-1}$; 10–200 mg Fe m^{-2} $year^{-1}$) compared with AISI 304 (0.18–0.57 mg Cr m^{-2} $year^{-1}$; 0.12–0.52 mg Ni m^{-2} $year^{-1}$; 10–140 mg Fe m^{-2} $year^{-1}$). Similar findings were obtained for AISI 304 exposed at unsheltered marine conditions (Brest, France) during 1 year with first year release rates of 0.4 mg Cr m^{-2} $year^{-1}$ and 36 mg Fe m^{-2} $year^{-1}$. Release rates of Ni and Fe from the stainless steel grades were orders of magnitude lower compared with their pure metals. Complementary exposures to artificial rain revealed that chromium and nickel release rates are highly pH dependent and also dependent on surface conditions. A freshly abraded stainless steel surface rapidly forms a passive layer with a chromium content only slightly higher than the bulk content, whereas the chromium content of a, for example, pickled (acid cleaned) and aged surface is significantly higher. Large variations in the extent of released chromium and nickel (and of iron) are hence expected. Release rates of chromium and nickel between 0.4 and 5 mg m^{-2} $year^{-1}$ have been reported upon exposure of grade 304 to artificial rainwater of different pH, compared with a pickled and aged surface that showed rates less than 0.7 mg m^{-2} $year^{-1}$ for the same conditions.

The extent of released metals is not possible to predict either from the surface oxide composition or the bulk alloy content, Figure 13.13.

13.6.2 Application Aspects

Modern architectural applications predominantly use the austenitic stainless steel grades AISI 304 (e.g., in the Gateway Arch, St. Louis, United States) and AISI 316 (e.g., in Petronas Twin Tower, Kuala Lumpur, Malaysia), even though higher alloyed (duplex, e.g., AISI 2205 in Millennium Footbridge, York, United Kingdom) grades are used at specific conditions. The Chrysler Building and the Empire State Building in New York, United States were both built using AISI 302, a grade with more carbon compared with AISI 304. AISI 304 is nowadays mainly used in rural and urban environments, whereas AISI 316 predominantly is the recommended grade for industrial and marine conditions, however not at sites close to the sea in tropical environments (high humidity, high daytime temperatures) where duplex steels predominantly are to be used.

Grade selection for a certain environment should be based on information of site-specific corrosivity conditions and extent of rain that is able to wash the surface from soiling and corrosive deposits, which thus hamper the occurrence of local corrosion processes and surface tarnishing (often yellowish appearance). The extent of surface soiling and corrosion resistance are both closely related to the surface finish and the surface roughness. The European Standard EN10088 recommends a surface roughness less than 0.5 μm for architectural applications where regular maintenance (cleaning) is implausible. However, regular cleaning procedures are recommended to maintain the desired aesthetic appearance and sustain a high corrosion resistance. This is in particular crucial for sheltered surfaces and for surfaces exposed in environments where the rain quantity and intensity are not sufficiently high. Surface soiling and retention of particles and deposits are facilitated on a rough surface compared with a smooth surface and may result in a reduced corrosion resistance with time due to the occurrence of localized corrosion at these areas, which allow retention of water and corrosive components such as chlorides. The importance of deposits on corrosion initiation has, for example, been shown in long-term (4–30 years) outdoor exposures of AISI 316 in the United Kingdom and during 12-year studies at different sites in China. Even a minor corrosion attack may influence the aesthetic appearance but still maintain a sufficient corrosion resistance.

Investigations by Asami and Hashimoto, Japan, have shown the importance of surface finish on the initial chromium content of the passive surface oxide and on the corrosion performance of different stainless steels at atmospheric marine conditions (1.5 years). Industrially abraded surfaces revealed the highest amount of chromium in the surface oxide (due to heat generated by the abrasion process), compared with bright-annealed (annealing in hydrogen gas at high temperature) and acid-pickled surfaces. The corrosion performance of these surfaces at the marine sites, evaluated as pitting depth, correlated with the chromium content of the surface oxide. Large variations in the initial chromium content, depending on surface finish, have also been shown by Wallinder et al., Sweden. However, these differences diminished with time. After 7 months of unsheltered urban and marine conditions did all AISI 304 stainless steel surfaces reveal relatively similar amounts of chromium in the surface oxide. A slight correlation was observed between the corrosion resistance (polarization resistance measurements) and the chromium content of the passive film at the urban site, however not at the marine site.

FURTHER READING

Copper-Based Architectural Alloys

S. Goidanich, J. Brunk, G. Herting, M.A. Arenas, and I. Odnevall Wallinder, Atmospheric corrosion of brass in outdoor applications. Patina evolution, metal release and aesthetic appearance at urban exposure conditions, *Science of the Total Environment, 412–413*, 46–57, 2011.

S. Goidanich, S. Jafarzadeh, L. Toniolo, and I. Odnevall Wallinder, Effect of wax based antigraffiti on patina composition and dissolution during 4 years of outdoor exposure, *Journal of Cultural Heritage, 11*, 288–296 2010.

S. Goidanich, I. Odnevall Wallinder, G. Herting, and C. Leygraf, Corrosion-induced metal release from copper-based alloys compared to their pure elements, *Corrosion Engineering, Science and Technology, 43*(2), 134, 2008.

R. Holm and E. Mattsson, Atmospheric corrosion tests of copper and copper alloys in Sweden—16-year results, in *Atmospheric Corrosion of Metals*, ASTM STP 767, eds. S.W. Dean Jr, E.C. Rhea, ASTM, Philadelphia, pp. 85–104, 1982.

E. Mattsson, Corrosion of copper and brass—practical experience in relation to basic data, *British Corrosion Journal, 15*, 6–13, 1980.

I. Odnevall Wallinder, T. Korpinen, R. Sundberg, and C. Leygraf, Atmospheric corrosion of naturally and pre-patinated copper roofs in Singapore and Stockholm—runoff rates and corrosion product formation, in *Outdoor and Indoor Atmospheric Corrosion*, ASTM STP 1421, ed. H.E. Townsend, ASTM, West Conshohocken, pp. 230–244, 2002.

I. Odnevall Wallinder, X. Zhang, S. Goidanich, N. Le Bozec, G. Herting, and C. Leygraf, Corrosion and runoff rates of Cu and three Cu-alloys in marine environments with increasing chloride deposition rate, *Science of the Total Environment, 472*, 681–694, 2014.

R. Piccichi, A.C. Ramos, M.H. Mendon, and I.T.E. Fonseca, Influence of the environment on the atmospheric corrosion of bronze, *Journal of Applied Electrochemistry, 34*, 989–995, 2004.

J. Sandberg, I. Odnevall Wallinder, C. Leygraf, and N. Le Bozéc, Corrosion-induced copper runoff from naturally and pre-patinated copper in a marine environment, *Corrosion Science, 48*(12), 4316–4338, 2006.

M.T. Sougrati, S. Jouen, B. Hannoyer, and A. Barbier, A study of copper and copper alloys runoff in urban atmosphere, in *Proceedings of the International Conference Copper '06*, ed. J.-M. Welter, Wiley-VCH Verlag GmbH, Weinheim, pp. 130–136, 2007.

X. Zhang, I. Odnevall Wallinder, and C. Leygraf, Mechanistic studies of corrosion product flaking on copper and copper-based alloys in marine environments, *Corrosion Science, 85*, 15–25, 2014.

Aluminum–Zinc Alloys

C.I. Elsner, P.R. Sere, and A.R. Di Sarli, Atmospheric corrosion of painted galvanized and 55% Al–Zn steel sheets: results of 12 years of exposure, *International Journal of Corrosion, 2012*, 16, 2012.

I. Odnevall Wallinder, W. He, P.-E. Augustsson, and C. Leygraf. Characterization of black rust staining of unpassivated 55% Al–Zn alloy coatings. Effect of temperature, pH and wet storage, *Corrosion Science, 41*, 2229–2249, 1999.

E. Palma, J. M. Puente, and M. Morcillo, The atmospheric corrosion mechanism of 55% Al–Zn coating on steel, *Corrosion Science, 40*(1), 61–68, 1998.

E.D. Persson, D. Thierry, and N. Le Bozec, Corrosion product formation on Zn55Al coated steel upon exposure in a marine atmosphere, *Corrosion Science, 53*, 720–726, 2011.

P. Qiu, C. Leygraf, and I. Odnevall Wallinder, Evolution of corrosion products and metal release from Galvalume coatings on steel during short and long-term atmospheric exposures, *Materials Chemistry and Physics, 133*, 419–428, 2012.

A. Ramus Moreira, Z. Panossian, P.L. Camargo, M. Ferreira Moreira, I.C. da Silva, and J.E. Ribeiro de Carvalho, Zn/55Al coating microstructure and corrosion mechanism, *Corrosion Science, 48*, 564–576, 2006.

H.E. Townsend and J.C. Zoccola, Atmospheric corrosion resistance of 55% Al–Zn coated sheet steel: 13-year test results, *Materials Performance, 18*, 13–20, 1979.

J.C. Zoccola, H.E. Townsend, A.R. Borzillo, and J.B. Horton, Atmospheric corrosion resistance of aluminum–zinc alloy-coated steel, in *Atmospheric Factors Affecting the Corrosion of Engineering Materials*, ASTM STP 646, ed. S.K. Coburn, ASTM, Philadelphia, pp. 165–184, 1978.

Weathering Steel

T. Kamimura, S. Hara, H. Miyuki, M. Yamashita, and H. Uchida, Composition and protective ability of rust layer formed on weathering steel exposed to various environments, *Corrosion Science, 48*, 2799–2812, 2006.

M. Morcillo, B. Chico, I. Díaz, H. Cano, and D. de la Fuente, Atmospheric corrosion data of weathering steels: a review, *Corrosion Science, 77*, 6–24, 2013.

M. Morcillo, I. Díaz, B. Chico, H. Cano, and D. de la Fuente, Weathering steels: from empirical development to scientific design, *Corrosion Science, 83*, 6–31, 2014.

M. Stratmann, K. Bohnenkamp, and T. Ramchandran, The influence of copper upon the atmospheric corrosion of iron, *Corrosion Science, 27*, 905–926, 1987.

M. Yamashita, H. Miyuki, Y. Matsuda, H. Nagano, and T. Misawa, The long term growth of the protective rust layer formed on weathering steel by atmospheric corrosion during a quarter of a century, *Corrosion Science, 36*(2), 283–299, 1994.

Stainless Steel

K. Asami and K. Hashimoto, Importance of initial surface film in the degradation of stainless steels by atmospheric corrosion, *Corrosion Science, 45*, 2263–2283, 2003.

G. Herting, I. Odnevall Wallinder, and C. Leygraf, A comparison of release rates of Cr, Ni and Fe from stainless steel alloys and the pure metals exposed to simulated rain events, *Journal of the Electrochemical Society, 152*(1), B23–B29, 2005.

J.R. Kearns, M.J. Johnson, and P.J. Pavlik, The corrosion of stainless steels in the atmosphere, in *Degradation of Metals in the Atmosphere*, ASTM STP 965, eds. S.W. Dean and T.S. Lee, ASTM, Philadelphia, pp. 35–51, 1988.

C. Liang and W. Hou, Twelve year atmospheric exposure study of stainless steel in China, in *Outdoor Atmospheric Corrosion*, ASTM STP 1421, ed. H.E. Townsend, ASTM, West Conshohocken, pp. 358–367, 2002.

I. Odnevall Wallinder, S. Bertling, D. Berggren Kleja, and C. Leygraf, Corrosion-induced release and environmental interaction of chromium, nickel and iron from stainless steel, *Water, Air, and Soil Pollution, 170*, 17–35, 2006.

I. Odnevall Wallinder, J. Lu, S. Bertling, and C. Leygraf, Release rates of chromium and nickel from 304 and 316 stainless steel during urban atmospheric exposure—a combined field and laboratory study, *Corrosion Science, 44*, 2303–2319, 2002.

D. Wallinder, I. Odnevall Wallinder, and C. Leygraf, Influence of surface treatment of 304L stainless steel on atmospheric corrosion resistance in an urban and a marine environment, *Corrosion, 59*(3), 220–227, 2003.

14

APPLIED ATMOSPHERIC CORROSION: UNESCO CULTURAL HERITAGE SITES

14.1 INTRODUCTION

Air pollution degrades cultural artifacts, including metals, stones, glasses, textiles, dyes, and leather. Such objects are exposed both outdoors and indoors in environments where relative humidity, temperature, and the concentration of corrosive species vary considerably. The environmental attributes under which decay is minimized differ with material and may, in fact, contradict each other. Whereas low relative humidity is important in preventing metallic artifacts from corroding, it promotes the degradation of fibers and embrittlement of moisture-containing materials, such as dyes, wood, leather, paper, and textiles. As briefly discussed in Section 7.6, no one set of environmental conditions fulfils the requirements of all types of materials. Because the subject of this book is the corrosion of metallic materials, it would lead too far to give examples for all materials. Stone materials are frequently occurring in cultural heritage buildings, and therefore one nonmetallic material (limestone) is included herein as an exception, for comparison purposes.

What distinguishes cultural artifacts from most other objects is their inherent value, from which it follows that decayed parts should be preserved rather than replaced. This naturally puts much more emphasis on conservation treatments of cultural objects than on objects that can be replaced more easily. Conservation refers to measures and actions aimed at safeguarding cultural heritage while respecting its significance, including its accessibility to present and future generations (EN 15898, Conservation of cultural property—Main general terms and definitions). There are two different approaches: prevention of decay through control of the environment and consolidation through treatment in order to arrest decay and stabilize the artifact against further deterioration. Restoration is actions applied to a stable or stabilized object aimed at facilitating its appreciation, understanding, and/or use while

Atmospheric Corrosion, Second Edition. Christofer Leygraf, Inger Odnevall Wallinder, Johan Tidblad and Thomas Graedel.

respecting its significance and the materials and techniques used. The renewal of material components when conservation is insufficient is termed renovation.

The aesthetic and artistic nature of artifacts, and their need for preservation, requires that corrosion scientists and conservators work as teams if optimum results are to be obtained. Conservators are generally very skilled at restoration and the maintenance of artistic integrity, while corrosion scientists, working from the perspective afforded by knowledge of physical and chemical processes, can often propose ways in which the corrosivity of an environment may be reduced or by which artifacts may be shielded from corrosive species. In all cases, treatments are more effective if done regularly; a neglected and seriously corroded artifact can seldom be restored to anything much like its original condition. The wide variety of materials, environments, and conservation approaches means that there will never be a general "cleaning and conservation treatment" suitable for all artifacts. Rather, experience, collaboration, and flexibility are required attributes for those who deal with degraded cultural artifacts.

In an effort to limit the information presented, two case studies used for illustration on how to assess corrosion rates; stock of materials at risk, that is, total exposed surface area of a material; and cost associated to corrosion for cultural heritage are presented here representing different environments and types of materials. To make the case studies livelier to the general reader, they have been selected from sites currently on the UNESCO list of world heritage sites. UNESCO or the United Nations Educational, Scientific and Cultural Organization is involved in a diversity of activities, but one of the most well known in the field of materials is the World Heritage List, which constitutes a mix of natural, cultural, and in-between sites considered to be of "outstanding universal value." The emphasis here is on environmental characteristics, on corrosion, and on related processes, but this is of course only one of several reasons why a site might be endangered, and a full risk analysis of a particular UNESCO site must take all these into account.

14.2 DESCRIPTION OF SELECTED SITES

The two sites included in these case studies are the Klementinum National Library in Prague, Czech Republic, and Acropolis in Athens, Greece. The Klementinum is situated in the middle of the city of Prague exposed to high traffic pollution, while the Acropolis is on a hill quite far away from the main roads of Athens. Compared to many other cities in Europe, Prague has a relatively large proportion of exposed metals, while the monuments at the Acropolis hill are made of stone. Therefore these two cases together will cover a broad spectrum of effects and situations.

14.2.1 The Klementinum, Prague, Czech Republic

The large complex of the Klementinum is situated next to the Charles Bridge in the middle of the historical center of Prague. In consists of several irregular buildings in three courtyards covering an area of about 2 ha. It consists of halls, classrooms,

FIGURE 14.1 Overview of the Klementinum area, Prague, Czech Republic.

churches, chapels, and the Astronomical Tower. It was formerly a Jesuit college with a history dating to one of the chapels dedicated to Saint Clement in the eleventh century. The Klementinum is presently home to the National Library of the Czech Republic.

Being in the city center, the Klementinum is surrounded by streets, the Krizovnicka street to the west of the complex, one of the most traffic-intense streets in Prague close to the Charles Bridge, and the Platnerska street to the north of the complex, which is slightly less traffic intensive.

The external materials of buildings and structures of the Klementinum are primarily stone, wooden, and rendered materials. However, for illustrative purposes calculations and discussions in the following for this site are given for metallic materials, which are otherwise relatively frequently exposed in this city.

An overview of the Klementinum including surrounding areas is given in Figure 14.1.

14.2.2 Acropolis, Athens, Greece

The Acropolis is in the middle of Athens, but as it is situated on a hill looking down to the city, it is not directly subject to local traffic pollution. The Acropolis hill includes a variety of monuments from different periods in time from fifth century B.C. to Roman monuments, for example, the Erechtheion, the temple of Athena Nike and the Theatre of Dionysus. The most well-known monument, which will be used as our case study, is the Parthenon. This monument was built between 447 and 438 B.C. and the sculptural decorations completed in 432 B.C. It is almost exclusively made of Pentelic marble and will therefore be used as an example on calculation and discussions of stone materials in general.

The Acropolis and the Parthenon are shown in Figure 14.2.

FIGURE 14.2 Overview of the Acropolis and Parthenon, Athens, Greece.

14.3 ESTIMATION OF CORROSION RATES

Only after a proper estimation of corrosion rates, it is possible to assess costs of damage in general and costs due to pollution in particular. It is quite frequent that the material used is unique to the regional area or even to the studied object, especially when it comes to stone materials. Local materials have often been used since transportation is easier and material composition and structure can vary even within a single quarry. It is likely that there will be limited information on the degradation mechanisms and corrosion rates of the exact material that is used. Therefore a common approach when assessing corrosion rates, risks, and economic loss due to corrosion is to use the concept of indicator materials. These materials have the benefit of being exposed worldwide in a variety of situations so that experience has been gathered about the degradation mechanism and corrosion rates in different environments while being sufficiently similar to the investigated material. For stone materials the most frequently used indicator material is Portland limestone, and for metals common indicator materials are carbon steel, zinc, and in some cases copper. Portland limestone is a carbonate stone with porosity slightly more than 20% (see Appendix D). It is an oolitic limestone from the Portland Beds of Upper Jurassic age (136–160 million years old), and the quarries are located on the Isle of Portland in Dorset, England. It is a popular material for architects and builders, and an important monument made of this stone is St. Paul's Cathedral in London.

Corrosion rates can be determined by exposure of sample materials on racks or estimated by using dose–response functions, as described in Chapter 8. Exposure data are available for Athens and Prague, but at other locations in the cities, and can only be used for comparison. For these sites, the use of dose–response functions (Section 8.4) is therefore the appropriate method.

14.3.1 Corrosion of Carbon Steel and Zinc in Prague and around the Klementinum

Corrosion of metals in Prague in general was highest in the period between 1970 and 1990 with a corrosivity C4, corresponding to 50–80 μm year^{-1} for carbon steel and 2.1–4.2 μm year^{-1} for zinc. Since then corrosion has decreased due to a significant decrease in levels of air pollution, and as a result corrosivity in Prague has since the turn of the century been C2 for carbon steel (<25 μm year^{-1}) and C2/C3 for zinc (about 0.7 μm year^{-1}). Figure 14.3 shows this trend of decreasing corrosion in Prague obtained by exposures of standard specimens. These values are just approximate general values for the entire city, but the actual corrosion rate varies depending on the local pollution situation, especially around the Klementinum, which is subject to the influence of local traffic sources.

Dose–response functions were presented in Chapter 8. For carbon steel in a multi-pollutant situation, the following dose–response function has been obtained:

$$R_{\mathrm{cst}} = 6.5 + 0.178[\mathrm{SO_2}]^{0.6}\,\mathrm{RH}_{60}e^{f(T)} + 0.166\mathrm{Rain}[\mathrm{H^+}] + 0.076\mathrm{PM}_{10} \quad (14.1)$$

where

R_{cst} is the corrosion rate of carbon steel (μm year^{-1})

$[SO_2]$ is the SO_2 concentration (μg m^{-3})

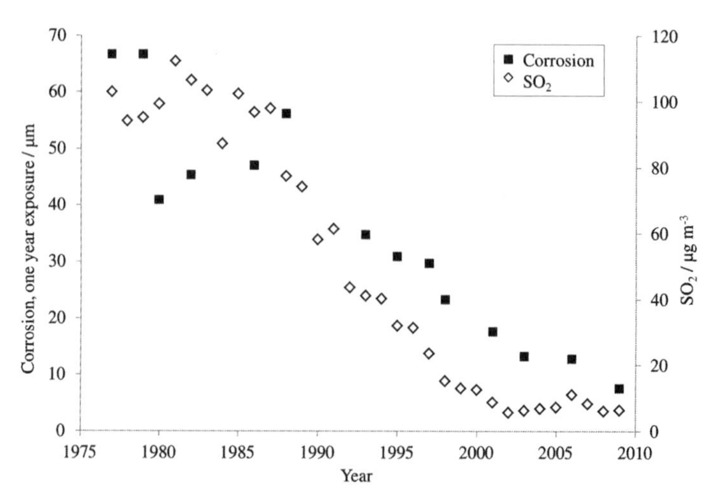

FIGURE 14.3 Trend of carbon steel corrosion and SO_2 gaseous concentrations in Prague, Czech Republic. (K. Kreislova, SVUOM, unpublished data.)

RH is the relative humidity (%)

RH_{60} is equal to RH at 60% when RH > 60, otherwise equal to 0

T is the temperature (°C)

$f(T)$ is equal to $0.15(T-10)$ when $T<10$, otherwise equal to $-0.054(T-10)$

Rain is equal to the annual amount of precipitation (mm)

$[H^+]$ is equal to the H^+ concentration in precipitation ($mg\,L^{-1}$)

PM_{10} is equal to the concentration of particles <10 μm ($\mu g\,m^{-3}$)

Values of SO_2 and PM_{10} concentration vary considerably around the Klementinum area, but the other parameters are relatively constant. For typical Prague values of RH = 75%, T = 10°C, rain = 500 mm, and pH = 6 the equation can be simplified to (the effect of acid rain only contributes to 0.1 μm year^{-1} at this pH level)

$$R_{cst} = 6.6 + 2.67[SO_2]^{0.6} + 0.076PM_{10} \quad (14.2)$$

Data of SO_2 and PM_{10} is in this area typically in the range 5–10 $\mu g\,m^{-3}$ and 20–30 $\mu g\,m^{-3}$, respectively, which results in a corrosion rate in the range 15–20 μm year^{-1} when applying Equation 14.2.

14.3.2 Corrosion of Portland Limestone in Athens

Surface recession of Portland limestone in central Athens was estimated around the turn of the century with measured values of about 10 μm year^{-1}, and these values have since then been similar or slightly less. The dose–response function for Portland limestone is

$$\begin{aligned} R_{por} &= 4.0 + (0.0059[SO_2] + 0.0078[HNO_3])RH_{60} \\ &+ \cdots 0.054Rain[H^+] + 0.0258PM_{10} \end{aligned} \quad (14.3)$$

Inserting values of environmental parameters in Equation 14.3 gives a surface recession in the Acropolis area of 6 μm year^{-1}, which not surprisingly is less than in the city center.

14.4 ESTIMATION OF CORROSION COSTS

Corrosion is not the only effect that air pollution and climate can have on materials. The effect of extreme temperatures and temperature–humidity cycles has a significant influence on the durability of stone materials. Furthermore the deposition of particulate matter not only increases the risk of corrosion but also results in discoloration and soiling of materials. It is important to realize that when a decision of a conservation or renovation measure is taken, it is based on the total condition of the object, which can be the result of corrosion and of other effects or, more likely, a combination of effects that cannot be completely separated. The examples presented here are therefore simplified and should not be quoted as official statements on actual costs for the selected cases but rather illustrations of principles on how to calculate costs for corrosion.

When estimating corrosion costs there are two fundamentally different costs associated with material damage, expenditures associated with conservation, restoration, or renovation and amenity loss. The first is rather simple to understand conceptually but requires quite a lot of detailed data. The amenity loss is more eluding and associated with the value of the object/building and the impairment of its aesthetical qualities as corrosion progresses. A direct consequence of serious amenity loss can, for example, be decreased in the amount of tourists visiting the object.

14.4.1 Estimating Renovation Costs and Stock of Materials at Risk

Before even starting to estimate the cost it should be realized that from a heritage manager's point of view facing an immediate restoration task, it is quite simple to just ask for quotations and then select the most attractive offer. However, when there is a need to estimate possible future costs for a maintenance plan or to estimate costs for certain types of objects over a larger geographical area, other methods are needed. This is especially true if there is a need to estimate the cost attributed to pollution and single pollutants and to prevent further damage through air quality policy through cost–benefit analysis or other methods (see later). The basic formula for calculating corrosion costs is

$$K = \frac{k \cdot S}{L} \tag{14.4}$$

where

K is the cost for maintenance or renovation (€ year^{-1})

k is the cost for a particular maintenance or renovation action (€ m^{-2})

L is the maintenance or renovation interval between actions (year)

S is the total surface area of the material (m^2)

The cost for a particular action varies quite a lot depending on material and is generally higher for objects of cultural heritage compared to other materials due to higher demands. For illustration purposes we use a value of 400 € m^{-2} corresponding to conservation of limestone/marble.

Just as in economic theory, there is an underlying assumption of human beings in general taking rational decisions based on the available information. Estimation of maintenance intervals is based on the assumption that the person in charge of renovation takes rational decisions. It would be reasonable to assume that the owner strives to minimize the cost over a longer period of time favoring small, more frequent interventions when possible instead of postponing action when consequences are serious. Consider, for example, the replacement of a roof made of galvanized steel sheet with a thickness of 30 μm situated in an area resulting in a corrosion attack for zinc of 0.7 μm year^{-1}. If the maintenance interval was set to 40 years resulting in

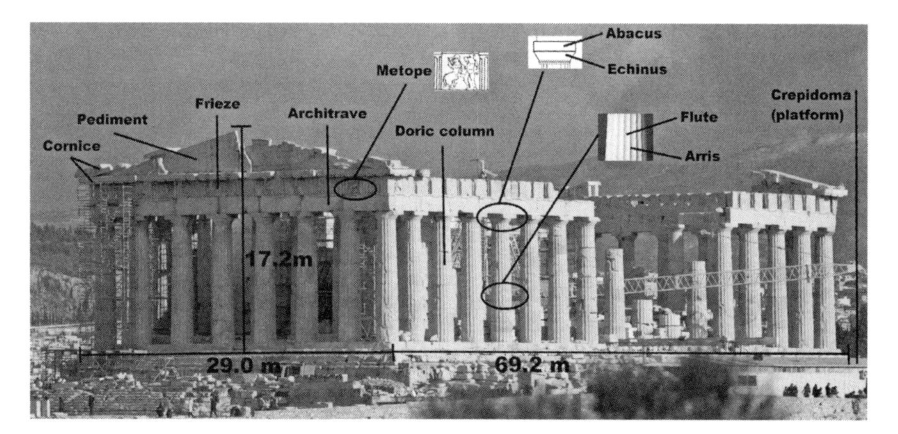

FIGURE 14.4 Different structural elements and measured distances outlining the procedure for estimating stock of materials at risk on an individual building (Parthenon).

an average corrosion of $40 \times 0.7 = 28\,\mu m$, there would be a serious risk of loss of protection and consequently structural integrity resulting in, for example, moisture penetration into the building with significant costs that could easily be avoided by choosing a slightly shorter and more rational maintenance interval. This is a conservative estimate, since the corrosion of zinc usually decreases with time. A criteria-based definition of the maintenance interval as that resulting in a specific corrosion attack, in this example by choosing a total attack of $20\,\mu m$ with a $10\,\mu m$ safety margin, makes it possible to use time-dependent dose–response functions to estimate rational maintenance intervals by "calculating backward" and replacing the exposure time with the maintenance interval.

Estimating the stock at risk is in principle simple but since the required data is in most cases not directly included in any type of existing registers, it is not easy to obtain this data at reasonable cost. Therefore several methods for estimating stocks at risk have been developed depending on the geographical scale from individual buildings, city districts, entire cities, countries, and to the continental level.

On the individual building level direct measurements are most appropriate as they permit an evaluation without any theoretical modeling or generalization by direct observation, counting, and measurement. This procedure is illustrated in Figure 14.4 for Parthenon resulting in a surface area of $3800\,m^2$. Combining this value with a conservation cost of $400\,€\,m^{-2}$ results in a total cost of each renovation of 1.5 million €, which then has to be divided by the maintenance interval. This is further developed in connection with the discussion on air quality policy in the following.

The next step in the generalization process is to use a so-called identikit. This is a generic building type devised to represent a typical building within a specific region. Example of an identikit could be "a typical gothic cathedral in Europe." To have an identikit it is not sufficient to only give this label, but a detailed description of the exposed materials and respective surface area needs to be given as illustrated in Table 14.1 giving an identikit for "a typical semi-detached house in UK."

TABLE 14.1 Identikit for a Typical Semidetached House in United Kingdom Used for Estimating Stock of Materials at Risk

Predominant Materials/Elements	Model (%)	Quantity (m^2 $building^{-1}$)
Flat roofs		*12.8*
Felt (flat roofs on garages)	50	6.4
Asbestos cement (on garages)	50	6.4
Pitched roof		*54*
Clay tiles	25	13.5
Slate tiles	50	27
Concrete tiles	25	13.5
External wall		*84*
Brickwork (bare)	58	48.7
Brickwork (painted)	10	8.4
Render	30	25.2
Timber framing or cladding	2	1.7
Windows and doors		*9*
Joinery	70	6.3
Aluminum/metal	30	2.7
Roof drainage		*3*
Plastic (gutters/pipes)	71	2.1
Cast iron (gutters/pipes)	29	0.9
Total external envelope		*165*

Adapted with permission from T. Yates, R. Butlin, and J. Medhurst, Assessing the economic benefits of reduced SO_2 concentrations for buildings in the UK, in *Economic Evaluation of Air Pollution Damage to Materials*, Proceedings of a UNECE Workshop, eds. V. Kucera, D. Pearce, and Y.-W. Brodin, Report 4761, Swedish Environmental Protection Agency, Stockholm, Sweden, pp. 162–185, 1996.
The identikit represents a weighted average of different architectural styles of semidetached buildings in an area. It shows, for example, three different types of roofing material, which would never be present simultaneously on a single building.

When assessing stocks at risk on a larger scale, entire cities or countries, it becomes no longer practical to use self-created identikits for all types of buildings. The approach instead is often to take what data is available in building registers and other sources and then to generalize by combining this data with data from individual inspections and identikits. This process is referred to as "grossing up."

The last step in the generalization is to use census data, that is, to assume that materials are exposed where people live and to associate each person with a specified amount of materials. This is illustrated in Figure 14.5.

14.4.2 Estimating Amenity Loss

In order to estimate the value of an object, it is in principle necessary to be able to sell it, but since the objects we are trying to assess are not directly on the market, it is only possible to use indirect methods to assess these indirect costs associated with the

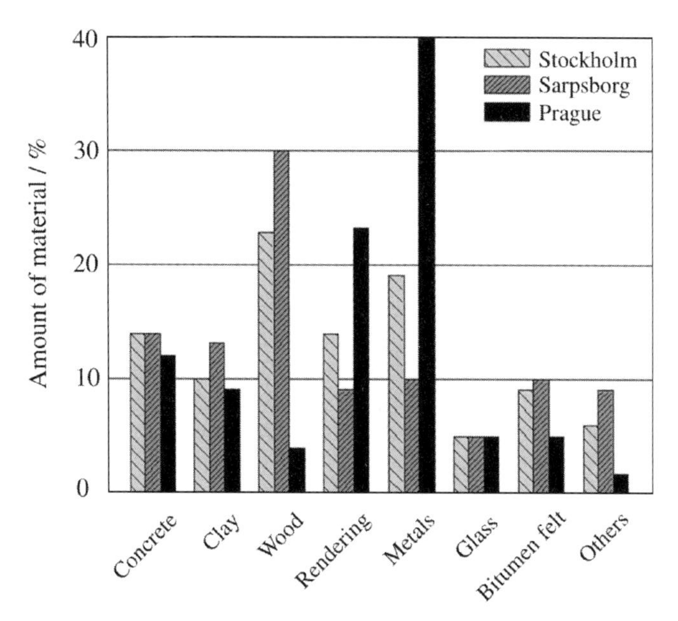

FIGURE 14.5 Example of the amount of material associated to a person (census data) for Stockholm, Sweden (total 132 $m^2 capita^{-1}$); Sarpsborg, Norway (165 $m^2 capita^{-1}$); and Prague, Czech Republic (83 $m^2 capita^{-1}$). (Adapted with permission from V. Kucera, J. Henriksen, D. Knotkova, and C. Sjöström, Model for calculations of corrosion cost caused by air pollution and its application in three cities, in *Progress in the Understanding and Prevention of Corrosion*, eds. J.M. Costa and A.D. Mercer, Institute of Materials for the Sociedad Española de Quimíca Industrial, Barcelona, Spain, vol. 1, pp. 24–32, 1993.)

amenity loss. Quite a range of approaches have been developed including but not limited to revealed preference techniques, experimental market techniques, and surrogate market methods. It would take too long to describe them all in great detail, but the contingent valuation method is a technique that requires mentioning. In this method the willingness to pay (WTP) or willingness to avoid (WTA) is obtained by directly asking people how much they are willing to spend or pay to avoid a certain event. An example of such a question could be "how much are you willing to pay in order to avoid the Parthenon being destroyed." As all questionnaire methods, this method of course has its weaknesses. The answer not only depends on the value of the object but also the available income of the questioned person.

Another, quite elegant way of estimating the amenity loss stems from a theoretical argumentation of rational decision-making as illustrated in Figure 14.6. By assuming that the decision on when to take an action to repair takes into account all costs, not only direct costs as in the example earlier for the zinc roof but also the amenity costs, it can be shown that the cost associated with amenity loss is equal to the direct costs for renovation or maintenance. While this method also has its drawbacks, it results in a very easy way to calculate the total cost of corrosion as twice that of the direct renovation cost.

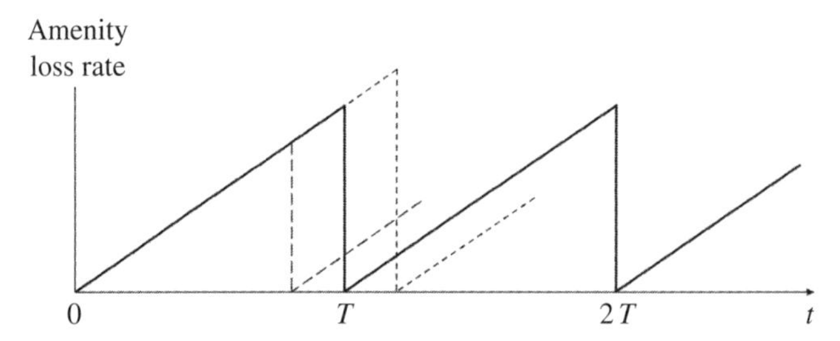

FIGURE 14.6 Example of time development of direct and indirect costs due to corrosion assuming a linear increase of loss with time. The loss rate can also have a nonlinear time dependence, either decreasing or increasing with time depending on the corrosion kinetics and the visual appearance. T indicates the renovation time and dashed lines renovation periods shorter or longer than T. (Adapted with permission from Elsevier; A. Rabl, Air pollution and buildings: an estimation of damage costs in France, *Environmental Impact Assessment Review, 19,* 361–385, 1999.)

14.5 PREVENTING FURTHER DAMAGE THROUGH AIR QUALITY POLICY

As implied from Figure 14.3 the pollution situation in Europe changed dramatically during the second half of the last century. The driving force for this improvement has not been the protection of materials and cultural heritage but the protection of human health. However, the framework for implementing this policy has been the Convention on Long-range Transboundary Air Pollution where an effect-based approach has been used throughout the process taking into account all possible effects to human health ecosystems and materials. Therefore strategies for assessing air pollution effects from a policy-oriented point of view have been developed. These fall into two main categories, calculating costs related to air pollution and assessment of tolerable corrosion rates.

14.5.1 Calculation of Air Pollution Costs

Equation 14.4 gives the total corrosion cost. The lifetime of materials and cultural heritage decreases due to pollution, but corrosion proceeds even in the absence of pollution and therefore can this equation not be used directly for calculating the costs attributed to pollution. What is possible, however, is to calculate differential costs where two different scenarios or states are considered, for example, by comparing the actual polluted situation with a similar situation but with clean air

$$\Delta K = k \cdot S\left(\frac{1}{L_{\mathrm{p}}} - \frac{1}{L_{\mathrm{c}}}\right) \quad (14.5)$$

where

ΔK differential cost due to pollution (€ $year^{-1}$)

L_p maintenance interval in polluted areas (year)

L_c maintenance interval in clean areas (year)

Consider again the example of Parthenon where $k \cdot S$ was estimated to 1.5 million €. If we assume a total corrosion attack of 100 µm before action is taken corresponding to an aged monument, the maintenance interval in the actual situation with a corrosion rate of 6 µm $year^{-1}$ would be $L_p = 100/6 = 17$ years. If instead the corrosion rate could be reduced to 4 µm by reducing pollution significantly, maintenance interval would be prolonged to $L_c = 100/4 = 25$ years resulting in a cost saving of 30 k€ $year^{-1}$. To complete a total cost–benefit analysis, this number should be added to all other benefits of reducing pollution and then be compared with the cost associated with the implementation.

14.5.2 Assessment of Tolerable Corrosion Rates

Due to the difficulties of assessing stock of materials at risk indicated previously, it is not always possible to make an estimation of corrosion costs, especially for objects of cultural heritage. Instead, the concept of tolerable corrosion rates has been developed. This is evaluated based on practical concerns first and foremost taking into account what maintenance measures need to be taken and how often in order to conserve the object. If a corrosion rate would result in necessary maintenance action every year, it is certainly not tolerable, and efforts need to be taken to improve the pollution situation. This improvement can be made by addressing different pollution sources. Taking equation 14.2 as a basis for describing the corrosion in the Klementinum area, improvements could be made by addressing sources of SO_2, PM_{10}, or both as is illustrated in Figure 14.7.

Finally, as a comparison with practical experiences of restoration, the Acropolis restoration project is described briefly. An extensive restoration and conservation program has taken place on the Acropolis' monuments since 1975 for treating the structural problems of the monuments and the degradation of their surfaces. The project may not be completed until 2020, and nearly $90 million will be needed to finish the work.

The project aims to restore damage caused to the Acropolis by three different means: mechanical (damage caused by earthquakes, explosions, bombardments, fires, freezing, plunders, alterations, and misplacements connected with errors of previous restorations), chemical (erosion suffered by the marble mainly as a result of acid rain and atmospheric pollution), and biological (corrosion caused by lichens, molds, bird droppings, plant roots, etc.).

The Parthenon restoration project included the dismantling, block by block, and repairing nearly every piece of the Parthenon (each of the Parthenon's 70,000 pieces is unique and fits in only one place), removal of all the inappropriate materials placed during previous restorations, and the reintegration of scattered ancient material

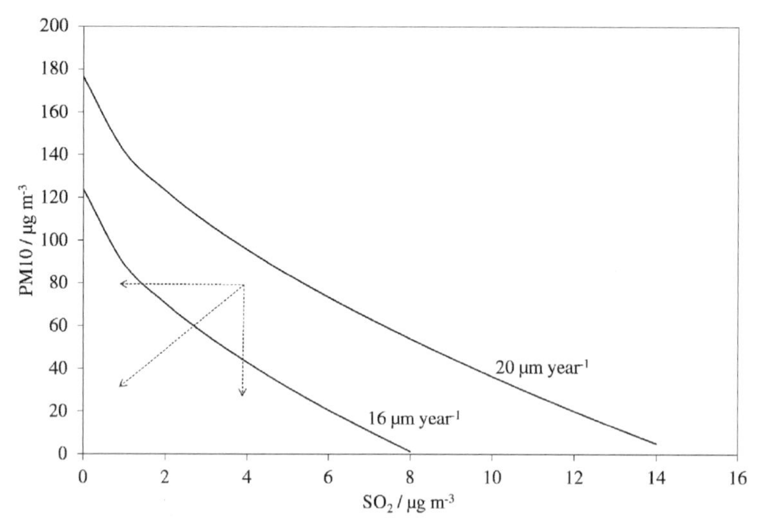

FIGURE 14.7 PM10 versus SO_2 with isolines of carbon steel corrosion (20 and 16 μm $year^{-1}$) calculated based on equation 14.2 illustrating different options for improving a pollution situation with the aim to decrease corrosion below a certain point (reducing PM10, SO_2, or both).

spread around the Acropolis. The monument was reassembled using the old stones, new stone infills (in Pentelic marble as the original) replacing missing stonework, and materials that are compatible with the old structure.

Damage caused by modern atmospheric pollution was repaired by cleaning the surfaces of sculptured architectural members (e.g., the Parthenon frieze, the high-relief metopes) with a dual-wavelength laser, with pulsed beams of ultrared and ultraviolet radiation. With this cleaning method, soot deposits and black crust were removed to the desired depth without damage to the underlying layer of marble and preserving the features of ancient sculpting, traces of the ancient stonecutter's tools, and the colored layers on the surface.

The Acropolis restoration project is the biggest renovation ever conducted on a monument in the world. Surely, the cost related to air pollution is only a small percentage of the total costs of the project (we estimated the total costs due to the atmospheric corrosion of 1.5 million € according to a conservation cost of 400 € m^{-2} and a surface area for Parthenon of 3800 m^2). However, while major restoration work will never be needed on the Acropolis for generations to come, degradation due to air pollution will continue in the future if appropriate countermeasures will not be taken.

FURTHER READING

UNESCO Sites

UNESCO, Convention Concerning the Protection of the World Cultural and Natural Heritage, adopted by the General Conference at its Seventeenth Session, Paris, November 16, 1972. Available at: http://whc.unesco.org/en/conventiontext/ (accessed on January 20, 2016).

Corrosion and Conservation

P. Dillmann, D. Watkinson, E. Angelini, and A. Adriaens, *Corrosion and Conservation of Cultural Heritage Metallic Artefacts*, European Federation of Corrosion (EFC) Series, vol. *65*, Woodhead Publishing Ltd, Cambridge, UK, 2013.

Effect of Air Pollution on Cultural Heritage

V. Kucera and S. Fitz, Direct and indirect air pollution effects on materials including cultural monuments, *Water, Air, and Soil Pollution, 85*(1), 153–165, 1995.

C. Tzanis, C. Varotsos, J. Christodoulakis, J. Tidblad, M. Ferm, A. Ionescu, R.-A. Lefevre, K. Theodorakopoulou, and K. Kreislova, On the corrosion and soiling effects on materials by air pollution in Athens, Greece, *Atmospheric Chemistry and Physics, 11*, 12039–12048, 2011.

C. Varotsos, C. Tzanis, and A. Cracknell, The enhanced deterioration of the cultural heritage monuments due to air pollution, *Environmental Science and Pollution Research, 16*(5), 590–592, 2009.

J. Watt, J. Tidblad, V. Kucera, and R. Hamilton, *The Effects of Air Pollution on Cultural Heritage*, Springer, New York, 2009.

Degradation of Stone and Other Building Materials

P. Brimblecombe, ed., *The Effects of Air Pollution on the Built Environment*, Imperial College Press, London, 2003.

T.E. Graedel, Mechanisms for the atmospheric corrosion of carbonate stone, *Journal of the Electrochemical Society, 147*, 1006–1009, 2000.

Estimation of Corrosion Costs

A. Rabl, Air pollution and buildings: an estimation of damage costs in France, *Environmental Impact Assessment Review, 19*, 361–385, 1999.

15

SCENARIOS FOR ATMOSPHERIC CORROSION IN THE TWENTY-FIRST CENTURY

Can one hope to generate accurate predictions of corrosion rates for the twenty-first century? Unfortunately not, because atmospheric corrosion rates depend on the atmospheric concentrations of corrosive agents, and those in turn depend on the activities of our technological society: how rapidly different countries and industrial sectors develop, what technologies are employed, how much attention is paid to pollution prevention, and so forth. Nonetheless, there is a good qualitative picture, at least, of the relationships between corrosive agents and materials, and the fluxes to the atmosphere of those corrosive agents are known reasonably well. It turns out that although it is not possible to predict, it is possible to construct scenarios of different futures that provide food for thought concerning the corrosion rates of the future.

Scenarios are detailed, carefully constructed stories that describe plausible alternative futures. They are not predictions but rather descriptions of possibilities. Their value is that they permit scientists, managers, policy makers, or other interested individuals to explore the consequences of different potential development paths. Scenarios have been used, for example, to guide corporate planning, to explore the impacts of climate change, and to investigate alternative paths of global development. In this concluding chapter, they are used to present possible future trends in atmospheric corrosion.

The term "scenario" is also often, in conflict with the definition earlier, used to denote a set of calculated values of climate and pollution parameters for a certain area and for a certain point in time, which would then be an extrapolated result of the scenario. The extrapolation itself is performed with advanced computer models involving emission, transport, chemical transformations in the atmosphere, and deposition. Direct anthropogenic emissions of SO_2 are almost entirely the result of the combustion of sulfur-containing fossil fuels and the smelting of metal

Atmospheric Corrosion, Second Edition. Christofer Leygraf, Inger Odnevall Wallinder, Johan Tidblad and Thomas Graedel.

sulfide ores. In addition, sulfides such as H_2S can either directly serve as a corrosion agent or oxidize to SO_2. Major emitting sectors include coal combustion, especially in large electric and heat power plants, and combustion of residential fuel oil. In deriving an emission inventory, the approach is generally to quantify fossil fuel combustion emissions from a combination of national inventory information and energy use statistics. Smelter emissions are identified and included on a facility by facility basis. Once emission rates are established, a suitable spatially discretized model of atmospheric chemistry is needed to convert emissions into concentrations, because it is the latter that is utilized in the dose–response functions (see Section 8.4). It is beyond the scope of this discussion to describe such models in detail. However, it is important to recognize that there are three main components. The first is the source rate of the species of interest, given by the emission inventories discussed previously. The second is a formulation of atmospheric motions and species transport, which is derived from meteorological information. The third is a calculation of loss rates, both chemical (i.e., reactions with other atmospheric species) and physical (i.e., losses to surfaces such as buildings, statuary, or soil).

The twentieth century and preceding centuries have been periods with significant changes in pollution, and changes are likely to continue into the twenty-first century. Climate change is today an even more effective driver of policy change, and we are just beginning to see the consequences. Seeking cobenefits of climate change and pollution policies is now on the agenda, but as far as scenarios are concerned, they are still mostly focused on one or the other. Corrosion is affected by both climate and pollution, and the developed dose–response functions provide an excellent tool when analyzing the effects of different scenarios. The chapter is primarily focused on effects on metals, but the underlying pollution and other environmental parameters used in the scenarios that are responsible for chemical attack on metals also affect a variety of other materials including, for example, stone materials. Before looking into the future, however, much can be learned from a historical perspective.

15.1 ATMOSPHERIC CORROSION IN THE RECENT MILLENIUM

When trying to describe corrosion in the distant past, one faces problems similar to those for the far future, summarized as lack of reliable measured corrosion data. This means that assessment of the past also needs to rely on the use of dose–response functions. Furthermore, environmental data are also often lacking. An exception is the city of London, where data has been derived for an extended period in time. Figure 15.1 shows calculated zinc corrosion in London based on the dose–response function for the SO_2-dominating situation (see Section 8.4). According to these data and calculations, the effect of SO_2 and the high corrosion rates associated with this pollutant provide an important but transient period in history. As will be seen, however, the effect of SO_2 is still important in many parts of the world and will continue to be so in some places well into the twenty-first century.

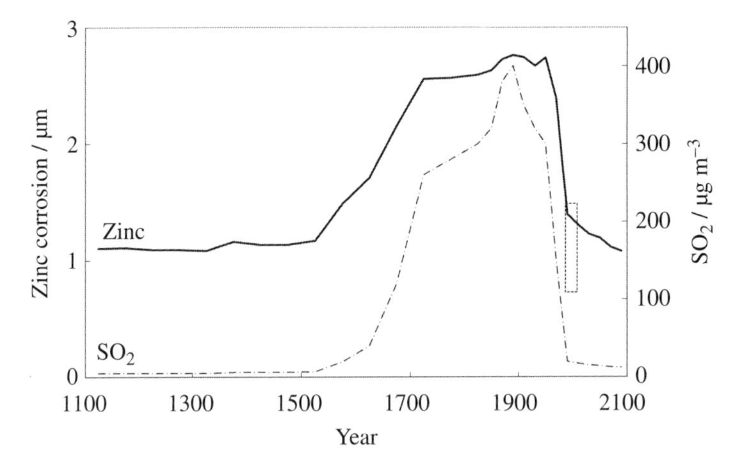

FIGURE 15.1 Zinc corrosion, 1 year of exposure, estimated for London, United Kingdom, and the corresponding SO_2 concentration on which the zinc corrosion was based. The dashed rectangular box indicates measured values of zinc corrosion. (Reproduced with permission from Elsevier; J. Tidblad, Atmospheric corrosion of heritage metallic artefacts: processes and prevention, in *Corrosion and Conservation of Cultural Heritage Metallic Artefacts*, eds. P. Dillmann, D. Watkinson, E. Angelini, and A. Adriaens, vol. 65, Woodhead Publishing Ltd, Cambridge, UK, pp. 37–52, 2013.)

15.2 ATMOSPHERIC CORROSION IN THE TWENTIETH CENTURY AND TODAY

Trends as depicted for the city of London have been established in recent years for many sites in Europe, based on measured corrosion data. All of these show similar results, as illustrated in Figure 15.2 for carbon steel, namely, decreasing trends in corrosion as a result of decreasing trends in pollution. Data given in Figure 15.2 is based on averages of several stations from the ICP Materials program described previously (see Section 8.3). When looking at individual sites, however, it is obvious that even if the general trend is decreasing, the magnitude and rate of decrease are quite different at different sites.

This is also clear when looking outside Europe. The Swedish International Development Cooperation Agency (Sida)-funded program on Regional Air Pollution in Developing Countries (RAPIDC) organized corrosion exposures at more than 20 test sites in 14 countries in Asia and Africa during the period 2001–2010. One part of the exposures aimed at investigating trends in corrosion and pollution. Figure 15.3 shows results for carbon steel and SO_2. In this case there is no obvious trend in corrosion/pollution at all. There are some sites where the pollution-induced corrosion is very high, up to SO_2 concentrations of 100 µg m^{-3} and carbon steel corrosion of 100 µm for 1 year of exposure, corresponding to the situation in Sweden in the 1960s when corrosion was at its highest. Thus, even if SO_2-induced corrosion is nonexistent in several places in Europe, it is still an

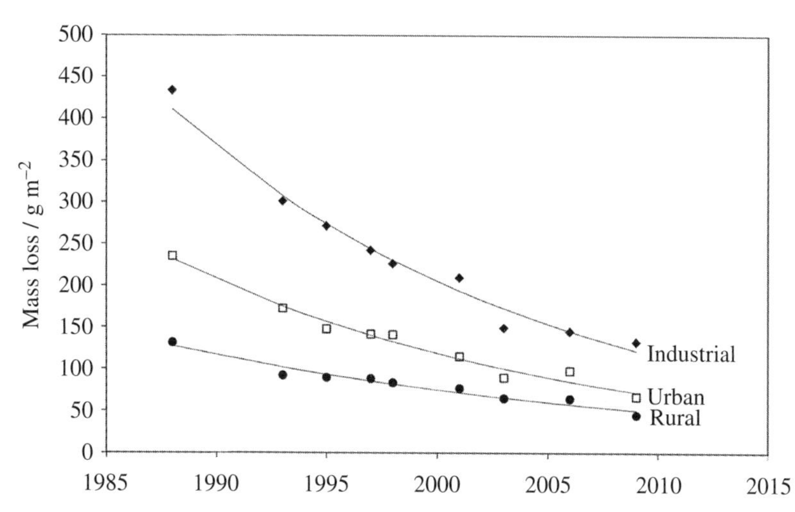

FIGURE 15.2 Carbon steel corrosion for industrial, urban and rural ICP Materials test sites. (Reproduced with permission from J. Tidblad, V. Kucera, M. Ferm, K. Kreislova, S. Brüggerhoff, S. Doytchinov, A. Screpanti, T. Grøntoft, T. Yates, D. de la Fuente, O. Roots, T. Lombardo, S. Simon, M. Faller, L. Kwiatkowski, J. Kobus, C. Varotsos, C. Tzanis, L. Krage, M. Schreiner, M. Melcher, I. Grancharov, and N. Karmanova, Effects of air pollution on materials and cultural heritage: ICP materials celebrates 25 years of research, *International Journal of Corrosion, 2012,* 16, 2012.)

important effect on a global scale and needs attention when constructing scenarios relevant for calculation of corrosion effects in the twenty-first century.

Before taking the step to use dose–response functions and data for different types of scenarios to estimate corrosion in the twenty-first century, a word of caution is appropriate. Even if the development of dose–response functions has been substantial in the last years, most of the underlying data have been obtained from sites located in temperate climates. Some data exist, as is evident from Figure 15.3, but the fact remains that most dose–response functions are biased toward predicting corrosion in a temperate climate. An example of a qualitative difference that has been observed is given in Figure 15.4.

At least for shorter exposures (1 year), the difference between zinc corrosion and copper corrosion in temperate and subtropical climates is striking. Corrosion rates of zinc, in absolute values, are higher than those of copper in Europe, but the opposite is true for Asia and Africa. The reason for this difference is not clear, mainly due to the lack of corrosion product characterization in these measurements. Sites in subtropical and tropical regions are characterized by higher temperatures and higher amounts of precipitation. This may cause higher formation rates at shorter exposure times in Asia and Africa, thereby causing the higher corrosion rates. In Figure 15.4 there is only one clear outlier, at Zn = 3.8, Cu = 1.3. This relates to a site in the copper belt area of Zambia, where the high zinc corrosion rate might be related to deposition of particulate copper. It should be added that the difference in corrosion rates of copper and zinc is less pronounced for longer exposure periods.

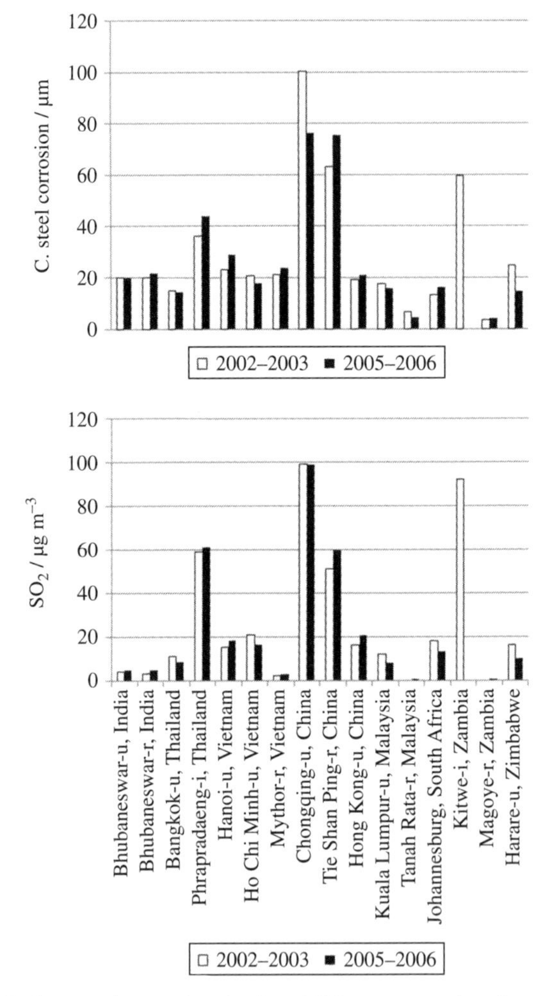

FIGURE 15.3 Carbon steel corrosion (top) and SO_2 concentration (bottom) for RAPIDC/corrosion sites in Asia and Africa.

15.3 ATMOSPHERIC CORROSION IN THE TWENTY-FIRST CENTURY: EFFECT OF CHANGES IN POLLUTION

As already indicated, SO_2 emissions in Europe are expected to decrease in the future. This is suggested by Figure 15.5, which shows past and projected SO_2 emissions until 2030. Emissions in Asia are expected to increase, emissions in Latin America and Africa are expected to remain at the current level, and emissions in the rest of the world are expected to decrease.

It is worth noting that even though the absolute magnitude of the decrease can vary substantially when comparing industrial, urban, and rural sites, the percentage decrease has been about the same. An example of these geographical differences is

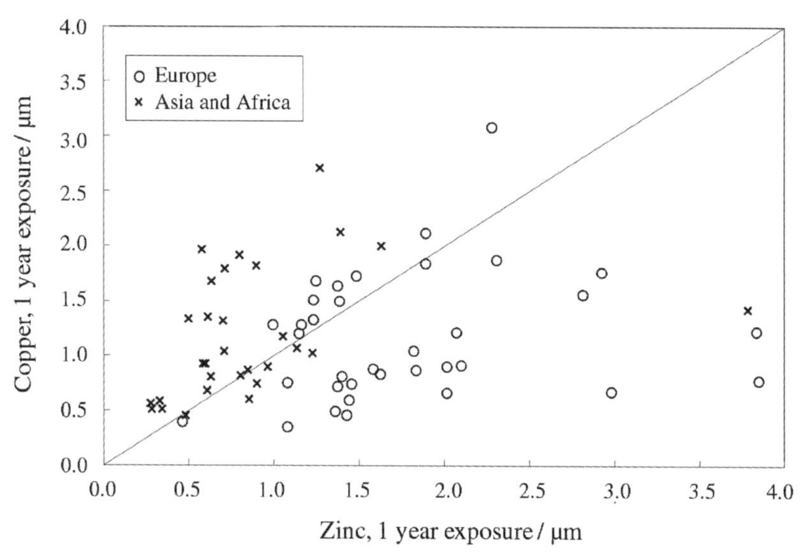

FIGURE 15.4 Copper corrosion versus zinc corrosion in Europe (unfilled circles), Asia, and Africa (crosses). (Reproduced with permission from John Wiley & Sons; J. Tidblad, K. Hicks, J. Kuylenstierna, B.B. Pradhan, P. Dangol, I. Mylvakanam, S.B. Feresu, and C. Lungu, Atmospheric corrosion effects of air pollution on materials and cultural property in Kathmandu, Nepal, *Materials and Corrosion*, doi: 10.1002/maco.201408043, 2015.)

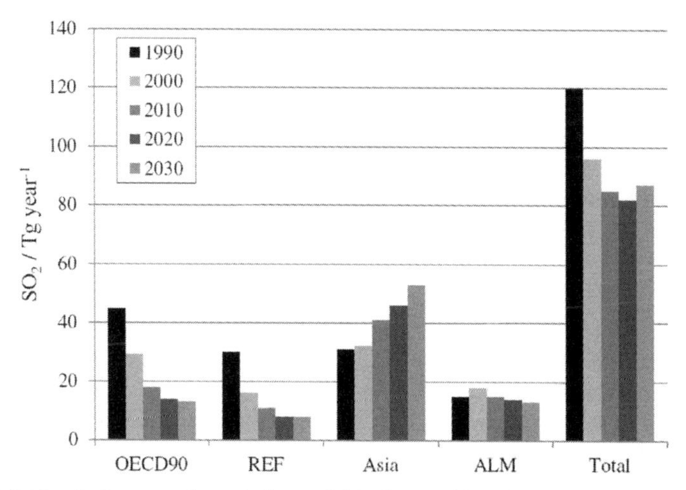

FIGURE 15.5 Anthropogenic emissions of SO_2 by world region (1990–2000–2010–2020–2030); OECD90 (North America Western Europe, Pacific OECD), REF (Central and Eastern Europe, Russia and Newly Independent States), Asia (Centrally Planned Asia, Other Pacific Asia, South Asia), and ALM (Latin America, Middle East, Africa). (Adapted with permission from Elsevier; J. Cofala, M. Amann, Z. Klimont, and K. Kupiainen, Scenarios of global emission of air pollutants and methane until 2030, *Atmospheric Environment, 41*, 8486–8499, 2007.)

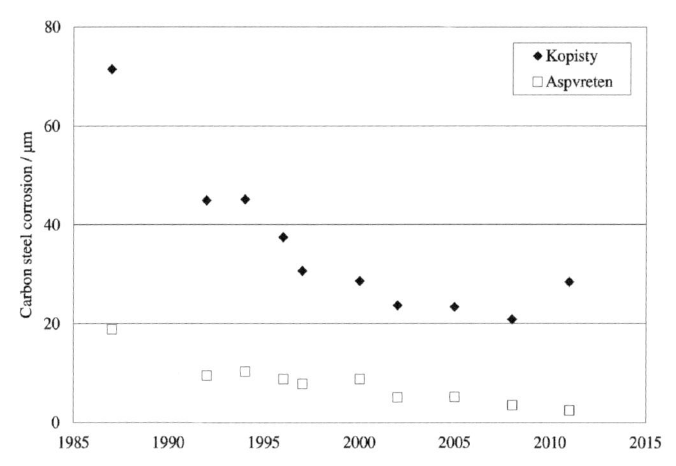

FIGURE 15.6 Trends in corrosion rate during the first year of exposure of carbon steel at two selected sites, one industrial (Kopisty, Czech Republic) and one rural (Aspvreten, Sweden). (Adapted with permission from J. Tidblad, T. Grøntoft, K. Kreislova, M. Faller, D. De la Fuente, T. Yates, and A. Verney-Carron, International cooperative programme on effects on materials, *Report 76 Trends in Pollution Corrosion and Soiling 1987–2012*, Swerea KIMAB, 2014.)

shown in Figure 15.6. The corrosion of carbon steel at a rural test site in Sweden (Aspvreten) was about 20 µm in 1987 for the first year of exposure, which is about the same value measured today (2014) at an industrial site in the Czech Republic (Kopisty). However, during this period there has always been a substantial difference in corrosion rates between the two sites. This is also an illustration that labels like "rural," "urban," and "industrial" are not particularly informative for judging the corrosion rate if not put in context.

The decrease in SO_2 pollution during the period 1985–2010 has been even more dramatic than during the early twentieth century. While corrosion rates have decreased to between 40 and 50% of their original value, SO_2 concentrations have decreased from 80 to 90%. This is manifested in the dose–response functions as nonlinear SO_2 terms. For example, if the SO_2 dependence in a dose–response function is given as $SO_2^{0.5}$, a decrease in SO_2 of 80% would result in a decrease in corrosion of 55%.

Another important aspect to consider when using scenario data to calculate corrosion effects is the importance of scale. When model results are obtained for large geographical areas, the smallest calculational unit (the "grid cell") is often too large to be of use for illustrating corrosion effects in dense urban or industrial areas. This is illustrated in Figure 15.7, which shows a comparison of calculated corrosion of carbon steel based on either measured (ICP Materials) or modeled (EMEP, the cooperative program for monitoring and evaluation of the long-range transmission of air pollutants in Europe) SO_2 values. For Kopisty (the industrial site), the modeled SO_2 values are too low, resulting in an underestimation of the corrosion rate, while for Aspvreten (the rural site), the modeled values are acceptable for calculating corrosion values on a large geographical scale. Current maps produced with large grid cells and covering

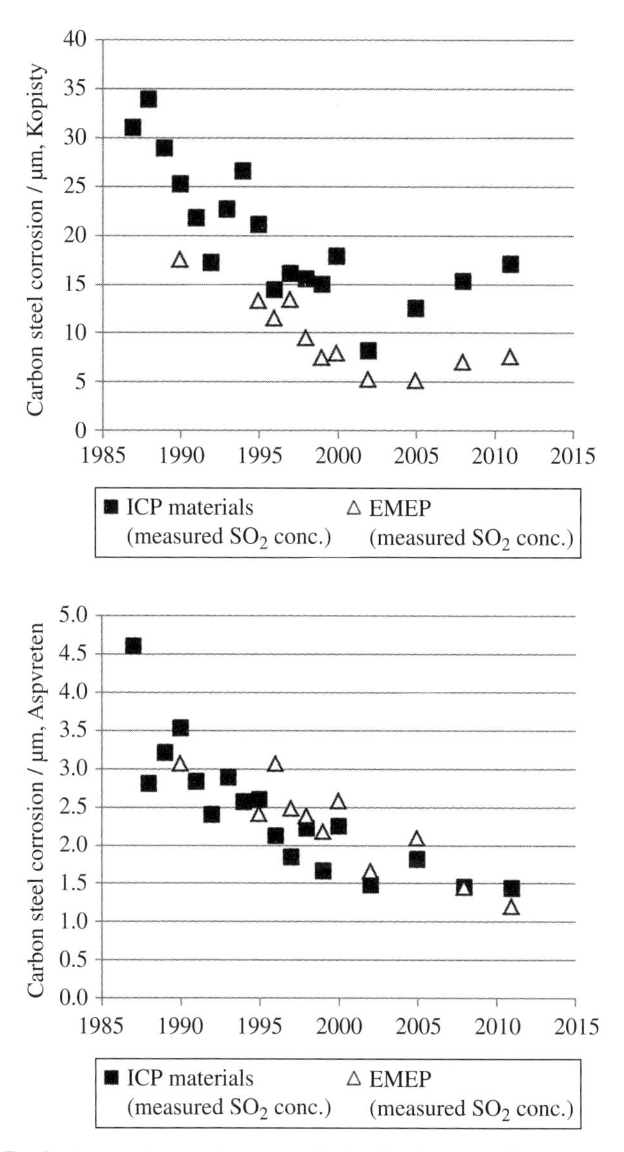

FIGURE 15.7 Carbon steel corrosion in μm during the first year of exposure based on measured SO_2 concentration (ICP Materials) at two selected sites, one industrial (top—Kopisty, Czech Republic) and one rural (bottom—Aspvreten, Sweden) compared to calculated SO_2 concentration (EMEP) at the same sites in a 50×50 km grid. Values are calculated by using only the SO_2 part of the dose–response function for the multipollutant situation (see Section 8.4). (Adapted with permission from J. Tidblad, T. Grøntoft, K. Kreislova, M. Faller, D. De la Fuente, T. Yates, and A. Verney-Carron, International cooperative programme on effects on materials, *Report 76 Trends in Pollution Corrosion and Soiling 1987–2012*, Swerea KIMAB, 2014.)

large geographical areas are not particularly suitable for showing future effects of pollution in urban and industrial areas. Climate changes, however, show less variation on a larger scale and can be studied under constant pollution scenarios. The future is promising in this respect. With increasing computing power, the EMEP program is now planning to change its grid size to grids of $0.1° \times 0.1°$ or even finer.

15.4 ATMOSPHERIC CORROSION IN THE TWENTY-FIRST CENTURY: EFFECT OF CHANGES IN CLIMATE

The first and most substantial results so far toward understanding the effect of climate change on degradation of materials were the product of the EU project "NOAH's ARK—Global climate change impact on built heritage" in the period 2004–2007. The project included assessment of a variety of materials and subjects, including pure climate-induced degradation effects like wet-frost events on stone materials and decay of wooden structures by fungal growth and the combined effects on climate and pollution on corrosion of metallic materials. The climate scenario used was the so-called IPCC A2 scenario, which describes a very heterogeneous world and projects the future for the periods 2010–2039 and 2070–2099. Dose–response functions for estimating corrosivity categories (see Section 8.4) were used to predict corrosion. Figure 8.9 and Section 8.4 provide a background for discussing and understanding the major effects of climate change on corrosion. When temperature increases, the corresponding change in corrosion in urban areas affected by SO_2 pollution can either increase or decrease, depending on whether the original temperature value is below or above 10°C. In coastal areas corrosion is expected to increase regardless of the initial temperature value.

Figure 15.8 shows estimated corrosivity categories for Europe in the future (2070–2099) based on the ISO dose–response functions of Chapter 8. In each grid cell the corrosivity for both carbon steel and zinc has been calculated. The higher of these two is then presented in the corresponding grid cell in the map. As a general statement inland areas are predicted to have a relatively low risk of corrosion (C2), while a higher risk is observed for coastal and near-coastal areas (C3–C5). This is in contrast to the situation today, in which corrosion in inland areas is more variable due to varying levels of pollution. Future corrosion in Europe will thus be dominated by the effects of sea salt rather than sulfur pollution (see also Figure 15.1) An exception to this conclusion, not shown in the map, is related to the use of deicing salts in road environments, as described in Chapter 12. In this case the effect of changing temperature values will be indirect, by moving the geographical zones where deicing salts are used today to new areas previously not affected by this parameter.

15.5 RESPONDING TO INCREASING RATES OF CORROSION

Atmospheric corrosion has to date been largely a defensive science, one of choosing materials with corrosion resistance as high as possible consistent with materials performance and cost, and hoping that the environment in which the materials were

FIGURE 15.8 Maximum corrosivity category C2–C5 (2070–2099) estimated from the ISO dose–response functions for carbon steel and zinc and temperature, relative humidity, SO_2, and chloride deposition data from the HadRM3 scenario. (Reproduced with permission from Elsevier; J. Tidblad, Atmospheric corrosion of metals in 2010–2029 and 2070–2099, *Atmospheric Environment, 55*, 1–6, 2012.)

used would be sufficiently benign that corrosion would be minimal. This empirical approach has yielded significant benefits but has not permitted much informed forward thinking. With dose–response functions now becoming available and scenarios being developed to represent a reasonable spectrum of possible environmental futures, one can begin to define a family of possible lifetime performance outcomes for specific materials. These outcomes differ with epoch and with geographic location. Clearly some future scenarios, some timescales, and some regions appear more problematic than others.

In general, the results indicate that the most rapid and severe changes in corrosion rates due to pollution seem likely to occur in Asia. In Europe pollution is expected to decrease further, but the effect on corrosion will not be as dramatic. Not only metals, which are the focus in this chapter, will be affected. Of special concern in Asia in the next several decades may be those corrosion-susceptible objects that are essentially irreplaceable—objects of art, particularly those in outdoor locations. Conservators of those objects—decorated buildings, statuary, monuments, and the like—might well be advised to employ the full spectrum of protective technology available, such as surface coatings. They should also become advocates for development policies that do not compromise the air quality in the local environment, for objects of art as well as objects of technology are among the products of society that may otherwise disappear. With climate change and increased temperatures, there will in addition be a general relative increase in corrosion in areas affected by chloride deposition.

This chapter has shown that corrosion rates will change substantially and differently at individual locations in the future. The underlying mechanisms for corrosion, the important parameters, and the corresponding protective measures that have been described in this book are, however, the same. Individuals responsible for planning and maintenance of infrastructure and heritage will face problems that are new to them, but that most likely have already been addressed in other parts of the world. Therefore, much is to be gained by international cooperation and transfer of knowledge within the scientific community, as well as within businesses focused on corrosion protection technologies. Much has been discovered in recent years so far as atmospheric corrosion science and technology is concerned, but as with all realms of knowledge, much remains to be accomplished. This book is an effort in that direction but certainly not the final word.

FURTHER READING

Corrosion and Air Pollution Trends

J. Tidblad, Atmospheric corrosion of heritage metallic artefacts: processes and prevention, in *Corrosion and Conservation of Cultural Heritage Metallic Artefacts*, eds. P. Dillmann, D. Watkinson, E. Angelini, and A. Adriens, vol. *65*, Woodhead Publishing Ltd, Cambridge, UK, pp. 37–52, 2013.

J. Tidblad, V. Kucera, M. Ferm, K. Kreislova, S. Brüggerhoff, S. Doytchinov, A. Screpanti, T. Grøntoft, T. Yates, D. de la Fuente, O. Roots, T. Lombardo, S. Simon, M. Faller, L. Kwiatkowski, J. Kobus, C. Varotsos, C. Tzanis, L. Krage, M. Schreiner, M. Melcher, I. Grancharov, and N. Karmanova, Effects of air pollution on materials and cultural heritage: ICP materials celebrates 25 years of research, *International Journal of Corrosion, 2012*, 16, 2012.

Air Quality, Emission Inventories, and Scenarios

J. Cofala, M. Amann, Z. Klimont, and K. Kupiainen, Scenarios of global emission of air pollutants and methane until 2030, *Atmospheric Environment, 41*, 8486–8499, 2007.

P.S. Monks et al. (+62 coauthors), Atmospheric composition change—global and regional air quality, *Atmospheric Environment, 43*, 5268–5350, 2009.

Corrosion Scenarios

C.M. Grossi, A. Bonazza, P. Brimblecombe, I. Harris, and C. Sabbioni, Predicting twenty-first century recession of architectural limestone in European cities, *Environmental Geology, 56*(3–4), 455–461, 2008.

J. Tidblad, Atmospheric corrosion of metals in 2010–2029 and 2070–2099, *Atmospheric Environment, 55*, 1–6, 2012.

APPENDIX A

EXPERIMENTAL TECHNIQUES IN ATMOSPHERIC CORROSION

A.1 INTRODUCTION

Progress in our understanding of atmospheric corrosion processes depends to a large extent on the proper choice and use of analytical techniques, capable of extracting information related to the corroding material surface. Many results presented in this book have been based on the use of such techniques. This appendix describes the most frequently used analytical techniques, as well as some techniques that are less frequently used in atmospheric corrosion but which possess the potential capability of becoming more important in future studies. Although atmospheric corrosion involves electrochemical processes, electrochemical techniques are generally limited to those for local probing of the corrosion potential variations along the surface.

Most techniques for analysis of corroded surfaces are based on bombardment of the sample with particles (the primary beam), for example, photons, electrons, or ions, and the subsequent measurement of the intensity, mass, energy, or direction of the particles that are emitted or transmitted as a result of the bombardment (the secondary beam). The particles of the primary and secondary beams are not necessarily of the same type. In what follows, the principle of each technique will be described in terms of the incoming probe particles and the outgoing signature particles. Depending on the principle, its suitability will be specified for *in situ* or only for *ex situ* studies.

Many analytical techniques are based on the use of electrons or ions as probing particles. Because of their strong interaction with the medium that they penetrate, such techniques must be performed in a dilute atmosphere corresponding to ultrahigh vacuum conditions (10^{-7} to 10^{-8} Pa). This requirement clearly results in a disturbance of the solid–liquid and the liquid–gas interfaces, and therefore the analyzed surface under such conditions is far from being representative of a corroding surface. Photons, on the other hand, possess much less interaction with the medium they

Atmospheric Corrosion, Second Edition. Christofer Leygraf, Inger Odnevall Wallinder, Johan Tidblad and Thomas Graedel.

penetrate. Hence, analytical techniques that use photons as probing particles can be performed at atmospheric pressures. This permits, at least under favorable circumstances, the corroded surface to be analyzed under *in situ* conditions.

Electrons and ions can more easily be focused to a very small beam than photons can. From this follows that analytical techniques based on electrons or ions as the primary beam in general possess higher spatial resolution. Because of their strong interaction with matter, these techniques also may exhibit higher surface sensitivity than those using photons as probing particles.

The techniques described can be grouped in different ways, depending on physical principles, type of information gained, *in situ* or *ex situ* capability, etc. We have chosen to group the techniques according to type of information they can provide: mass change, surface topography, surface elemental composition, chemical state, corrosion product phases, corrosion potential, and atmospheric composition. These discussions constitute the remainder of this appendix.

A.2 TECHNIQUES FOR DETECTING MASS CHANGE

A.2.1 Microgravimetry

Mass loss measurements by means of microgravimetry are obtained after removal of the corrosion products. The data consists of the difference between the sample mass before and after exposure. It is this method that provides the most direct measure of the corrosion effect. It is applicable as long as there is a reliable chemical stripping treatment for removing the corrosion products that have been formed. Currently, the method can measure a mass loss detection limit of around $10\,mg\,m^{-2}$ ($=1\,\mu g\,cm^{-2}$). This value can be compared with typical mass loss values of copper after 1 year of exposure in a very mild indoor environment, around $50\,mg\,m^{-2}$. Hence, the detection limit of mass loss measurements corresponds to a few months of exposure in the mildest possible indoor environment. In more aggressive environments, the detection limit permits shorter exposure times to be used.

Mass increase measurements by means of microgravimetry require no removal of corrosion products and can be used on virtually any material. The detection limit with a high-precision instrument is typically a few milligram per square meter. Thus, somewhat smaller mass gains can be measured compared with mass losses because of the absence of the removal procedure in mass gain measurements. The major drawback is that the measured mass increase can be caused not only by corrosion effects but also by other surface-related phenomena such as humidity sorption and dust deposition. Measurements of mass gain, but not of mass loss, can though be performed *in situ*.

This book contains several examples of results based on microgravimetry measurements; see, for instance, Figures 7.5, 7.6, 7.7, 7.8, 7.9, and 8.7.

A.2.2 Resistance Sensor Measurements

A resistance sensor is based on a change in the ohmic resistance of thin metal films upon atmospheric corrosion. The advantage is that the corrosion effects of any metal can be followed *in situ* by monitoring the change in ohmic resistance caused by the change of the cross section of the foil. The detection limit is a few milligram per square meter under favorable circumstances, that is, when the ohmic resistance of the corrosion products is high and the corrosion attack over the metal surface is uniform. Comparison with other techniques has shown, however, that the resistance sensor can overestimate the corrosion effect because of nonuniform corrosion attack. Measurements can be performed *in situ* (see Fig. 7.10).

A.2.3 Electrolytic Cathodic Reduction

The technique is based on the cathodic reduction of corrosion products in an electrochemical cell using an electrolyte that will have only minor chemical interaction with the corroded surface. Measurements are normally performed by applying a constant cathodic current while monitoring the electrode potential as a function of time. The detection limit is around $5\,mg\,m^{-2}$ under favorable conditions. Electrolytic cathodic reduction has mostly been used for estimates of mass of corrosion products formed on copper and silver. Difficulties in ambiguous interpretation of the electrode potential make the technique somewhat imprecise. This is often the case for copper, more seldom for silver. The technique can only be applied *ex situ*.

A.2.4 Quartz Crystal Microbalance Measurements

The material to be studied is deposited onto a piezoelectric quartz crystal that oscillates at its resonance frequency. A change in mass of the oscillating device results in a change in resonance frequency, which can be measured with high precision. The quartz crystal microbalance is one of the most mass-sensitive techniques used in atmospheric corrosion and can be applied to any material, which can be deposited onto quartz. The detection limit is of the order of $0.1\,mg\,m^{-2}$, that is, nearly two orders of magnitude lower than mass loss measurements. It can be used to study mass changes of metals during indoor atmospheric corrosion with a time resolution of less than an hour. The major disadvantage is the same as for mass increase measurements by microgravimetry, that is, the difficulty in distinguishing corrosion effects from other surface effects that can cause mass changes. The technique has been used for *in situ* studies, both outdoors and indoors. Examples of applications are seen in Figure 7.4 (indoors) and Figures 8.2, 8.3, and 8.4 (outdoors).

A.3 TECHNIQUES FOR ANALYZING SURFACE TOPOGRAPHY

A.3.1 Scanning Electron Microscopy

In scanning electron microscopy a highly focused electron beam impinges on a solid surface and gives rise to a backscattered electron beam, which is collected and amplified. By scanning the incoming electron beam and simultaneously analyzing the low-energy part of the backscattered beam (so-called secondary electrons), topographical information on the solid surface can be gained with high lateral resolution. Scanning electron microscopy has become a frequently used imaging technique for studying both inorganic and organic solid materials. Its versatility can be increased by simultaneous analysis of either characteristic X-rays (Section A.4.3) or Auger electrons (Section A.4.2) emitted from the surface through excitation of the incoming electron beam. In both cases information can be gained about the elemental composition of the sample surface. The highest magnification is of the order of 10^5, corresponding to a lateral resolution of around 5 nm. The technique can be used for many purposes, including the study of the morphology, grain size, and elemental composition of corrosion products (the latter through simultaneous analysis of characteristic X-rays) or the spatial distribution of corrosion attacks. In an environmental scanning electron microscope, imaging of the sample is obtained at elevated atmospheric pressure (>20 Torr). This environment retains moisture on the material surface, thereby permitting *in situ* analysis during atmospheric corrosion. Examples of the application of the technique in atmospheric corrosion are seen in Figures 2.9 and 9.2.

A.3.2 Atomic Force Microscopy

In atomic force microscopy the morphology of a surface is imaged by monitoring the forces between a cantilever and the surface to be analyzed. When the distance between the tip of the cantilever and the surface is close enough, there is an attractive (van der Waals) force that results in a bending of the cantilever. The deflection can be monitored in different ways. By analyzing the deflection as a function of position over the surface, an image of the surface is obtained with very high resolution, of the order of 1 nm or better. As opposed to scanning tunneling microscopy (see Section A.3.3), both conducting and nonconducting materials can be studied. During ongoing atmospheric corrosion in laboratory exposures, atomic force microscopy has been shown to reveal topographic *in situ* information down to a resolution of the order of a few tenths of nanometer. This type of measurements requires that the surface is well polished so that features of reaction products can be distinguished from features of the original surface. An example of such an application is seen in Figure 2.7.

A.3.3 Scanning Tunneling Microscopy

In this technique, the morphology of a surface is monitored by scanning a tip over it. The distance from the tip to the surface is measured through analysis of the tunneling current between the tip and the surface. Under favorable conditions the technique has a resolution of 1 nm or better and permits individual atoms to be imaged. The sample has to be a good conductor; thus, thicker corrosion products with semiconducting or insulating properties cannot be studied with this approach. As with atomic force microscopy, scanning tunneling microscopy can be used under vacuum conditions or in environments at atmospheric pressure. The technique therefore has the potential capability for *in situ* studies of the initial stages of atmospheric corrosion before the reaction products formed hinder the tunneling current between tip and sample. Under atmospheric corrosion conditions, the spatial resolution is no better than 1 nm.

A.4 TECHNIQUES FOR ANALYZING SURFACE COMPOSITION

A.4.1 X-ray Photoelectron Spectroscopy

In X-ray photoelectron spectroscopy, the material surface is irradiated by a monochromatic (i.e., single energy) beam of X-rays. The intensity of photoemitted electrons is monitored as a function of their kinetic energy. For the low kinetic energy electrons (<2000 eV), the information originates from a very thin surface layer, around 0.5–3 nm in depth, corresponding to one to five monolayers of solid compounds. Since each element has a characteristic spectrum of emitted electrons, the overall spectrum provides an elemental analysis of the investigated surface region. All elements except hydrogen can be detected. The kinetic energies are slightly shifted if the oxidation state changes, so that X-ray photoelectron spectroscopy can distinguish metallic iron (Fe^0) from oxidized iron (Fe^{2+} or Fe^{3+}), for example. The measured area is normally a few square millimeter, but instruments with spatial resolution of around 10 μm are available. The detection limit is of the order of 0.1% for many elements but is considerably reduced if the surface is contaminated by hydrocarbons during sample handling. Data can be quantified with an uncertainty of around 10–20% of the absolute value. This permits single phases to be detected if only one or two phases are present in the corrosion products. In a multiphase situation, however, the identification of the phases present is ambiguous. X-ray photoelectron spectroscopy is normally restricted to solid surfaces, although special arrangements have been made to obtain spectra from liquid surfaces. Similar arrangements could be made to perform *in situ* measurements during ongoing atmospheric corrosion, but only *ex situ* measurements have been reported so far.

A.4.2 Auger Electron Spectroscopy

The characteristic process in Auger electron spectroscopy is the emission of Auger electrons, which are produced in a two-electron process. The primary beam consists of X-rays, electrons, or ions. The kinetic energies of emitted Auger electrons are in the same range as the emitted photoelectrons in X-ray photoelectron spectroscopy. From this it follows that the thickness of the analyzed area in the two techniques is similar, around 0.5–3 nm. Each element except the two lightest, hydrogen and helium, emits Auger electrons with characteristic energies; the technique therefore provides an elemental analysis of all elements except hydrogen and helium located in a thin surface layer. The potential for distinguishing different oxidation states from each other is less than in X-ray photoelectron spectroscopy, but the spatial resolution is superior, around 0.1 μm or better, provided the primary beam consists of electrons. This permits Auger electrons to be excited and analyzed from different parts of a corroded surface, a technique called scanning Auger microscopy. The versatility of the technique is further increased through so-called profile studies in which the elemental composition of a layer of corrosion products is analyzed as a function of depth from the surface through simultaneous ion sputtering. Detection limits and quantification possibilities in Auger electron spectroscopy are similar to X-ray photoelectron spectroscopy, as is the limited feasibility of performing *in situ* analysis.

A.4.3 Analysis of Characteristic X-Rays

When electrons with sufficient kinetic energy irradiate a solid material, X-rays are emitted in a broad range of different energies characteristic of each emitting element. These X-rays are used for elemental analysis in a similar way as are photoelectrons and Auger electrons. A major difference is that the X-rays originate from much deeper inside the material, of the order of 1 μm, as compared to the 0.5–3 nm, which is the information depth of X-ray photoelectron spectroscopy or Auger electron spectroscopy. Hence, the analysis of characteristic X-rays can be considered a bulk method rather than a surface method. There are two ways to detect the characteristic X-rays: through wavelength-dispersive spectrometers (WDS), in which each energy or wavelength of emitted X-rays is detected separately, or through energy-dispersive spectrometers (EDS), in which the whole X-ray spectrum is detected simultaneously. EDS detection results in higher sensitivity and in shorter analysis time than WDS detection (except for light elements where the reverse holds true) but in less energy resolution and thus in reduced possibilities to quantify the data. Hence, EDS detection is better used for identifying elements, whereas WDS detection is better used for quantifying them. All elements can be detected except those lighter than beryllium. No information on oxidation state can be extracted. The spatial resolution for detecting characteristic X-rays is of the order of 1 μm, and the detection limit is around 0.01% (WDS) and 0.5% (EDS). An example of such an application is displayed in Figure 9.4.

A.4.4 Glow Discharge Optical Spectroscopy

In this technique the sample acts as a cathode for a plasma that has been created by evacuating the sample chamber, filling it with low-pressure argon and applying a high voltage between the cathode and a surrounding anode. Positive ions from the plasma sputter the surface and remove excited atoms from it. During subsequent de-excitation there is an optical emission with characteristic lines for each element in the removed layer of material. Hence, in glow discharge optical spectroscopy, a depth profile is obtained in which the emission signal of removed atoms is monitored as a function of sputtering time. The cathode area is usually in the range between one and several tenths of square millimeter and the lateral resolution therefore poor. The sputtering process is rapid, which limits the depth resolution to around 1 nm. In terms of surface sensitivity, the method has an information depth that covers the range from typical surface analytical techniques, such as X-ray photoelectron spectroscopy, to bulk analytical techniques, such as analysis of characteristic X-rays. No information on oxidation state can be gained, and no *in situ* measurements are possible.

A.4.5 Infrared Reflection Absorption Spectroscopy

When infrared radiation interacts with any material, part of the radiation is absorbed at specific energies (or wavelengths) characteristic of vibrational motions of atoms in a molecule or vibrational motions of the molecule as a whole. The infrared region of interest is in the range from 2.5 to 50 μm, corresponding to the spectral region between 4000 and 200 cm^{-1}. Infrared techniques can be used in a variety of working modes, including transmission of infrared radiation through the material or reflection, diffuse or specular, of the radiation. Specular reflection is used in infrared reflection absorption spectroscopy and has been applied to obtain *in situ* information on thin layers (0.5 nm) of reaction products during initial stages of atmospheric corrosion (see Figures 6.6 and 6.9). The surface sensitivity is enhanced by having a high angle of incidence and p-polarized infrared light. The analyzed area is around 1 cm^2, so there is no lateral resolution. In an infrared microscope samples can be analyzed down to a size of the order of 10 μm. In this case, however, the surface sensitivity is more limited.

A.4.6 Sum Frequency Generation (SFG)

It uses two laser beams that overlap at the investigated surface of a material or the interface between two materials. An output beam is generated at a frequency of the sum of the two input beams. One of the beams has a constant wavelength in the visible range, while the other beam is tunable with a wavelength in the infrared range. Through tuning the infrared beam, the system can scan over resonances and obtain vibrational spectra of interfaces. This unique technique is monolayer sensitive and capable of providing molecular information from the gas–solid, gas–liquid, and liquid–solid interfaces.

A.5 TECHNIQUES FOR IDENTIFYING PHASES IN CORROSION PRODUCTS

A.5.1 X-Ray Diffraction

The technique is based on a monochromatic beam of X-rays directed onto the solid material to be investigated. If the material is crystalline, one observes diffracted X-ray beams in directions that fulfill the requirements of Bragg's law. This well-known law formulates a relationship between the wavelength of X-rays, the angle of diffraction, and the interplanar spacing of the crystalline material. An unknown phase in the corrosion products can be identified by comparing the X-ray pattern of the corrosion products with that of a powder diffraction data base of known phases. Several phases can be identified in a multiphase system, provided all are crystalline. The detection limit for a single phase is a few percent. The most frequent way of identifying phases in corrosion products is the powder method, which uses a randomly oriented powder to increase the range of angles of diffracted beams into a line pattern. The sampling depth is a few micrometer. Examples of results from phase identifications are displayed in Figures 9.8 and 9.11.

A.5.2 Confocal Raman Microscopy

Confocal Raman microscopy is based on an optical microscope with an excitation laser, a spectrometer or monochromator, and an optical sensitive detector that has been added to perform Raman microscopy measurements. The technique can be used to identify phases on a surface with a resolution better than 1 μm. Figure 2.4 illustrates one example of application.

A.5.3 Mössbauer Spectroscopy

In Mössbauer spectroscopy the sample is irradiated by photons of well-defined high energy produced by a radioactive source containing a specific nucleus. By slightly moving the X-ray source forward and backward with respect to the sample, the energy of radiation can be absorbed by the same nucleus, which is transformed to an excited state. The excitation energies of the nucleus depend on the surrounding electronic states, and the abundance of the specific nucleus is determined by measuring the emitted or transmitted beam. The technique is well suited for identifying or distinguishing crystalline or amorphous magnetic phases from each other, mainly iron-, cobalt-, and tin-containing compounds. An important application in atmospheric corrosion is the identification of different iron oxides or iron oxy-hydroxides in corrosion products on steel formed during atmospheric exposure (Section 9.4). When transmitted X-rays are analyzed, the method assesses the bulk material, as it has an information depth of around 15–20 μm. If backscattered X-rays or backscattered electrons are analyzed instead, the information depth can be reduced. The sensitivity is of the order 1%, and the technique possesses no spatial resolution.

A.5.4 Thermal Analysis

Thermal analysis encompasses a number of analytical techniques in which some physical parameter of the sample, such as its mass or dimensions, is determined as a function of temperature. The temperature may either be held constant or varied at a controlled rate. Thermogravimetry is one of the principal techniques of thermal analysis that has found applications in atmospheric corrosion research. It is based on the change in sample mass as a function of temperature or time, caused by chemical or physical transformations induced by the temperature change. Possible applications include phase identification in corrosion products of metals or stones and evaluation of the amount of moisture in exposed materials. The sensitivity for detecting specific compounds by thermogravimetry is usually in the range from 0.01 to 1% of the total sample mass. A related technique is thermal desorption spectroscopy. Here the sample is placed in a vacuum environment, and the gases emitted during thermal analysis are detected by means of a mass spectrometer. It should be emphasized that a single thermal analysis technique does not provide unambiguous information about a given sample but often needs complementary or supplementary information supplied by other methods. For example, thermogravimetry in combination with thermal desorption spectroscopy may be a useful analytical combination, since the species emitted during thermogravimetry at a given temperature can be identified by the latter technique.

A.6 TECHNIQUES FOR CORROSION ELECTRODE POTENTIAL

The utility of applying electrochemical techniques to atmospheric corrosion studies has been limited thus far. One main reason is the difficulty of reproducing the atmospheric exposure situation in an electrochemical experiment, in which the material surface normally is covered by an aqueous layer with a thickness that far exceeds that of the aqueous layer formed during atmospheric corrosion. Another principal reason is the measurement of the electrode potential, which traditionally has been accomplished by a reference electrode combined with a Luggin capillary. In the presence of a thin aqueous layer, such measurements will influence the layer thickness and therefore change the measurement itself. The Kelvin probe represents an instrumental development that has overcome some of these obstacles.

A.6.1 Kelvin Probe

The Kelvin probe for atmospheric corrosion studies permits measurement of the corrosion potential of a corroding metal surface without touching the surface covered by the thin aqueous layer. The Kelvin probe is a vibrating metal electrode that is separated from the metal to be investigated by an air gap. If the Kelvin probe and the investigated metal are electrically connected, their Fermi levels are equal. Because of the presence of a thin aqueous layer on the metal, there is a difference in Volta potential between the Kelvin probe and the metal. This difference can be measured through the alternating current that is generated between the vibrating Kelvin probe and the metal surface. It turns out that the measured Volta potential difference changes linearly with the corrosion potential of the sample. As a result, the Kelvin probe has found frequent applications in atmospheric corrosion for *in situ* studies of variations in corrosion potential over a heterogeneous metal surface, for example, a surface with nonuniform corrosion products caused by deposits of aerosol particles. Another important application has been to measure the local variation in corrosion potential of metals covered by an organic coating. Here the Kelvin probe is located external to the organic coating, and yet it can probe the corrosion potential changes at the metal/coating interface caused by changes such as partial delamination of the coating. This may result in enhanced metal corrosion and concomitant reduction in corrosion potential in delaminated areas. The lateral resolution for corrosion potential measurements is around 5 μm.

A.6.2 Scanning Kelvin Probe Force Microscopy (SKPFM)

This scanning probe method is based on atomic force microscopy, whereby the potential offset between a probe tip and a surface can be measured with the same principle as a macroscopic Kelvin probe. It enables the work function of surfaces to be measured with high lateral resolution, usually much less than 1 μm. The map of the work function can be related to many surface phenomena, including corrosion, and has been used to empirically deduce the relative nobility of different phases in microstructures of a surface.

A.6.3 Scanning Electrochemical Techniques

Newly developed electrochemical techniques, such as the scanning reference electrode, scanning vibrating electrode, localized electrochemical impedance spectroscopy, and methods based on atomic force microscopy or scanning tunneling microscopy, permit the potential or current distribution across a metal surface to be determined with a spatial resolution of around 10 μm or better. Although these techniques have mainly been used for metals exposed to a bulk electrolyte, some of them may eventually find applications related to atmospheric corrosion as well.

A.7 TECHNIQUES FOR MONITORING ATMOSPHERIC CORROSIVE SPECIES

Atmospheric corrosion research can also involve the measurement of gaseous, particulate, and waterborne corrosive agents, both during laboratory and field exposures (see, for instance, Figs. 7.1 and 7.2). There exist two different modes of sampling, periodic or continuous. In periodic sampling the air can be sucked through an absorption unit, usually a filter impregnated with chemicals that selectively absorb a particular gaseous component of the atmosphere. In continuous sampling the gas is probed with a calibrated analytical beam of radiation. The following paragraphs describe periodic and continuous measuring procedures for different gas corrosive agents.

A.7.1 SO_2

In periodic sampling the absorption can be accomplished in a paper filter impregnated with potassium hydroxide and glycerol. Afterward, all sulfur is oxidized to sulfate with hydrogen peroxide and analyzed by means of ion chromatography. The detection limit is below 1 ppbv. An alternative procedure, the sulfation plate method, uses the affinity of lead oxide (PbO) to react with SO_2 to form lead sulfate ($PbSO_4$). A small disk of PbO is facing the ground during exposure, and it is assumed that the PbO surface mainly is exposed to SO_2 and not to any significant amounts of sulfate aerosols. After a fixed exposure period, usually 30 days, the exposed area is analyzed with respect to sulfate. In continuous monitoring the SO_2-containing atmosphere is subjected to a UV beam at a given wavelength. The resulting fluorescence is detected and converted to SO_2- concentration, with a detection limit of around 1 ppbv.

A.7.2 NO_2

In periodic sampling the absorption solution can be a mixture of triethanolamine, *o*-methoxyphenol, and sodium metabisulfite. The resulting solution is analyzed spectrophotometrically at 550 nm after adding sulfanilamide and ammonium-8-anilino-1-naphthalene-sulfonate (ANSA). The detection limit is below 1 ppbv. Continuous monitoring is based on chemiluminescence. Here the NO_2 is first reduced to NO and then mixed with O_3 to produce NO_2 in an electronically excited state. The transition of NO_2 to its ground state is accompanied by photon emission that is converted to NO_2 concentration, with a detection limit of around 1 ppbv.

A.7.3 H_2S and Other Reduced Sulfur Compounds

In periodic sampling the absorption can be accomplished by a paper filter impregnated with silver nitrate. The filter is analyzed by means of X-ray fluorescence. The detection limit is below 1 ppbv. Continuous monitoring is based on conversion of reduced sulfur compounds to SO_2 and subsequent fluorescence measurements.

A.7.4 Cl_2

Cl_2 is analyzed following absorption into a paper filter impregnated with *o*-toluidine. This filter is not specific for Cl_2 alone. It may react with other chlorine-containing compounds as well, for example, HCl that is a much more common gaseous corrodent than Cl_2. The filter is immediately analyzed by means of X-ray fluorescence. The detection limit is below 1 ppbv. No suitable techniques are available for continuous monitoring.

A.7.5 NH_3

NH_3 is collected by absorption in a paper filter impregnated with oxalic acid. The filter is extracted in water and analyzed spectrophotometrically at 660 nm following the indophenol method. The detection limit is below 1 ppbv. Continuous measurements are based on chemiluminescence.

A.7.6 NH_4^+

Ammonium in aerosols can be collected by means of a paper filter. Afterward, the filter is extracted in water and analyzed according to the indophenol method. The detection limit is 0.1 $\mu g\,m^{-3}$.

A.7.7 Cl^-

Chloride in aerosols can be collected by means of a paper filter, extracted in water, and analyzed by means of ion chromatography. The detection limit is 0.1 $\mu g\,m^{-3}$. An alternative procedure is the wet candle method. The "candle" is a moistened gauze wick wrapped around a glass tube. When the wind blows past the candle, chlorides from the atmosphere are picked up and their amount determined after fixed exposure periods, usually 30 days.

A.8 SUMMARY

A complete description of the processes involved in atmospheric corrosion requires that all three phases involved, the solid, the liquid, and the gaseous, need to be characterized. There is obviously no single analytical technique able to fulfill such a requirement. Nevertheless, a number of complementary techniques exist that are capable of providing vital information on the phases and their interfacial domains. The intention of this brief compilation has been not only to present techniques used in atmospheric corrosion research today but also techniques that might be used in future studies. The limits of instrumentation for surface analysis, electrochemical analysis, and gaseous analysis are constantly being pushed toward finer spatial resolution, higher sensitivity, and increased versatility. Future progress in our understanding of atmospheric corrosion will depend not only on further instrumental development but also on our imagination in the use of these instruments.

FURTHER READING

A.T. Kuhn, ed., *Techniques in Electrochemistry, Corrosion and Metal Finishing: A Handbook*, John Wiley & Sons, Inc., Chichester, 1987.

P. Marcus and F.B. Mansfeld, eds., *Analytical Methods in Corrosion Science and Engineering*, CRC Press, Boca Raton, 2005.

APPENDIX B

COMPUTER MODELS OF ATMOSPHERIC CORROSION

B.1 FORMULATING COMPUTER MODELS

As in nearly every field of science, atmospheric corrosion advances as a result of a symbiotic relationship among theory, experiment, and computer models. It is in the construction of a model that it is revealed how well and how completely a physical and/or chemical process is understood. It was mentioned in Chapter 1 that the field of atmospheric corrosion has historically progressed much more slowly than many others. A principal reason for this slow pace has been the very low level of effort devoted over the years to the construction of models, although another has certainly been the difficulty of representing the complex physics and chemistry of the corrosion process.

A model begins with a review of the scientific concepts that are involved in a physical process of interest. When such a review has progressed to the point where the principal ingredients of the process have been identified, a conceptual model can be constructed. Such a model describes the way in which the process occurs and identifies components that may not be well understood or complicated factors that may need to be simplified. The purpose of the conceptual model is to paint a word picture of what takes place during the occurrence of the process.

The second stage of model building is the act of taking the conceptual model and expressing it in a mathematical description. Each part of the process is rendered, so that the description is transformed from a concept or series of concepts into a quantitative representation. It is not uncommon at this stage to discover that a feature that can be described in a conceptual way cannot readily be described quantitatively because sufficient experimental data are not available. Two actions then commonly ensue. One is the estimation of the data from theory or from data

Atmospheric Corrosion, Second Edition. Christofer Leygraf, Inger Odnevall Wallinder, Johan Tidblad and Thomas Graedel.

dealing with analogous systems. The next is the inauguration of laboratory or field experiments to supply the missing information.

When a system can be represented mathematically, even if only in a preliminary fashion, it may be embodied in a computer model. In such a model, all components of the process are described by equations and the resulting equations solved electronically. Modern computers are capable of dealing with large systems of equations, and so the descriptions may be quite detailed, but machine computation capabilities often limit the degree to which the model can adequately describe the physical system.

In practice, computer models, theoretical constructs, and experiments form a triad of approaches to the understanding of a physical system. Models cannot be formulated until there is a sufficient theoretical and experimental picture. After that occurs, and after model results emerge, the results often stimulate new thinking and new investigations. These activities produce new data, which can be fed back into the models and the results studied. Once the agreement between model results and data is good enough, the model can be used in predictive fashion. In this way, science progresses by a series of iterative steps.

Computer models can readily be justified on the following grounds:

- They contribute to understanding and insight.
- They aid in the interpretation of experimental data.
- They aid in the testing of hypotheses.
- They enable predictions to be made concerning the susceptibility of a given system to corrosion.

B.2 THE STATUS OF COMPUTER MODELS OF ATMOSPHERIC CORROSION

The earliest chemical process models potentially applicable to atmospheric corrosion were those generated for the study of equilibria in geochemical systems. Several generations of these models have proven extremely useful for systems sufficiently isolated from changing environments that they have time to reach or approach equilibrium. Such systems represent a first approximation to atmospheric corrosion conditions, but the normal variations that occur in temperature, humidity, solar radiation, and other factors render equilibrium models inadequate to express processes that take place in a more dynamic environment.

The next stage of model development involved the application of chemical kinetics to environmental systems of various kinds. Models of gas-phase atmospheric chemistry were the first to be formulated, followed by treatments of liquid-phase atmospheric chemistry. In their original manifestations, these models treated only a single phase, parameterizing any necessary processes in related phases. ("Parameterizing" is the practice of approximating complex processes with one or two ensemble parameters; it is utilized where the details of the process are uncertain or where a computation is made more tractable but no less reliable by parameterization.)

In the middle of the 1990s, atmospheric corrosion models were developed, treating a number of phases explicitly. These models combined the chemistry of the equilibrium models with the kinetic approaches used in other environmental systems. In these models the reactions within two or more phases were formulated in detail. This step requires advanced computational approaches as well as advanced chemical approaches, because reactions in different phases generally proceed at very different speeds (a situation termed "stiffness"), and the computer solution of such a problem requires special formulation of the numerical solution equations. More or less concurrently, electrochemists have added detailed chemical treatments to their detailed electrode process models, though the ability to control the variables in a laboratory or industrial setting has enabled them to retain equilibrium approaches to the calculations.

Thus, most of the components are available to permit the development of comprehensive kinetic models of atmospheric corrosion. Models have treated the multiphase chemistry, the electrochemistry, and the challenge of evolving layers of water and corrosion products. It appears that useful model development now requires considerably more laboratory data in order to put constraints on rates of reaction, phase change, and transport. As with much of science, highly interactive laboratory studies, field experiments, and modeling activities will generate the most rapid progress. It is likely that sophisticated computer models capable of reproducing and revealing much of atmospheric corrosion chemistry can be in place within a decade.

B.3 AN OVERVIEW OF CHEMICAL MODEL FORMULATION

Atmospheric corrosion is fundamentally a process in which the surface of a material exposed to the atmosphere is chemically transformed. For any chemical reaction, we can write the generalized transformation as

$$aA + bB \rightarrow cC + dD \tag{B.1}$$

where the capital letters refer to reactants and products and the small letters to the appropriate stoichiometric coefficients. All reactions have two constraints: mass is conserved (i.e., the number of atoms of each element is the same on each side of the reaction), and charge is conserved (i.e., there is no gain or loss of electrons). The rate of an elementary reaction such as (B.1) is given by

$$R_1 = k_{1f}[A]^a[B]^b \tag{B.2}$$

where the square brackets indicate concentrations and k_{1f} is an empirically measured quantity known as the rate constant or reaction rate coefficient.

In some cases, Reaction B.1 is irreversible. More often, however, the reaction can occur also in the reverse direction, that is,

$$cC + dD \rightarrow aA + bB \tag{B.3}$$

with a rate given by

$$R_3 = k_{1r}[C]^c[D]^d \tag{B.4}$$

In the simplest exemplification of a corrosion system, Reactions B.1 and B.3 comprise the entire set of transformations. In such a system, if left undisturbed for a long time period with respect to transformations, the forward and reverse rates will approach equilibrium:

$$k_{1f}[A]^a[B]^b = k_{1r}[C]^c[D]^d \tag{B.5}$$

An equilibrium constant is then defined by

$$K_{eq} = \frac{k_{1f}}{k_{1r}} = \frac{[C]^c[D]^d}{[A]^a[B]^b} \tag{B.6}$$

Given the value of K_{eq} and any three of the constituent concentrations, one can then solve for the remaining concentration.

The simple system described earlier can be created and studied in the laboratory. In real-world atmospheric corrosion, however, many additional factors must be considered. One is the presence of additional constituents and thus of additional chemical transformation possibilities. Suppose, for example, that species E and F are also present. This allows for reactions such as:

$$aA + eE \rightarrow cC + fF \tag{B.7}$$

$$dD + eE \rightarrow bB + fF \tag{B.8}$$

Each of these added reactions has its own rate and equilibrium constant. A chemical equilibrium computational model then has the task of taking initial concentrations of a number of materials, say, $[A]$, $[B]$, $[D]$, and $[E]$, together with thermodynamic information (e.g., the K values) and solving for the set of concentrations of all species that satisfies the data.

Equilibria in many atmospheric corrosion systems are very slow, while conditions can change rapidly. As a result, equilibrium models are often less appropriate than kinetic models in which each reaction is followed independently. The input to such a model consists not only of initial constituent concentrations and rate parameters but also of information on changes in one or more associated conditions with time. This may be formulated as

$$\frac{d[A]}{dt} = Y_A - Z_A + X_A \tag{B.9}$$

where Y_A = the rate of production of species A as a consequence of chemical reactions within the regime. The rate is computed by summing the rates of those reactions that produce species A. If all the reactions are second order, for example,

$$Y_A = \Sigma[r_m][r_n]k_{m,n} \tag{B.10}$$

where r_m and r_n are the reactants and $k_{m,n}$ is the reaction rate coefficient.

Z_A = the rate of loss of species A as a consequence of chemical reactions within the regime. The rate is computed by summing the rates of those reactions that consume species A. If all the reactions are second order, for example,

$$Z_A = \Sigma[r_A][r_j]k_{A,j} \tag{B.11}$$

where r_j is the reactant and $k_{A,j}$ is the reaction rate coefficient.

X_A = the sum of source and sink rates for species A, incorporating entrainment and detrainment of the species across the vertical boundaries of the regime. For the gas regime, for example, this term is the sum of the entrainment and detrainment across the top of the regime boundary under consideration and the entrainment and detrainment across the gas–liquid interface at the bottom boundary of the regime. Factors incorporated into this term for the liquid regime include the dissolution and precipitation of minerals, the imposition of an external electrical circuit, and the thickening and thinning of adsorbed water layers. In the solid phase, surface morphology and its changes are major considerations. No matter what the regime, physical processes that transport constituents are crucial to satisfactory model formulation. In the remainder of this appendix, we discuss some of the aforementioned aspects in more detail.

B.4 THE TRANSPORT OF REACTANTS

In a well-mixed single regime system, considerations of the transport of reactants are, by definition, unimportant. Such a system does not occur in atmospheric corrosion, however, because in a corroding system there are several regimes (as described in Chapter 3) and ubiquitous concentration gradients. The two key transport topics become (i) the movement of reactive or potentially reactive constituents from one regime to another and (ii) the rate of presentation of the mobile constituents to a unit surface area. In the general formulation of the diffusion equation, the flux J (with typical units of molecules or ions per unit area per unit time) is given by

$$J = -D_i \frac{d[C_i]}{dx} \tag{B.12}$$

where D_i is the diffusion constant (units of area per unit time) for species i in the medium within which transport is occurring.

The diffusion equation can be formulated in a number of different ways to make it appropriate to the regime for which a computer model is being constructed. Consider, for example, application to the atmosphere. Except near a major source of a constituent of interest, atmospheric species are generally considered to be well mixed, and atmospheric chemistry models thus decouple transport and chemistry. An important consideration, however, is the deposition of atmospheric constituents to surfaces. To represent that process, Equation B.12 is often written as

$$J = [C_i]V_{\text{dep}} \tag{B.13}$$

where V_{dep}, the deposition velocity (units of length per unit time), is a value derived from measurement that incorporates the gaseous diffusion constant and any surface resistances that establish a concentration gradient.

For the simulation of chemical reactions in aqueous solution, the well-mixed assumption is again generally appropriate. Thus, Equation B.12 is used but with diffusion constants appropriate to the liquid phase rather than to the gaseous.

If the transiting medium is the solid phase, transport must proceed through connected pores, down grain boundaries or cracks, or along some other structural heterogeneity. In such a case the diffusion constant in Equation B.12 tends not to be very well defined, and poorly supported estimates are often required.

If the transiting species is electrically charged, as is the case with many constituents of interest in atmospheric corrosion, other considerations enter in. Picture, for example, a surface on which a film of corrosion products has formed. Since corrosion is a process in which the oxidation state of the corroding material changes, ions will, in general, move from the solid phase into the overlying aqueous solution and vice versa. This movement requires that they transit a corrosion product membrane that contains a fixed charge that may aid or impede the ion transport. The ion flux is given by

$$J = \frac{2\left(d[C_i]/dx\right)^2 D_i}{h_D[X]} \tag{B.14}$$

where h_D is the thickness of the membrane and $[X]$ is the concentration of fixed charge in the membrane. Since corrosion product membranes are not spatially homogeneous and are difficult to handle in laboratory experiments, there are few data for D_i or $[X]$ values.

A final consideration, which has historically been central to electrochemistry, concerns ionic transport in the presence of an imposed electric field. This is treated by adding a second term to the diffusion equation to give

$$J = D_i\left\{\frac{d[C_i]}{dx} - \frac{Z_i e}{kT}[C_i]E_x\right\} \tag{B.15}$$

where $Z_i e$ is the electronic charge on the transiting ion, k is Boltzmann's constant, and E_x is the electric field. Larson (see "Further Reading") discusses the second term in Equation B.15 in some detail.

B.5 PHYSICOCHEMICAL PROCESSES

Atmospheric corrosion takes place when surfaces are wet, or at least moist. Aqueous reaction processes are thus central. As has been discussed earlier in this book (e.g., Chapters 2 and 7), a related feature of the process is that the amount of water on a corroding surface is seldom constant for very long; more commonly, it increases or decreases as a function of relative humidity (or, if exposed outdoors, of the rate and type of precipitation). Thus, while a snapshot of corrosion processes can be acquired

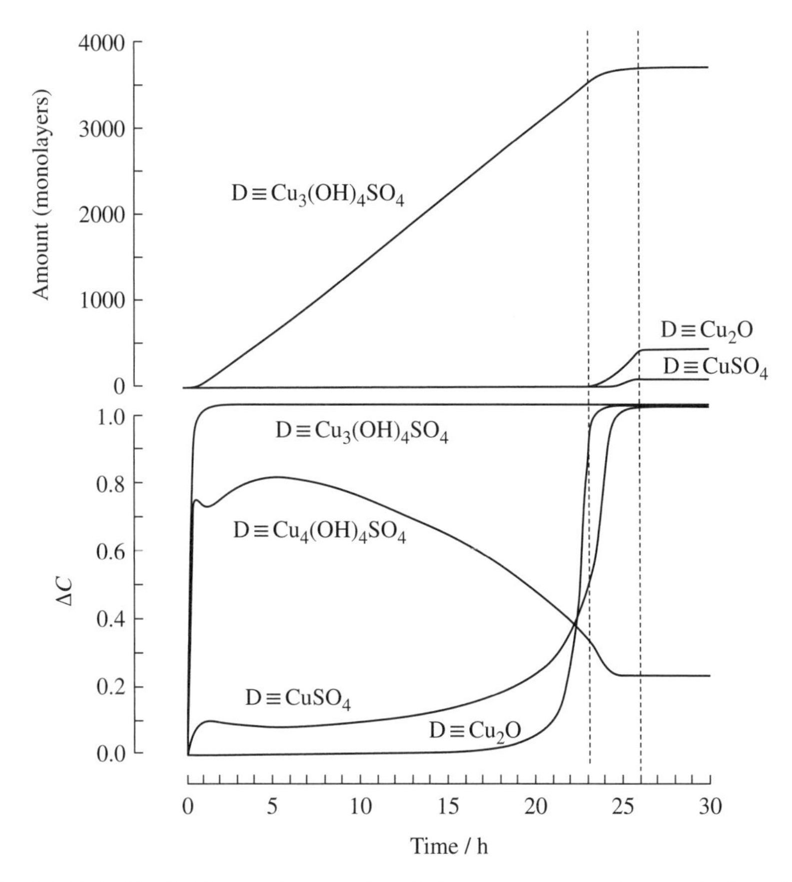

FIGURE B.1 (Top) Computed amounts of the copper corrosion products antlerite [$Cu_3(OH)_4SO_4$], cuprite [Cu_2O], and chalcocyanite [$CuSO_4$] for a model in which a 7 μm surface film of water evaporated at a constant rate over a 30 h period. The D≡ notation indicates a compound that has precipitated from the liquid phase. The vertical dashed lines correspond to the starting and ending times for the precipitation of cuprite. (Bottom) The degree of saturation ΔC in the aqueous surface film for antlerite, cuprite, chalcocyanite, and brochantite [$Cu_4SO_4(OH)_6$]. (Reproduced with permission from The Electrochemical Society; J. Tidblad and T.E. Graedel, GILDES model studies of aqueous chemistry. IV. Initial $(NH_4)_2SO_4$-induced atmospheric corrosion of copper, *Journal of the Electrochemical Society, 144,* 2666–2675, 1997.)

through a model in which the surface water layer is stable, a more realistic model would allow the water layer (and the corresponding chemistry) to vary over time.

As the amount of water on the surface changes, the aqueous concentrations of constituents in solution change as well. If the water is evaporating rather than condensing, some of those constituents may eventually exceed their saturation values and begin to precipitate. The process can be modeled if an evaporation scenario is incorporated, constituent concentrations are periodically recomputed, and precipitation rates are followed. An example for the corrosion of copper appears in Figure B.1. Antlerite, $Cu_3SO_4OH_4$, with very low solubility, forms in the presence of a thick water layer and precipitates immediately. Other species, much more soluble, do not precipitate until the layer has become quite thin and the loss of water has rendered their concentrations much higher.

The precipitation of corrosion products implies, of course, a growing layer of the products atop the copper. In such cases the model must keep track of the rate of that growth, because the layer thickness and morphology will influence ion transport and thus the subsequent corrosion rates. A similar computational requirement applies to the opposite situation in which the chemical evolution of the aqueous solution or the deposition of added water onto the surface promotes dissolution of the solid phase and the enhancement of transport.

B.6 ELECTROCHEMICAL PROCESSES

Electrochemistry enters into atmospheric corrosion in either of two situations: (i) an inherent potential difference exists between two parts of a corroding system, or (ii) an applied potential difference has been created between two parts of a corroding system. In actuality, since corrosion is an electrochemical process, it is always appropriate to incorporate electrochemistry into a model of atmospheric corrosion. As with other facets of computer model building, the critical questions asked include "How important is a full electrochemical treatment to satisfactory simulation?" and "How difficult conceptually and computationally must the electrochemical treatment be?"

The general corrosion reaction can be written as

$$M_{(s)} \xrightarrow{r_1} M^{n+}_{(l)} + ne^- \xrightarrow{r_2} \xrightarrow{r_3} \xrightarrow{r_4} P_{(l)} \text{ or } P_{(s)} \tag{B.16}$$

where r_i are the reactions in the reaction chain and $P_{(l)}$ and $P_{(s)}$ are corrosion products. The steps in this sequence that are explicitly electrochemical are the first (where the corroding molecule is oxidized) and the last (where it is reduced). Consider the schematic diagram of a parallel-plate electrolytic cell, shown in Figure B.2a. Species are oxidized at the anode and reduced at the cathode. The rate of charges entering or leaving the liquid phase is given by a Butler–Volmer equation (see Chapter 3) and ion transport as specified before. Standard aqueous-phase chemical transformations occur in the core of the cell.

An example of an electrochemical process utilizing such a cell is shown in Figure B.2b for a model designed to reproduce the synthesis of propylene oxide from a propylene-saturated bromide electrode. With the electrochemical equations, chemical equations, and transport equations formulated to address this process at steady state, results were obtained as shown in Figure B.1. Among the more interesting aspects of the results are the sharp concentration transitions that are predicted to occur at the boundaries between the diffusion layers and the well-mixed core of the cell.

The example of Figure B.2 does not include any reactions involving the electrode material itself, so is not a corrosion model. To convert it into one, however, one need

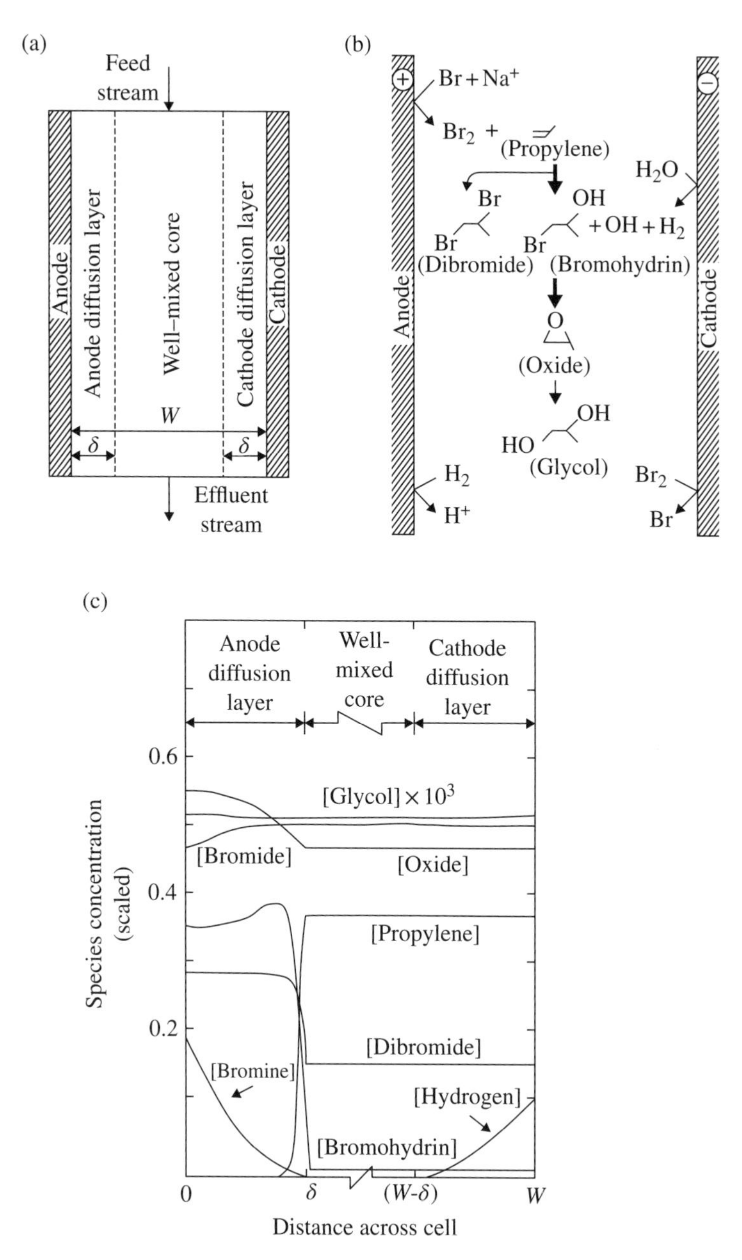

FIGURE B.2 (a) Schematic diagram of a parallel-plate electrolytic cell. (b) The model of reaction chemistry within the cell. Desired reactions are indicated by heavy arrows, side reactions by light arrows. (c) Computed spatial variations of chemical species across the cell. (Reproduced with permission from The Electrochemical Society; R.C. Alkire and J.D. Lisius, Incorporation of complex reaction sequences in engineering models of electrolytic cells, *Journal of the Electrochemical Society, 132,* 1879–1888, 1985.)

only add appropriate electrode reactions, open the system to the atmosphere, and permit constituent interchange among solid, liquid, and gaseous regimes. While conceptually straightforward, such studies are computationally very difficult, and no fully satisfactory examples have yet been produced.

FURTHER READING

General

H. Simillion, O. Dolgikh, H. Terryn, and J. Deconinch, Atmospheric corrosion modeling, *Corrosion Reviews, 32,* 73–100, 2014.

Chemical Models

L.A. Farrow, T.E. Graedel, and C. Leygraf, GILDES model studies of aqueous chemistry: II. The corrosion of zinc in gaseous exposure chambers, *Corrosion Science, 38,* 2181–2199, 1996.

W.W Nazaroff and G.R. Cass, Mathematical modeling of chemically reactive pollutants in indoor air, *Environmental Science and Technology, 20*(9), 924–934, 1986.

N.L. Plummer, D.L. Parkhurst, and D.C. Thorstenson, Development of reaction models for ground-water systems, *Geochimica et Cosmochimica Acta, 47,* 665–686, 1983.

J. Tidblad and T.E. Graedel, Gildes model studies of aqueous chemistry. III. Initial SO_2-induced atmospheric corrosion of copper, *Corrosion Science, 38,* 2201–2224, 1996.

C.J. Weschler, M.L. Mandich, and T.E. Graedel, Speciation, photosensitivity, and reactions of transition metal ions in atmospheric droplets, *Journal of Geophysical Research, 91,* 5189–5204, 1986.

Physicochemical Processes

B.B. Brady and L.R. Martin, Use of surface chemkin to model multiphase atmospheric chemistry: application to nitrogen tetroxide spills, *Atmospheric Environment, 29*(6), 715–726, 1995.

C.W. Liu and T.N. Narasimhan, Redox-controlled multiple species reactive transport, 1. Model development, *Water Resources Research, 25*(5), 869–882, 1989.

J.H. Payer, G. Ball, B.I. Rickett, and H.S. Kim, Role of transport properties in corrosion product growth, *Materials Science and Engineering, A198,* 91–102, 1995.

J. Tidblad and T.E. Graedel, GILDES model studies of aqueous chemistry. IV. Initial $(NH_4)_2SO_4$-induced atmospheric corrosion of copper, *Journal of the Electrochemical Society, 144,* 2666–2675, 1997a.

J. Tidblad and T.E. Graedel, GILDES model studies of aqueous chemistry. V. Initial SO_2-induced atmospheric corrosion of nickel, *Journal of the Electrochemical Society, 144,* 2676–2683, 1997b.

J.C. Walton, Mathematical modeling of large-scale nonuniform corrosion: coupling of corrosion, transport, and geochemical processes in nuclear waste isolation, *Nuclear and Chemical Waste Management, 8*(2), 143–156, 1988.

Electrochemical Models

M. Ramasubramanian, B.N. Popov, R.E. White, and K.S. Chen, A mathematical model for electroless copper deposition on planar substrates, *Journal of the Electrochemical Society, 146*, 111–116, 1999.

F.E. Sloan and J.B. Talbot, Evolution of perhydroxyl ions on graphite/epoxy cathodes, *Journal of the Electrochemical Society, 144,* 4146–4151, 1997.

J.C. Walton, Mathematical modeling of mass transport and chemical reaction in crevice and pitting corrosion, *Corrosion Science, 30*(8/9), 915–928, 1990.

Comprehensive Models

R.C. Alkire and J.D. Lisius, Incorporation of complex reaction sequences in engineering models of electrolytic cells, *Journal of the Electrochemical Society, 132,* 1879–1888, 1985.

R. Goody, J. Anderson, and G. North, Testing climate models: an approach, *Bulletin of the American Meteorological Society, 79,* 2541–2549, 1998.

T.E. Graedel, Gildes model studies of aqueous chemistry. I. Formulation and potential applications of the multi-regime model, *Corrosion Science, 38,* 2153–2180, 1996.

R.S. Larson, A physical and mathematical model for the atmospheric sulfidation of copper by hydrogen sulfide, *Journal of the Electrochemical Society, 149,* B40–B46, 2002.

APPENDIX C

THE ATMOSPHERIC CORROSION CHEMISTRY OF ALUMINUM

C.1 INTRODUCTION

Aluminum is regularly exposed to the indoor and outdoor atmosphere, both inadvertently and by direct intent. It is widely used in the telecommunications and electronics industries, where its electrical conductivity and processability make it the material of choice in a wide variety of applications. Aluminum also sees extensive outdoor use in applications that include siding, window trim, and automotive decorative elements.

The corrosion of aluminum in products exposed to the indoor or outdoor atmosphere is generally modest but can sometimes be a constraint on use. A major hazard is related to the interaction of aluminum with chloride (Cl^-) ions, which can cause the corrosion of thin film aluminum on microelectronic circuits during or following manufacture. In some outdoor applications, especially those connected with marine environments, the use of aluminum is also strongly limited by the chloride sensitivity. In addition, sensitivity to sulfur dioxide (SO_2) is known to cause significant deterioration in some environments.

C.2 CORROSION LAYER FORMATION RATES

The corrosion of aluminum has historically been examined in relation to its outdoor use. All of the studies agree that upon initial exposure aluminum rapidly forms a thin film of aluminum oxide. Under continued exposure this film may grow and be transformed into various other atmospheric products.

The following ranges of corrosion rates for outdoor atmospheric exposure in different environments have been reported:

Atmospheric Corrosion, Second Edition. Christofer Leygraf, Inger Odnevall Wallinder, Johan Tidblad and Thomas Graedel.

Rural	0.0–0.1 μm year^{-1}
Urban	≈1 μm year^{-1}
Marine	0.4–0.6 μm year^{-1}

The laboratory observation that the aluminum sulfate salts require a few weeks to a few months to precipitate from solution is consistent with field data that 3 months or less is sufficient for the formation of pits visible to the naked eye, aluminum sulfates being among the principal constituents of the degraded surfaces.

Indoors, aluminum tarnishes slowly. Surface analyses demonstrate that indoor aluminum surfaces often accumulate adherent particles containing large concentrations of Cl^- and sulfate (SO_4^{2-}) ions. The indoor aluminum corrosion rate is affected by relative humidity, but corrosion only begins for relative humidities of about 70%. At or above that point, the aluminum corrosion products absorb sufficient moisture to stimulate chemical degradation processes.

C.3 THE MORPHOLOGY OF ATMOSPHERIC CORROSION LAYERS ON ALUMINUM

The structure of the oxide films that form upon the exposure of aluminum to the atmosphere has been examined in detail by ultramicrotomy. The films vary considerably in thickness after exposures of several weeks, the thickest portions reaching 100 nm depth and showing great lateral heterogeneity. Oxide film flaws of horizontal dimension 70–200 nm are seen on industrially processed aluminum and appear to be associated with metal ridges and troughs produced by mill rolling processes. Aluminum thin films used in microelectronics are not subject to mill flaws but commonly have small-scale imperfections that respond electrochemically to the high current densities typical of microcircuitry. Well-formed aluminum oxide/hydroxide films are stable in water at potentials more positive than −1.5 V (SCE). Since potentials typical of exposure to the atmosphere are of order +0.5 V (SCE), aluminum is expected to be stable in the absence of aggressive–corrosive species.

In the presence of water or water vapor, the outer skin of the oxide film on aluminum is converted to a hydrous form. The resulting film is often highly porous; it is clear that part of the hydration process represents a chemical transformation into hydroxide minerals, while another part is simply the penetration of water. This penetration renders the underlying thin barrier layer of oxide, which may be imperfect, susceptible to the full spectrum of complex chemistry of the aqueous surface layer. Anions in this aqueous solution are readily incorporated by the aluminum, almost certainly at flaws or dislocations of the oxide film.

Corrosion of aluminum appears to occur principally at flaws or grain boundaries. Aluminum corrosion manifests itself as deep pits rather than as uniform surface deterioration, and the points of chemical attack are more likely to be mechanical flaws in the material than flaws occurring naturally during the refining process. The pitting action traces the segregation of residual impurities during heat treatment, a pattern that introduces corrosion cells into the surface. Pitting is particularly characteristic of Cl^- attack but is also seen in the carboxylic acid (RCOOH) corrosion of aluminum.

C.4 CHEMICAL MECHANISMS OF CORROSION

C.4.1 Oxides and Hydroxides

Table C.1 indicates in its rightmost column those compounds of aluminum that have been conclusively or provisionally detected in corrosion layers resulting from exposure to natural environments. γ-Alumina [aluminum oxide] is the first

TABLE C.1 Minerals and Other, Mainly, Crystalline Substances Found on Corroded Aluminum

Substance[a]	Hey Index No.[b]	Crystal System	Formula	K_{sp} [c]
Oxides and hydroxides				
Aluminum oxide		Cubic	γ-Al_2O_3	
Akdalaite	7.6.8	Hexagonal	$Al_2O_3{\cdot}\frac{1}{4}H_2O$	$4.8{\cdot}10^{-13}$
Boehmite	7.6.2	Orthorhombic	γ-AlOOH	
Gibbsite	7.6.4	Monoclinic	$Al(OH)_3$	$3.7{\cdot}10^{-15}$
Bayerite	7.6.5	Monoclinic	$Al(OH)_3$	$1.5{\cdot}10^{-14}$
Tucanite		Amorphous (Monoclinic)	$Al(OH)_3{\cdot}\frac{1}{2}H_2O$	
Sulfates				
Aluminum sulfate hydrate		Amorphous	$Al_x(SO_4)_y{\cdot}(H_2O)_z$	
Aluminum sulfate		—	$Al_2(SO_4)_3{\cdot}4H_2O$	
Aluminum sulfate		—	$Al_2(SO_4)_3{\cdot}5H_2O$	
Aluminum sulfate		—	$Al_2(SO_4)_3{\cdot}16H_2O$	
Jurbanite			$Al(SO_4)(OH){\cdot}5H_2O$	$1.6{\cdot}10^{-18}$
Aluminite	25.6.5	Monoclinic	$Al_2(SO_4)(OH)_4{\cdot}7H_2O$	
—	—		$Al_3(SO_4)_2(OH)_5{\cdot}9H_2O$	
Mendozite	25.6.13	Monoclinic	$NaAl(SO_4)_2{\cdot}11H_2O$	
Chlorides				
Aluminum chloride		Monoclinic	$AlCl_3$	
Cadwaladerite	8.6.3	Amorphous	$AlCl(OH)_2{\cdot}4H_2O$	
Lesukite		Isometric-Hexoctahedral	$Al_2Cl(OH)_5{\cdot}2H_2O$	
Carbonates				
Dawsonite	11.7.3	Orthorhombic	$NaAlCO_3(OH)_2$	

[a] Included under some circumstances may be basic (b) or hydrated (h) compounds.
[b] A.M. Clark, *Hey's Mineral Index*, 3rd edition, Chapman and Hall, London, 1993.
[c] Solubility products are taken from a variety of sources, especially R.P. Frankenthal, in *Handbook of Analytical Chemistry*, ed. L. Meites, McGraw-Hill, New York, pp. 5–30, 1963 and D.R. Turner, M. Whitfield, and A.G. Dickson, The equilibrium speciation of dissolved components in freshwater and sea water at 25°C and 1 atm pressure, *Geochimican et Cosmochimica Acta, 45*, 855–881, 1981.

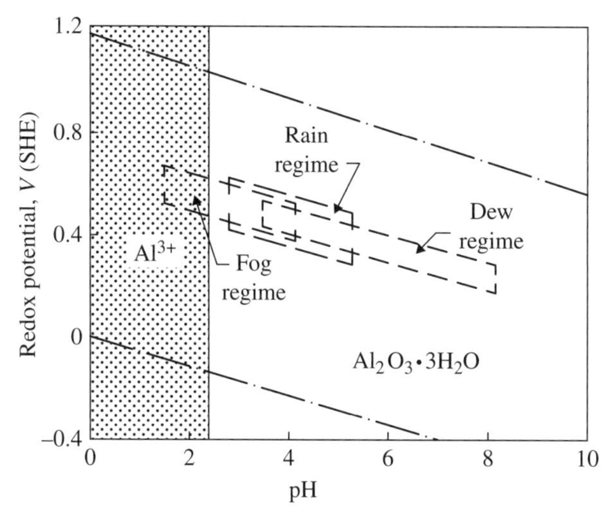

FIGURE C.1 Potential–pH diagram for the system Al—H_2O at 25°C, for a concentration of aluminum ionic species of 0.1 M. The approximate regimes for fog, rain, and dew are indicated.

component formed upon exposure to the atmosphere; it has an initial depth of 2–3 nm. For exposures underwater or for a few months in air, the oxide becomes covered by a thin layer of boehmite [γ-AlOOH]. The boehmite is subsequently covered by bayerite [$Al(OH)_3$, also written as $Al_2O_3{\cdot}3H_2O$]. Akdalaite [$Al_2O_3{\cdot}\frac{1}{4}H_2O$] and tucanite [$Al(OH)_3{\cdot}\frac{1}{2}H_2O$], which are hydrated variants of the oxide and hydroxide, are sometimes present as well.

The potential–pH diagram for the aluminum–water system is shown in Figure C.1. The stability of bayerite over a very wide range of acidities is readily apparent. It is not until pH 2.4 that complete dissolution of the oxide is expected in the simple aluminum–water system. It is an experimental observation, however, that some dissolution begins at pH 4. Solutions with acidities at this or higher levels include most rains and almost all fogs, evaporating solutions of deposited electrolyte, and the altered chemical conditions found in surface pits on aluminum. Even at pH values as low as 2.0, however, the degradation of the surface layer is still of the pitting type, with some oxide film remaining. In the case of dew, the typical solution is too basic to be expected to dissolve the surface oxide.

Another O—H species deserving mention is hydrogen peroxide (H_2O_2), which is always present in precipitation, is a strong oxidizer, and is a species identified as a participant in the atmospheric corrosion mechanisms of other metals. In addition, H_2O_2 is known to be formed naturally on fresh aluminum surfaces in the presence of water, a situation that would occur when the surface is exposed by acid etching. It does not appear, however, that H_2O_2 is likely to play any direct role in the atmospheric corrosion of aluminum, since it has very little activity toward the metal. (Aluminum is the metal called "most resistant" to hydrogen peroxide.) As will be seen in the following, however, H_2O_2 has roles other than direct action to play in the atmospheric metamorphosis of aluminum.

C.4.2 Sulfides and Sulfates

Sulfur compounds are important to the atmospheric corrosion chemistry of aluminum, since SO_4^{2-} is known to be incorporated into the corroding surface layers and since amorphous aluminum sulfate hydrate is observed to be the most abundant corrosion product on aluminum exposed to marine and industrial atmospheres. The dominant abundance of this noncrystalline product renders unproductive any attempt to use equilibrium diagrams to deduce chemical mechanisms for aluminum corrosion, since the amorphous aluminum sulfate hydrate is a metastable species. The amorphous material has a very long lifetime at atmospheric temperature and pressures and is formed in the presence of sulfuric acid at quite acidic pH conditions. Aluminite, a crystalline aluminum hydroxysulfate, has also been reported in atmospheric samples. Both of these constituents are produced by interactions with dissolved SO_4^{2-}, which is generated either in the gas phase or in the surface aqueous layer from the oxidation of SO_2. Indoors, SO_4^{2-} is the most abundant anion found incorporated into aluminum surface layers in electronic equipment buildings, where its accumulation rates range from 0.1 to 0.7 $\mu g\,cm^{-2}\,year^{-1}$.

Two factors favor the formation of the amorphous phase. The first is that the crystalline aluminum sulfate minerals that are capable of being formed naturally are composed of between 15 and 25 atoms. Assembling so many constituents properly is a statistically disfavorable process. The second factor of importance, and a factor that applies to outdoor formation of any mineral constituent on a corroding surface, is the tendency for solutions atop outdoor aluminum to undergo rapid temperature changes. A decrease in temperature results in an increase in the degree of saturation of the solution and an increase in the solution viscosity. If the liquid is cooled sufficiently rapidly, atomic rearrangements will not have enough time to lead to a crystalline formation and the solution is left in a frozen liquid state.

A schematic process diagram for the sulfur chemistry component of atmospheric aluminum corrosion is given in Figure C.2. This figure is the first of two that summarize the chemical processes involved in aluminum surface metamorphosis upon atmospheric exposure, indoors or out. It shows that any sulfur gas dissolving into the aqueous film will eventually be oxidized and/or ionized to bisulfite ion (HSO_3^-). The bisulfite ion is readily converted to SO_4^{2-} by O_3, H_2O_2, or transition metal ion impurities (largely iron or manganese). Once SO_4^{2-} is formed by solution chemistry or supplied by particles or precipitation, it is available to form the aluminum hydroxysulfates. The aluminum is made available only by dissolution or bypassing of the passive oxyhydroxide surface layer. Such dissolution can be expected in outdoor exposures if the aqueous surface layer is moderately or highly acidic. With the exception of the precipitation input, the processes shown in the figure apply to indoor surfaces as well as to those outdoors.

Although reduced sulfur compounds are present in the atmosphere, they do not play any significant role in aluminum's corrosion chemistry.

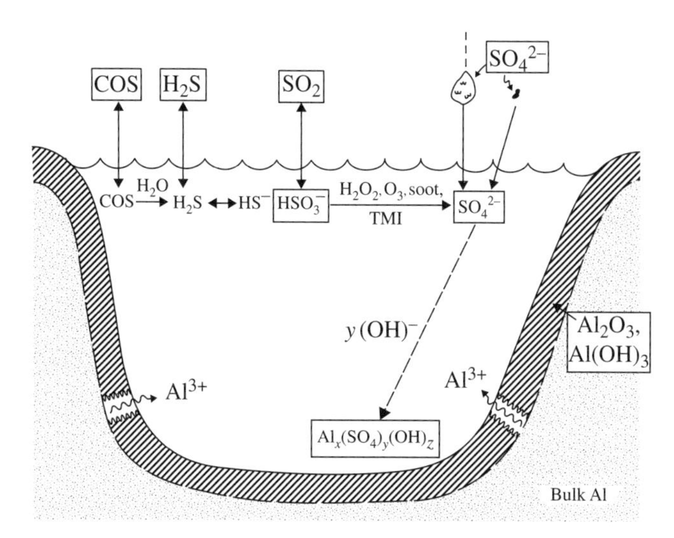

FIGURE C.2 A schematic representation of the sulfur chemistry involved in the atmospheric corrosion of aluminum. The atmosphere is at the top of this diagram, a surface water layer is in the center, and the corroding metal is at the bottom. For simplicity, the diagram pictures a pit or depression in the metal surface, although deposited hygroscopic particles, heavy dew, or rain might be capable of producing similar conditions anywhere on the surface. The wavy line indicates aluminum ions diffusing from the bulk solid, through the oxide layer, and into the aqueous solution. In this and the following figure, species whose presence in the indicated phase has been confirmed by field measurement are enclosed in a rectangular box; the presence of species not so enclosed is unconfirmed but is thought to be reasonable. Solvation or reaction processes that have been confirmed by laboratory studies are shown as solid arrows; those for which the mechanism is uncertain are shown as dotted arrows. Rain is indicated by the atmospheric water drop and airborne particles by the irregular small atmospheric object.

C.4.3 Chlorides

A significant cause of aluminum surface degradation upon exposure is Cl^-. The solubility of the resulting products containing chloride prevents large buildups of chloride surface species, but some Cl^- is definitely incorporated into the corrosion layer. No specific compound containing chlorine and aluminum has yet been identified on aluminum surfaces exposed to the outdoor atmosphere.

Indoors, Cl^- is readily detected on aluminum surfaces. Its accumulation rates, covering the range 0.01–0.13 μg cm^{-2} year^{-1}, are about a tenth of the surface accumulation rates of SO_4^{2-}. Among the specific products that have been identified is aluminum chloride [$AlCl_3$]. It has been proposed that the aluminum chloride is formed by a stepwise chlorination of aluminum hydroxide; this is consistent with the identification of the intermediate compound cadwaladerite [$AlCl(OH)_2{\cdot}4H_2O$] in laboratory experiments and the observation of equilibria among the several

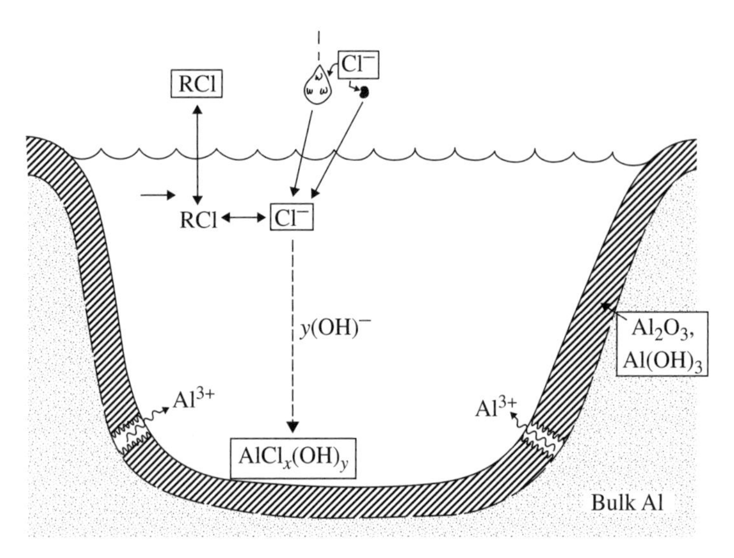

FIGURE C.3 A schematic representation of the chemical processes involved in the formation of components containing chlorine during the atmospheric corrosion of aluminum (see Fig. C.2 legend for explanatory details).

aluminum hydroxychloride species $Al(OH)_nCl_{3-n}$. The aluminum hydroxychlorides are observed to form in the presence as well as the absence of SO_4^{2-} and Cl^-. The reaction sequence is therefore

$$Al(OH)_3 + Cl^- \rightarrow Al(OH)_2Cl + OH^- \quad (C.1)$$

$$Al(OH)_2Cl + Cl^- \rightarrow Al(OH)Cl_2 + OH^- \quad (C.2)$$

$$Al(OH)Cl_2 + Cl^- \rightarrow AlCl_3 + OH^- \quad (C.3)$$

and is thought to proceed largely by competition between the OH^- and Cl^- for surface sites on the hydrated aluminum oxide surface. In Figure C.3 a schematic diagram for potential Cl^- deposition and reactions is shown. Chlorine enters the aqueous surface layer through the incorporation of gaseous HCl or an organic gas containing chlorine or through the deposition of sea salt, background aerosol particles, or precipitation. Once present, sequential combination reactions can lead to aluminum chloride or the hydroxychlorides. These constituents are soluble in weak acid solutions and so would be expected to be retained only if they can be isolated from the generally acidic aqueous surface layer. This relatively straightforward picture overlooks some complications suggested by laboratory experiments. For example, the presence of SO_4^{2-} in the chloride solutions enhances the rate of pitting, as does H_2O_2. Laboratory evidence suggests that a corroding aluminum surface has a variety of adsorption sites with different adsorption properties, only a minority of these sites being active for pitting corrosion. It is obvious that a complete chemical description of the corrosion of exposed aluminum in a mixed electrolyte solution will be a difficult and detailed task, involving the interactions of a number of electrolytic and neutral species.

C.4.4 Carbonates and Organic Compounds

There is no evidence that aluminum surfaces overlain by water and exposed to the atmosphere take part to any significant degree in carbonate or organic chemistry. Nonetheless, the occasional formation of compounds containing carbon is probably a part of aluminum corrosion layer development. The potential products include carbonates, formates, and acetates.

C.4.5 Other Components

The deposition of atmospheric particles onto aluminum surfaces is a reasonable presumption, but again the outdoor data are very limited. They show that the corrosion layers resulting from industrial exposures contain quartz and clays; samples from marine environments contain, in addition, sodium chloride and numerous marine microorganisms. No detailed chemical analyses of any of these constituents were attempted.

Indoors, small amounts of ammonium, calcium, sodium, potassium, and magnesium have been measured in aluminum surface layers as a consequence of the deposition of small airborne particles onto the metal surfaces.

C.4.6 Chemical Process Summary

The major anions in surface water layers on aluminum exposed to the outdoor atmosphere are Cl^-, SO_4^{2-}, and NO_3^- (and probably CO_3^{2-} and $COOH^-$). All of these are known to adsorb onto the aluminum oxide surface from electrolyte solutions, although not very rapidly.

Inside equipment buildings, a significant amount of information is available concerning the rates of deposition of ion to aluminum surfaces. The ranges of measured rates are given in Table C.2 for three anions: SO_4^{2-}, Cl^-, and NO_3^-. The last column of Table C.2 relates the accumulation rates to the surface density of the initial oxide/hydroxide surface layer. As can be seen, there is approximately one anion supplied per surface alumina molecule per year. Since aluminum generally undergoes pitting corrosion involving only a small fraction of the surface (less than a tenth),

TABLE C.2 Rates of Accumulation of Ions from the Atmosphere on Indoor Aluminum Surfaces

Ion	Accumulation Rate ($\mu g\,cm^{-2}\,year^{-1}$)	Reactive Constituent Ratio Rate (Ions Per Surface Al_2O_3 Molecule Per Year (Central Value))
SO_4^{2-}	0.1–0.7	0.2
NO_3^-	0.03–0.27	0.2
Cl^-	0.01–0.13	0.4

it follows that the anion supply is ample for reactions to occur and that the aluminum corrosion rate will not be constrained by a shortfall of reactants. Outdoors, where the greater concentrations of reactive gases and aerosol particles and the presence of precipitation provide sources of ions of much higher magnitude than those indoors, anion supply will probably never be a limitation to the rate of aluminum corrosion.

Oxide thinning (or partial dissolution of the oxide layer) occurs for aluminum if the associated electrolyte solution is sufficiently acidic or basic. Although the anions associated with the hydrogen ion are present in the solution, it is the pH that determines the dissolution rate of the oxide, as the dissolution is proton-induced. In at least three circumstances, high acidities can occur on the surface of aluminum exposed to the atmosphere: by the adsorption of highly acidic fogs, by the evaporation of surface electrolytic solutions produced by rainfall or water adsorption, and by the generation of highly acidic conditions in pits and crevices. The penetration of the reactive ion through the surface oxide/hydroxide layer to the metal surface is thought to be the rate-determining step in the aluminum corrosion process, at least in the case of small, relatively mobile ions such as chloride.

The presence of compounds of aluminum including sulfate and chloride is evidence for chemical reactions being central to the corrosion process, rather than merely the proton-induced dissolution of the aluminum oxide. The rate of reaction between oxygenated aluminum and SO_4^{2-} is similar to that for oxygenated aluminum and Cl^-, therefore, if $[SO_4^{2}] \gg [Cl^-]$, as will often be the case in both the indoor and outdoor environments, the rate of aluminum dissolution by SO_4^{2-} interaction will be much greater than that due to Cl^-. The predominance of sulfates over chlorides is, in fact, observed on outdoor aluminum specimens. The reactions may occur either in the hydrated oxide lattice in the solid phase or in the associated solution in the liquid phase. Whichever regime predominates, the process for forming stable, long-lasting corrosion products is slow. Laboratory studies show that several days or weeks are required for the formation of the aluminum salts; such times might be diminished under atmospheric conditions by preexisting seed crystals of the salts from earlier solution–evaporation cycles.

Once the oxide film on aluminum is sufficiently thinned by dissolution or passed by the adsorption of solution into pores or by ion transport down defects, relatively rapid corrosion of the underlying metal can occur. The corrosion that is initiated, largely near or beneath flaws in the original oxide layer, will tend to occur in pits in which the solution differs substantially from the bulk. This chemical difference usually encompasses highly acidic pH values, with associated dissolution of the proximate aluminum. Much of the aluminum that is liberated combines with the abundantly available SO_4^{2-} to form the poorly soluble amorphous aluminum sulfate hydrate, which is the main constituent of the corrosion layers.

C.5 SUMMARY

Compared with many metals, aluminum corrodes rather slowly upon exposure to the atmosphere. Its rate of corrosion is enhanced in marine areas, however (largely by elevated levels of Cl^-), and enhanced still more in urban areas (largely by elevated levels of SO_4^{2-}). The initial surface layer on aluminum prior to interaction with atmospheric corrosive species is an oxide/hydroxide; it is porous and freely sorbs water (and thus any ions present in the water as well). The subsequent corrosion occurs preferentially at flaws or grain boundaries. Three processes that have been assessed in connection with chemical reactions in the aqueous layer atop the exposed aluminum surface may be summarized as follows: (i) high acidities are shown to be common outdoors but not inside; they provide a supply of aluminum ions to the surface water layers as a consequence of the dissolution of the Al_2O_3 lattice; (ii) formation of insoluble amorphous aluminum sulfate is favored outdoors over that of crystalline aluminum hydroxysulfate because of the chemical complexity of the naturally formed crystalline compounds and the freezing out of solution constituents by repeated wet–dry cycles; and (iii) corrosion products containing chloride are formed by Cl^- replacement of hydroxide moieties in the aluminum oxyhydroxide surface layer, followed by solubilization and runoff of the hydroxychloride mixed salt.

Sulfate species are the most abundant indoor and outdoor constituents of the surface corrosion layers; in the outdoor environment they have been shown to arise primarily from SO_2 incorporation and from precipitation. Indoors, the SO_4^{2-} is present largely as a consequence of deposited fine particles. Chloride species are the next most common corrosion product; the indoor sources of chloride are reactive gases and particles, and the outdoor sources are reactive gases, airborne particles, and precipitation. Anion budget estimates suggest that the supply of ions is not rate limiting, but rather the rate of penetration of the ions through the surface oxide/hydroxide layer.

Details concerning the information in this appendix may be found in T.E. Graedel, Corrosion mechanisms for aluminum exposed to the atmosphere, *Journal of the Electrochemical Society, 136*, 204C–212C, 1989. A more comprehensive description of evolution of different corrosion products formed on aluminum, as a result of atmospheric exposures, is given in Chapters 9 and 13 together with suggestions of further reading.

APPENDIX D

THE ATMOSPHERIC CORROSION CHEMISTRY OF CARBONATE STONE

D.1 INTRODUCTION

Calcium is not exposed to the atmosphere as a pure element, but as a compound, most often a carbonate. Two forms are in common use, limestone and marble, the latter being a metamorphosed form of the former. Limestone has traditionally been a common building material, though is now largely supplanted by concrete. The corrosion of the limestone exteriors and decorative features of many historic structures remains of interest, as does the corrosion of marble statuary, which is most often but not always kept indoors.

The mechanisms involved in the corrosion of carbonate stone remain poorly determined despite considerable effort over the years. The physical system is complicated in the case of stone by porosity, biological interactions, and physical degradation, and the combined influences of these factors have proven difficult to disentangle. Nonetheless, a small number of carefully controlled laboratory experiments have combined with detailed field research to yield significant information about the processes involved.

D.2 CORROSION LAYER FORMATION RATES

In the case of many metals, the corrosion layers that form are highly insoluble, and formation rates are determined by aggressive physical or chemical removal of the corrosion products following field exposure. The situation for calcium carbonate is almost the exact opposite; the corrosion products that form are highly soluble and tend to be lost during field exposure. The data thus gives recession rates, that is, rates

Atmospheric Corrosion, Second Edition. Christofer Leygraf, Inger Odnevall Wallinder, Johan Tidblad and Thomas Graedel.

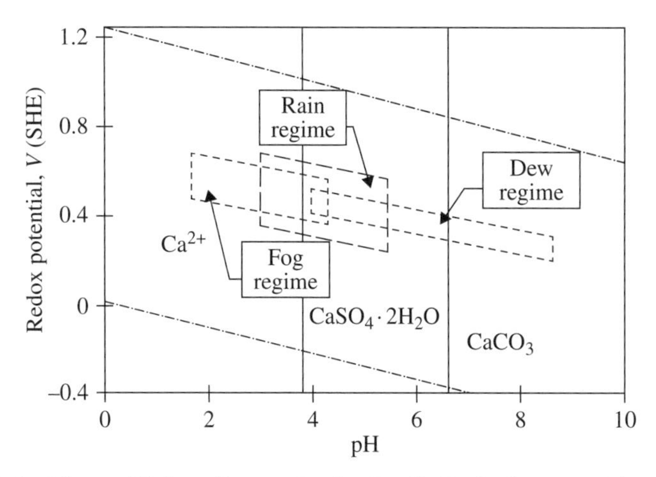

FIGURE D.1 The equilibrium diagram for the Ca—C—O—H—S system. The contours are for different SO_4^{2-} activities in aqueous solution. (Adapted with permission from H.H. Schmidt, ed., *Equilibrium Diagrams for Minerals at Low Temperature and Pressure*, The Geological Club of Harvard, Cambridge, MA, 1962.)

of loss of the exposed surface. In the most complete study conducted at various sites within the continental United States, the average rates were determined to be:

Marble	0.5–50 μm $year^{-1}$
Limestone	10–150 μm $year^{-1}$

D.3 MORPHOLOGY OF ATMOSPHERIC CORROSION ON CARBONATE STONE

Unlike metals, carbonate stone is highly porous, the total porosity being about 0.5–1% for marble and 15–20% (or even as much as 45%) for limestone. Water and any chemical species dissolved in the water thus have ready access to the interior of the material, especially in the case of limestone.

The dominant corrosion product on carbonate stone is gypsum, $CaSO_4 \cdot 2H_2O$. As Figure D.1 shows, gypsum is the stable product in mildly acidic aqueous systems involving carbonate and sulfate ions in combination with calcium. Compounds with sulfur in a reduced state are not thermodynamically favored. The equilibrium situation says nothing about the speed of the transformation, of course, but the transformation turns out to be quite rapid. In an experiment designed to study this situation, tiny particles of calcite were placed on electron microscope slides and exposed several groups of them to the atmosphere for different 3-day periods. The results are shown in Figure D.2. It is apparent that calcite conversion to gypsum can occur within a few days under these conditions, though the rates vary with each exposure situation.

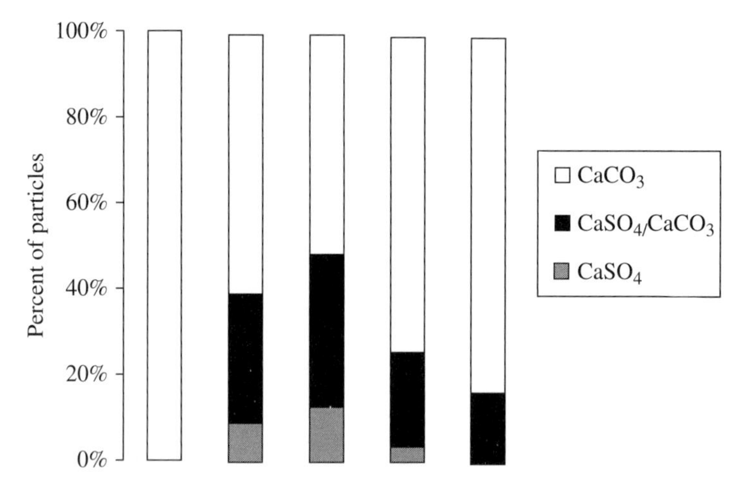

FIGURE D.2 The conversion of calcite to gypsum on atmospheric particles. Samples of calcite ($CaCO_3$) were exposed to the atmosphere for 2–3 day periods at Ithaca, NY. The bars represent (left to right) a control (not exposed to the atmosphere), and exposures on May 10–13, May 24–27, May 28–30, and May 30–31 (all 1986). (Data from T.J. Butler, Composition of particles deposited to an inert surface at Ithaca, New York, *Atmospheric Environment,* 22, 895–900, 1988.)

Similar results are seen for much larger samples of limestone as well, with significant amounts of gypsum formed after 2-month exposures in central London.

A complication with the study of stone corrosion is that the porosity permits freeze/thaw cycles to occur in high-latitude climates in winter. In these situations, the expansion of water undergoing freezing generates internal pressure that can rupture the adhesive materials holding the stone together. The interrelationships of freeze and thaw cycles with chemical deterioration have not yet been well studied.

Another complication with the study of carbonate stone corrosion is that the rough surfaces and pores provide convenient homes for a variety of organisms: bacteria, algae, fungi, and lichens, for example. The organisms commonly release acid metabolites that contribute to the chemical deterioration of the stone.

Finally, carbonate stone corrosion occurs in the presence of soot deposited on the rough stone surfaces. The soot not only degrades the visual appearance of the stone but also enhances the retention of water and provides the potential for catalytic enhancement of corrosion processes.

D.4 CHEMICAL MECHANISMS OF CARBONATE STONE CORROSION

D.4.1 Introduction

Ions or molecules that have been detected or suggested as potential participants in the corrosion chemistry of carbonate stone were used to direct a survey of the mineralogical and chemical literature for possible mineral constituents. The resulting list

TABLE D.1 Carbonate-Containing Minerals and Other Crystalline Substances with Relevance to the Atmospheric Corrosion of Carbonate Stone[a]

Substance	Hey Index No.	Crystal System	Formula	Detection[b]
Carbonates				
Calcite	11.4.1	Rhombohedral	$CaCO_3$	Yes
Aragonite	11.4.2	Rhombohedral	$CaCO_3$	Yes (P)
Dolomite	11.4.6	Trigonal	$CaMg(CO_2)_2$	
Ankerite	11.13.6	Rhombohedral	$Ca(MgFe)(CO_3)_2$	Yes
Sulfites				
Calcium sulfite hemihydrate	—	—	$CaSO_3{\cdot}\frac{1}{2}H_2O$	
Hannebachite	27.1.1	Orthorhombic	$CaSO_3{\cdot}H_2O$	
Calcium sulfite dihydrite	—	—	$CaSO_3{\cdot}2H_2O$	
Sulfates				
Anhydrite	25.4.1	Orthorhombic	$CaSO_4$	
Bassanite	25.4.2	Orthorhombic	$CaSO_4{\cdot}\frac{1}{2}H_2O$	Yes (P)
Gypsum	25.4.3	Monoclinic	$CaSO_4{\cdot}2H_2O$	Yes
Nitrates				
Nitrocalcite	13.5	Monoclinic	$Ca(NO_3)_2{\cdot}4H_2O$	Yes
Oxalates				
Whewellite	31.1.5	Monoclinic	$Ca(C_2O_4){\cdot}H_2O$	Yes
Weddellite	31.1.6	Tetragonal	$Ca(C_2O_4){\cdot}2H_2O$	Yes

[a] Included under some circumstances may be basic (b) or hydrated (h) compounds.
[b] A.M. Clark, *Hey's Mineral Index*, 3rd edition, Chapman and Hall, London, 1993.
(P) indicates detection on atmospheric carbonate particles; the same species presumably are formed also on large carbonate stones.

of candidate species is given in Table D.1. It contains 13 entries, some of which have no mineralogical names and are therefore not known to be produced naturally by geochemical processes.

The final column in the table indicates whether a specific mineral has been found in atmospheric corrosion layers on carbonate stone. The initial surface, almost entirely of $CaCO_3$, calcite, most often has a coating of one or both of the calcium oxalates in unpolluted areas and of gypsum in highly polluted ones. One study claims the presence of $Ca(NO_3)_2{\cdot}4H_2O$, nitrocalcite.

D.4.2 Absorbed Water

The atmospheric corrosion of carbonate stone is strongly influenced by the presence or absence of water. The amount of water absorbed into or upon the stone varies with the stone itself, with relative humidity, and with temperature but is always substantial.

Unlike the case of metals, where water can be pictured as absorbed atop a solid surface, incorporated water on carbonate stone is largely contained within the pores.

The critical role of water in contact with carbonate stone is to provide a medium into which the calcium carbonate can dissolve. The stone forms an equilibrium with the CO_2 from the air that is dissolved in the water:

$$CaCO_3(s) + H_2CO_3 \Leftrightarrow Ca^{2+} + 2HCO_3^- \quad (D.1)$$

In the absence of other dissolved species, the equilibrium is strongly to the left. Nonetheless, it is this natural process termed karst dissolution that has been the active agent in limestone dissolution over geologic time.

In the case of some of the materials that corrode upon exposure to natural environments, the mechanisms and rates of oxidation of the mobilized ions from one valence state to another are items of interest as possible limiting steps in the degradation process. Such is not the case with calcium, which has only the +2 valence state normally available to it. As a consequence, the chemical degradation mechanisms need consider only questions of reaction and dissolution of the Ca^{2+} ion and not of the ion's oxidation or reduction.

D.4.3 Carbonates

Of the principal calcium carbonate minerals, dolomite is the most stable, calcite is less so, and any aragonite, $CaCO_3$, laid down as sediment is transformed into calcite over time. The carbonates are rather insoluble in neutral or basic solution but dissolve with increasing ease as in-contact water solutions are acidified.

D.4.4 Sulfites and Sulfates

Sulfur dioxide is universally regarded as the most damaging species in the atmospheric corrosion of carbonate stone, because of its common presence in urban areas as a consequence of the combustion of sulfur-containing fossil fuels, together with its ability to generate sulfuric acid. When initially dissolved in solution, however, SO_2 forms H_2SO_3, a weak acid, in which the sulfur atom is in the +4 (S^{IV}) oxidation state. In order for the strong acid H_2SO_4 to be generated, the sulfur atom must be oxidized to +6 (S^{VI}). This process is the most potentially limiting chemical step in the atmospheric corrosion of carbonate stone.

In the presence of SO_2 but the absence of effective oxidizing species, sulfur remains as S^{IV}. In laboratory experiments by Johansson and coworkers at the Chalmers University of Technology, Gothenburg, exposure of calcareous stones to SO_2 alone in humid air produced only calcium sulfite corrosion products. With an oxidizer present, however, S^{VI} readily resulted.

In aqueous solutions exposed to the atmosphere, there are many potential routes for S^{IV} oxidation. Hydrogen peroxide and ozone scavenged from the atmosphere are both very effective oxidizers, the former at low pH and the latter at higher. Transition metal ions, especially iron and manganese, are very effective in catalyzing the

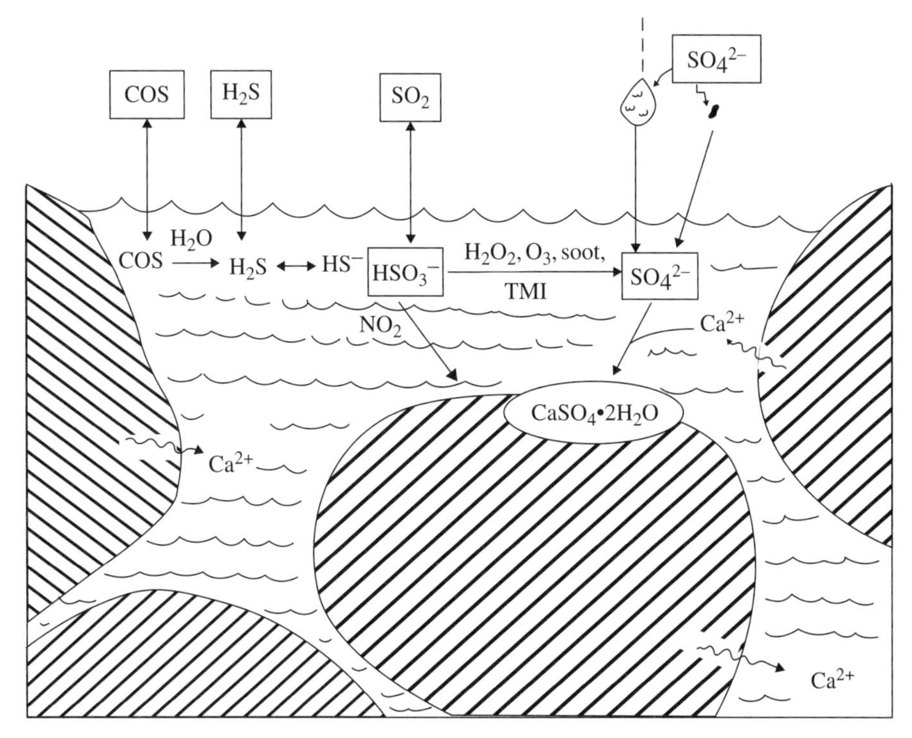

FIGURE D.3 A schematic representation of the processes potentially involved in the sulfur chemistry of the atmospheric corrosion of carbonate stone. The atmosphere is at the top of this diagram, and surface water has filled the pores between the stone particles. The wavy line indicates Ca^{2+} ions diffusing into the aqueous solution from the bulk solid. Species shown within the rectangular boxes are present as solution constituents: that in the oval is present as a precipitate. TMI indicates transition metal ions: they and soot catalyze the S^{IV} to S^{VI} oxidation. Rain is indicated by the atmospheric waterdrop and airborne particles by the irregular small atmospheric object.

oxidation, as is soot and fly ash. All of these constituents are commonly present in building stones as impurities or surface deposits. An additional, highly effective, oxidant is NO_2, which transforms S^{IV} to S^{VI} by what appears to be a surface-mediated reaction.

The overall process can thus be represented as follows:

$$SO_2(g) \overset{\text{rapid}}{\leftrightarrow} SO_2(l) \overset{\text{slow}}{\leftrightarrow} CaSO_3{\cdot}xH_2O \leftrightarrow CaSO_4{\cdot}2H_2O(s) \tag{D.2}$$

(with a bypass arrow from $SO_2(g)$ directly to $CaSO_4{\cdot}2H_2O(s)$)

where the chain involving sulfite represents $S^{IV} \rightarrow S^{VI}$ transformation in solution and the route that bypasses the sulfite stage represents surface-mediated transformation. The processes marked with an asterisk are slow in the absence of an efficient oxidizer and fast in its presence. Figure D.3 is a schematic representation of the sulfur chemistry of carbonate stone, in which the solution oxidation and surface-mediated oxidation are clearly distinguished. The sulfur atom is incorporated into the solid phase as part of the gypsum, $CaSO_4{\cdot}2H_2O$, corrosion product. More important, however, is the increased acidity of the resulting solution. Under such conditions, the carbonate stones undergo rapid dissolution.

D.4.5 Nitrates

There has been a single report of the detection of a nitrate mineral on calcium. Given the relatively high solubility of such a species (Table D.1), this report must be regarded as provisional. Nonetheless, NO_3^- is certainly present in deposited particles and raindrops, in which nitric acid serves as a source of protons.

D.4.6 Oxalates

Oxalates (combinations of calcium with the anion of a weak acid) are the only organic corrosion products consistently identified on carbonate stone. They appear to predominate in regions where air pollution is low. There are at least three sources that can supply oxalate ions to the stone surface: secretions from algae and lichens, degradation of surface pigments or protective compounds, and atmospheric particles and hydrometeors. Although the evidence is not unambiguous, microorganism secretions are very likely the dominant source.

The chemical processes involved in calcium oxalate formation appear straightforward. In the presence of water on the surface of the stone or in its pores, any oxalate ions present will readily enter into aqueous solution. They will then combine with the available calcium ions. The crystalline form that results will be dependent on the degree of agitation during the precipitation process. The sequence is shown schematically in Figure D.4.

An interesting aspect of carbonate stone corrosion is that oxalates and sulfates tend not to occur together. At any rate, their concentrations are not proportional to each other. A suggested explanation is that algae and lichens, apparently the primary source of the oxalate ions, are highly sensitive to sulfur compounds. High pollution levels would thus be incompatible with the presence of high concentrations of oxalate-generating microorganisms.

D.5 SUMMARY

The corrosion of carbonate stone is chemically straightforward, but the overall process is made quite complex by several factors ultimately related to the porosity of the material: the presence of microorganisms, the water reservoirs, and the tendency of the stone to "breathe." In addition, carbonate stone is not chemically pure, and the transition metals and noncarbonate minerals that are invariably present complicate the picture, as does calcium carbonate's solubility in acidified solutions.

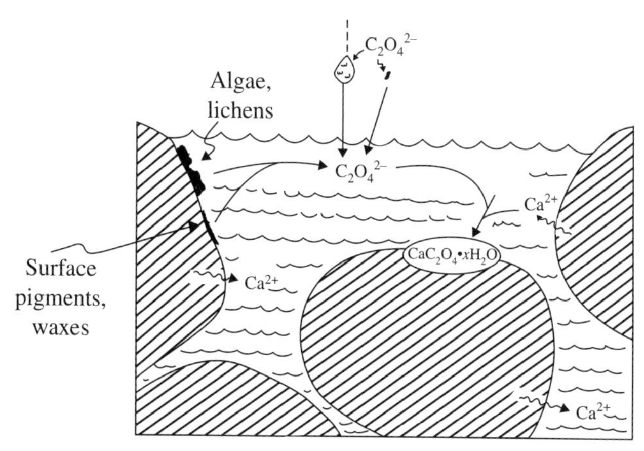

FIGURE D.4 A schematic representation of the processes potentially involved in the organic chemistry of the atmospheric corrosion of carbonate stone (see Fig. D.3 legend for explanatory details).

Notwithstanding the foregoing, the conceptual corrosion picture is rather simple. In the presence of significant amounts of atmospheric SO_2 and either high humidity or some other water source, the dissolved SO_2 is oxidized to SO_4^{2-} in solution, and the calcite is then transformed into gypsum. The latter, being quite soluble, can be lost to runoff. Alternatively, the stone may undergo spallation as the weakened crystalline structure is subjected to cycles of wetting and drying and thawing and freezing.

Should atmospheric sulfur be minimal or absent, microorganisms take up residence on and in the stone and generate oxalate ions. The resulting calcium oxalates then form a weakly protective surface layer.

Attempts to mitigate the corrosion of carbonate stone take two forms: applying protective agents to repel water and reactive compounds and impregnating the pores of the stone with inert materials. Some successes have resulted from these approaches, but a number of problems remain. For the preservation of existing structures, continued research on protective materials and their use is needed. For new structures, carbonate stone should be chosen only with full knowledge of its physical and chemical properties, its susceptibility to corrosion, and the necessity of protective treatments initially and throughout its lifetime.

Details concerning the information in this appendix may be found in Chapter 14, further reading, degradation of stone and other materials.

APPENDIX E

THE ATMOSPHERIC CORROSION CHEMISTRY OF COPPER

E.1 INTRODUCTION

When exposed to the atmosphere, copper (and alloys of copper) eventually forms a thin layer of corrosion, generally brownish green or greenish blue. This layer is designated the patina, a name derived from the green crust often found on ancient Roman dishes, or patens. Copper patinas are generally regarded as aesthetically pleasing, and much of the use of copper in architecture and sculpture is based on patina being formed. Once established, patina tends to be extremely stable and to become a permanent part of the building or object to which it is attached. Any significant change in these patinas, which can occur under some conditions, is thus generally detrimental. In contrast, copper and its alloys are used in a number of applications where maintenance of the original surface characteristics is important: in electrical contacts and connectors, for example. In these applications the formation of any patina is undesirable, and the conditions under which patina formation occurs, as well as the formation rate, assume major importance.

Copper patinas are chemically and metallurgically complex structures. Their major chemical constituents are well known and are related to trace species found in the atmosphere. The patina components do not reflect the atmospheric composition directly, however, but clearly favor those atmospheric constituents whose products possess an optimum set of solubilities, crystal structures, chemical reactivities, and formation rates.

Atmospheric Corrosion, Second Edition. Christofer Leygraf, Inger Odnevall Wallinder, Johan Tidblad and Thomas Graedel.

E.2 CORROSION LAYER FORMATION RATES

The rate of copper corrosion is generally less than that for zinc or many steels but somewhat larger than that for aluminum. The following approximate ranges of corrosion rates in different environments have been determined:

Rural	$\approx 0.5\,\mu m\, year^{-1}$
Urban	$\approx 1\text{–}2\,\mu m\, year^{-1}$
Industrial	$<2.5\,\mu m\, year^{-1}$
Marine	$\approx 1\,\mu m\, year^{-1}$

Three (or perhaps four) stages in the evolution of the visual appearance of exposed copper from its initial state to that of final patination can be identified. New freshly cleaned copper, prior to exposure to an outdoor atmosphere, is a salmon-pink color (that of a new penny). After a few weeks of exposure, copper turns a dull brown shade (that of an older penny). This shade gradually deepens to become black or nearly so. Finally, the greenish-blue layer of terminal copper patina is formed.

The time scales for patina formation vary substantially with geographic location, and there is some suggestion that the rate of patination has increased over the past century. In the 1920s, in the Western Hemisphere, the time required to develop the green patina was at least 10 years, with two or three decades being more common. Toward the latter half of the twentieth century, patina formation within a few years became the norm in many locations. The change is exemplified by Copenhagen where the time scale for green patina formation dropped from 20 to 30 years in the 1930s to 8 years in the 1960s.

E.3 THE MORPHOLOGY OF NATURAL PATINAS ON COPPER

Natural copper patinas do not have a smooth, uniform appearance on copper, but are quite spatially heterogeneous. The surfaces show a highly porous structure common to patinas found in the atmosphere. This porosity can be easily demonstrated by the ability of copper patinas to absorb significant amounts of water.

When large surfaces of copper are exposed in configurations possessing a variety of inclinations or solar exposures, patination does not proceed uniformly. Instead, horizontal or inclined surfaces patinate more rapidly than vertical surfaces. Laboratory studies indicate that solar photons can accelerate copper corrosion in controlled environments, which might suggest that the inclination differences result from differing solar exposures. In this connection, it is known that horizontal surfaces have higher maximum temperatures than vertical surfaces; they thus have larger temperature variations and would be expected to offer more opportunities for the precipitation of corrosion products. However, careful examinations of roofing copper installed in identical fashion and inclination but at differing degrees of solar insolation show no indication that photons play a controlling role in the patination process. Thus, other factors, such as the ability of inclined surfaces to retain dew, appear capable of overriding any solar effects that may be present.

The topography of materials exposed to the atmosphere can sometimes have major influence on their degradation. In the case of copper, it is widely recognized as a disadvantage that the runoff of patina from the most rapidly patinating surfaces causes discoloration of masonry below, and it is clear that rundown from inclined copper surfaces helps to patinate adjacent vertical panels. The form of the copper in this runoff material is thought to be copper sulfate that formed prior to the development of brochantite.

The influence of the metallic structure of the underlying copper can be substantial. Laboratory studies show that untextured polycrystalline material initially sulfidizes much more rapidly than single crystal copper, probably because the grain boundaries accelerate nucleation. On exposed statuary, patina forms more rapidly on wrought materials than on cast materials.

Under normal conditions, the adherence of the green patina layer to the underlying copper is very high. It is clear, for example, that the patina is not disrupted by several successive 90° bends. The patina is also stable to rubbing with abrasive pads, either under dry conditions or in the presence of water or alcohol. However, the patina exfoliates readily if wetted with acetone and rubbed with abrasive pads. This finding suggests that the patina minerals may utilize acetone-soluble organic matter as a binder during cementation, a process that bears some similarities to the diagenesis (i.e., the formation and evolution) of sedimentary rocks incorporating both inorganic and organic components.

E.4 CHEMICAL MECHANISMS OF COPPER CORROSION

E.4.1 Introduction

Ions or molecules that have been identified or suggested as significant patina components were used to survey the mineralogical and chemical literature for potential mineral constituents. The resulting list of candidate species is given in Table E.1. The table contains 18 entries, some of which have no mineralogical names and are therefore not known to be produced naturally by geochemical processes. Included in the table are index numbers, information on crystal habit, formulas, and solubilities.

In considering the development of the natural patinated layer on copper, it is clear that brochantite [$Cu_4SO_4(OH)_6$] is the key substance that needs to be characterized, since it is almost universally found to be the major component of aged copper patinas and since it represents the terminal stage of the patination process. At a minimum, there are four ingredients necessary to the production of brochantite:

1. A supply of copper ions. The process by which the ions are supplied and the ion flux are established primarily by the metallurgical properties (crystal structure, defect density, etc.) of the underlying bulk copper.
2. An aqueous layer at the surface of the copper or of its corrosion products. This layer may be produced by adsorption of water vapor under high humidity conditions, or it may result from active precipitation.

TABLE E.1 Minerals and Other Crystalline Substances Found on Corroded Copper[a]

Substance	Hey Index No.	Crystal System	Formula	Solubility[+]
Metals, oxides, and hydroxides				
Copper	1.1	Cubic	Cu	iw,sa
Cuprite	7.3.1	Cubic	Cu_2O	iw,sa
Tenorite	7.3.2	Monoclinic	CuO	iw,sa
Spertiniite	7.3.4	Orthorhombic	$Cu(OH)_2$	iw,sa
Sulfides				
Chalcocite	3.1.1	Monoclinic	Cu_2S	iw,sa
Sulfates				
Chalcocyanite (hydrocyanite)	25.2.1	Orthorhombic	$CuSO_4$	sw,a
Chalcanthite	25.2.3	Trigonal	$CuSO_4{\cdot}5H_2O$	sw,a
Antlerite	25.2.6	Orthorhombic	$Cu_3SO_4(OH)_4$	iw,sa
Brochantite	25.2.7	Monoclinic	$Cu_4SO_4(OH)_6$	iw,sa
Langite	25.2.9	Orthorhombic	$Cu_4SO_4(OH)_6{\cdot}2H_2O$	iw,sa
Posnjakite	25.2.8	Monoclinic	$Cu_4SO_4(OH)_6{\cdot}H_2O$	iw,sa
Strandbergite*			$Cu_{2.5}SO_4(OH)_3{\cdot}H_2O$	
Chlorides				
Nantokite	8.2.1	Cubic	$CuCl$	iw,sa
Atacamite	8.2.4	Orthorhombic	$Cu_2Cl(OH)_3$	iw,sa
Paratacamite	8.2.5	Trigonal	$Cu_2Cl(OH)_3$	iw,sa
Botallackite	8.2.6	Monoclinic	$Cu_2Cl(OH)_3$	iw,sa
Carbonates				
Malachite	11.2.1	Monoclinic	$Cu_2(CO_3)(OH)_2$	iw,sa
Azurite	11.2.2	Monoclinic	$Cu_3(CO_3)_2(OH)_2$	iw,sa
Nitrates and nitrites				
Gerhardite	13.3	Orthorhombic	$Cu_2(NO_3)(OH)_3$	iw,sa
Organics				
Copper(II)formate		Orthorhombic	$Cu(HCO_2)_2$	sw,a
Copper(II)acetate			$Cu(CH_3CO_2)_2$	sw,a
Copper(II)oxalate			$Cu(C_2O_4){\cdot}xH_2O$	iw,sa

[a] Included under some circumstances may be basic (b) or hydrated (h) compounds.
[+] Solubilities in room temperature water (w) and strong acids (a) are given as insoluble or sparingly soluble (i) or soluble with or without decomposition.
* Not an official crystallographic name.

3. A source of sulfur. This source may be an atmospheric gas, atmospheric particles, or trace ions in precipitation.
4. An oxidizer. The oxidizer may be an atmospheric gas or it may be a component of precipitation.

Although an explanation of how brochantite is formed would satisfy a great deal of the uncertainty surrounding the natural development of copper patinas, the explanation would be incomplete without reference to the many other components commonly found there. At a minimum, any explanation of the presence and formation of these additional components must include the following:

1. A source of organic acids, either as gases in the atmosphere or as constituents of precipitation. The source might equally well involve precursors to the acids, provided those precursors could be readily transformed by surface chemical processes. Aliphatic aldehydes, known to be common in both gas and liquid phases in the atmosphere, seem the most likely precursors.
2. A source of chloride ions. This source could be gas-phase compounds such as HCl, but the demonstrated dominance of chloride corrosion in marine locations argues for a significant contribution from a particle source such as sea-salt aerosol.
3. A source of nitrate ions. This source could be either atmospheric gas-phase compounds or nitrate ions in precipitation.

E.4.2 Adsorbed Water

Copper exposed to the atmosphere, particularly outdoors, invariably has water atop its surface oxide layer. The most obvious example of this situation is during precipitation events, but several monolayers of water are adsorbed on the surface at moderate to high humidities even in the absence of active precipitation. Table E.2 gives rough estimates of some typical water thicknesses. Since, as soon as more than about three monolayers of water are present, the water layer possesses the properties of bulk water, from the standpoint of surface chemistry, moderate relative humidities result in enough surface water so that any reactions occur either in a thin liquid phase or at a liquid–solid interface. Even less water (such as might be present at rather low humidities) may be sufficient for liquid-phase chemical processes, since small amounts of water tend not to be evenly distributed but to coalesce to clusters, particularly if hygroscopic particles have been previously deposited on the metal surface.

TABLE E.2 Estimate of Water on Copper Surfaces under Different Conditions

Condition	Deposited Weight ($g\,m^{-2}$)	Water Thickness (nm)	Water Monolayers
Clean copper, 60% RH, 20°C	0.005	5	15
Clean copper, 90% RH, 20°C	0.001	10	27
Copper with deposited particles, RH_{crit}	0.01	10	[a]
Copper at 100% RH	1.0	10^3	$3 \cdot 10^3$
Copper covered by dew	10	10^4	$3 \cdot 10^4$
Copper wet from rain	100	10^5	$3 \cdot 10^5$

[a] Dependent on the amount and type of deposited particles.

E.4.3 Oxides

Cuprite [copper(I)oxide or Cu_2O] is the initial surface component formed on copper upon atmospheric exposure. It is readily detected by X-ray diffraction on field samples of copper, particularly for exposures of a few months or more. Cuprite is a high-symmetry cubic crystal with each metal atom having two close oxygen neighbors and each oxygen atom surrounded by a tetrahedron of copper atoms. Cuprite is insoluble in water and slightly soluble in acid.

E.4.4 Sulfur Chemistry

When copper is exposed to reduced sulfur gases in humid air in the laboratory, copper(I)sulfide, Cu_2S, is the principal constituent of the tarnish film. The formation of Cu_2S requires, of course, that copper be present in the Cu^+ state. Since Cu^+ is rapidly oxidized to Cu^{2+} in environments exposed to the atmosphere, the formation of copper(I) compounds is unlikely except under unusual conditions. A further constraint is the low anticipated concentrations of the most abundant reduced sulfur compounds—hydrogen sulfide, H_2S, and carbonyl sulfide, COS. In support of these considerations, sulfides are only rarely identified in copper patinas.

Corrosion of copper can occur by the action of oxidized as well as reduced sulfur, with atmospheric sulfur dioxide as the active agent. In this connection, the stability of common copper minerals that contain oxidized sulfur can be examined with the use of a diagram showing stability domains as functions of solution acidity and anion concentration. Such a diagram for the $Cu{-}SO_3{-}H_2O$ system is presented in Figure E.1. Here it is seen that the region in which brochantite, $Cu_4SO_4(OH)_6$, is stable is centered within the rain regime. Furthermore, as precipitation on a metal surface begins to evaporate, the ionic concentration will increase, the pH will decrease, and the system will move upward into a region of much wider brochantite stability and toward antlerite stability. In the case of fog, antlerite, $Cu_3SO_4(OH)_4$, is clearly the preferred crystalline form.

Brochantite is a mixed sulfate-hydroxide salt of copper, $Cu_4SO_4(OH)_6$. It is nearly always the most common component of the green patina that copper forms upon extended atmospheric exposure and is readily detected by X-ray diffraction, thus clearly establishing its crystalline nature. Antlerite is also a mixed sulfate-hydroxide salt of copper, $Cu_3SO_4(OH)_4$. It is not uncommon as a patina constituent, being reported in several studies on samples a few decades old. There is some evidence that antlerite occurs only in more highly polluted environments. A third sulfate mineral, posnjakite, $Cu_4SO_4(OH)_6{\cdot}H_2O$, is also found in copper patinas. Posnjakite, antlerite, and brochantite are structurally related in that they consist of corrugated sheets (see Section 9.2). Posnjakite may coexist with or undergo transformation to brochantite because of its structural similarity.

It should be noted that there are two limitations on the use of diagrams such as Figure E.1. The first is that the diagram represents chemical equilibrium conditions, which may not be met in the exposure of materials to the atmosphere. The second is that the diagram applies in a strict sense only to pure systems. In reality, a variety of

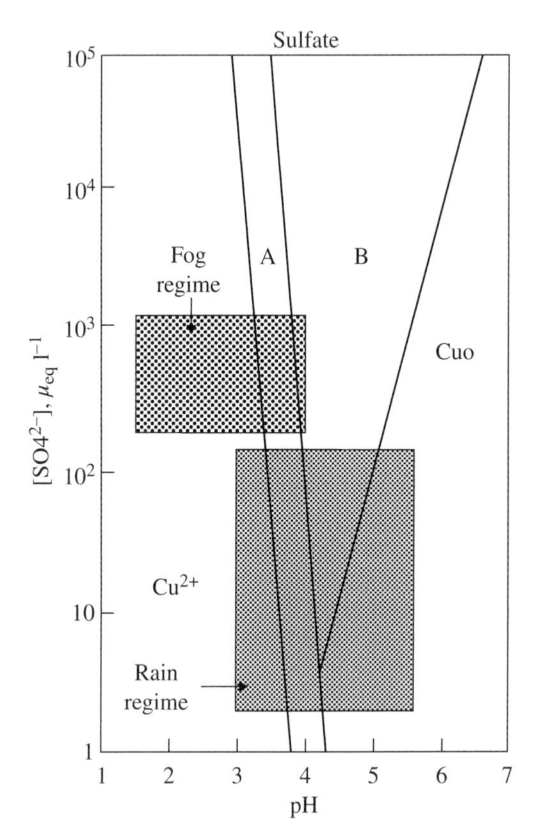

FIGURE E.1 Stability diagram for the system $Cu—SO_3—H_2O$. Typical ranges of SO_4^{2-} and pH in rain and fog are indicated. A and B indicate the regions in which antlerite, $Cu_3SO_4(OH)_4$, and brochantite, $Cu_4SO_4(OH)_6$, are stable.

additional ions and molecules are present, often at fairly high concentrations. The potential clearly exists for metastable compounds, complex species, and synergistic effects. As a result, the stability domain diagrams should be regarded as only crude guides to the chemical products expected as a consequence of patina formation.

If the previous information is combined with the perspective of atmospheric chemistry, one can construct the schematic diagram shown in Figure E.2. The diagram is the first of several attempting to link atmospheric and surface species with surface corrosion processes. In them, the species and processes known to occur in these systems are indicated, together with species and processes thought likely to be involved but at present unconfirmed. These diagrams have three regions: an atmospheric phase at the top, comprised of gases, atmospheric particles, and precipitation elements; a liquid-phase region in the center, which receives chemical species both from the atmosphere above and the region below; and a solid-phase region, which consists of the corroding copper and any patina layer already formed. Three gases appear in the gas-phase region: sulfur dioxide (SO_2), carbonyl sulfide (COS), and

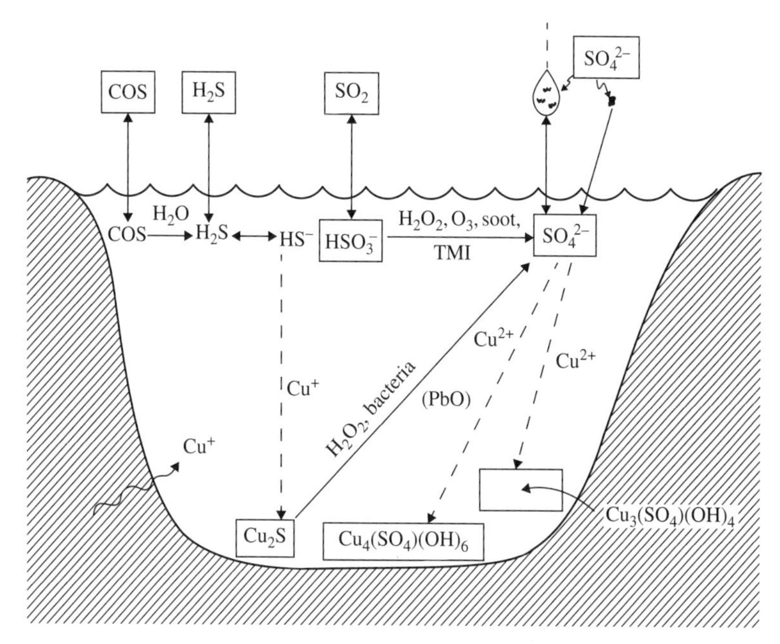

FIGURE E.2 A schematic representation of the processes involved in the formation of the sulfate components of copper patinas. The atmosphere is at the top of the diagram, a surface water layer in the center, and the corroding metal at the bottom. The patina formation is spatially homogeneous, and for simplicity the diagram pictures a depression in the metal surface, although deposited hygroscopic particles, heavy dew, or rain could produce similar conditions anywhere on the surface. The wavy line indicates copper ions. In this and several following figures, species whose presence in the indicated phase has been confirmed by field measurement are enclosed in a rectangular box; the presence of species not so enclosed seems reasonable but is unconfirmed. Solvation or reaction processes that have been confirmed by laboratory studies are shown as solid arrows; those for which the mechanism is uncertain are shown as dotted arrows. Rain is indicated by the atmospheric water drop and airborne particles by the irregular small atmospheric object. TMI refers to transition metal ions.

hydrogen sulfide (H_2S). All are implicated in corrosion processes and all are present in the gas phase in urban atmospheres. Sulfur enters the liquid phase as well by virtue of the presence of sulfate ions on particles and in precipitation. In each case, it is one of the major species present.

Within the liquid phase, sulfur progresses monotonically toward an increased ionization state. COS hydrolyzes to H_2S, and the resulting sulfide may form chalcocite if the reduced sulfur species is sufficiently abundant. In the right portion of the diagram, dissolved SO_2 hydrolyzes to bisulfite ion, which is readily converted to sulfate ion by ozone, hydrogen peroxide, or transition metal ions, all of which are expected in aqueous surface films on copper. Once sulfate is formed or incorporated into the aqueous layer, it is available to form brochantite (or perhaps, if the solution is sufficiently acidic, antlerite) by catalytic processes perhaps involving PbO. In most cases, posnjakite, $Cu_4SO_4(OH)_6 \cdot H_2O$, acts as an unstable intermediate in the brochantite-forming process.

E.4.5 Chlorine Chemistry

The stability diagram for the chloride–copper–water system appears in Figure E.3. The chloride mineral atacamite, $Cu_2Cl(OH)_3$, is stable only at chloride concentrations near the top of the range of concentrations found in rain. The situation would be enhanced by evaporation, since higher concentrations of chloride ion would move the system into the stable atacamite region. Atacamite is soluble in weak acid, so is not commonly found in highly polluted regions. In patina formed near the sea, however, atacamite is similar to or greater than brochantite in abundance.

In Figure E.4 a schematic diagram for the chloride deposition and reactions is shown. This evolution scheme includes the formation of $Cu_2Cl(OH)_3$ (atacamite or paratacamite) as end product and proceeds through CuCl (nantokite) as a precursor. Chlorine enters the liquid-phase region through the incorporation of gaseous HCl or through the deposition of sea-salt particles or precipitation. Through the reaction of cuprous ions and chloride ions, CuCl is formed and then acts as a seed crystal for the

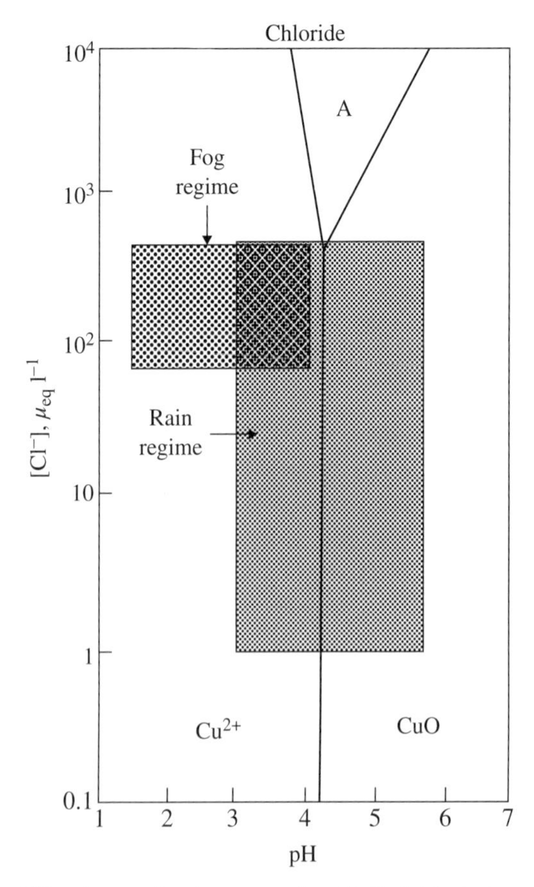

FIGURE E.3 Stability diagram for the system Cu—Cl—H_2O. Typical ranges of Cl^- and pH in rain and fog are indicated. A indicates the region in which atacamite, $Cu_2Cl(OH)_3$, is stable.

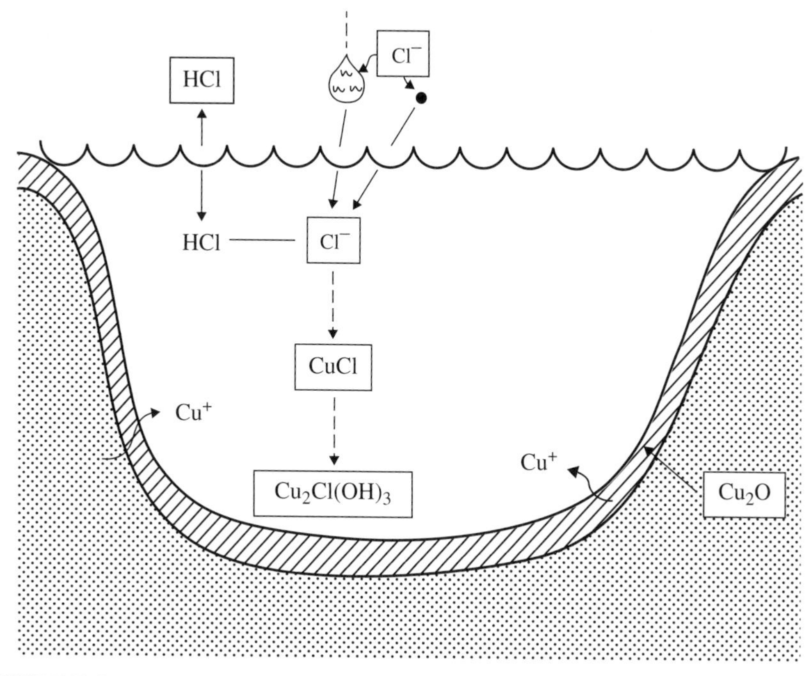

FIGURE E.4 A schematic representation of the processes involved in the formation of the chloride-rich corrosion products of copper patina (see Fig. E.2 legend for explanatory details).

formation of $Cu_2Cl(OH)_3$ via many subsequent dissolution-ion pairing-precipitation steps. Once present, straightforward reactions lead to the formation of atacamite. Because of the solubility of atacamite in acid solution, however, it is likely to be present only if it is physically isolated from the aqueous surface layer. An alternate suggestion is that the presence of relatively high concentrations of the strong acid H_2SO_4 permits the sulfate ion to displace the chloride ion from the crystalline solid. The resulting HCl can then be lost by evaporation.

E.4.6 Carbon Chemistry

The stability data for the carbonate–copper–water system shows that the concentrations of dissolved CO_2 are too low for malachite, $Cu_2(CO_3)(OH)_2$, to be stable, even at a concentration enhancement of an order of magnitude. As a result, malachite is not a favored precipitate in atmospheric systems and, in fact, has never been unambiguously observed. The organic chemistry of the patina is much more interesting, especially indoors, where copper carboxylates are found to be major constituents in the corrosion process. As seen in Figure E.5, the atmosphere is a source of both aldehydes and organic acids, in the gas phase and in precipitation. In aqueous atmospheric systems, aldehydes are readily transformed to acids by transition metal ions, OH· radicals, and other reactants. The acids are moderately to poorly soluble in water, so they will tend to precipitate promptly onto the underlying patina, the

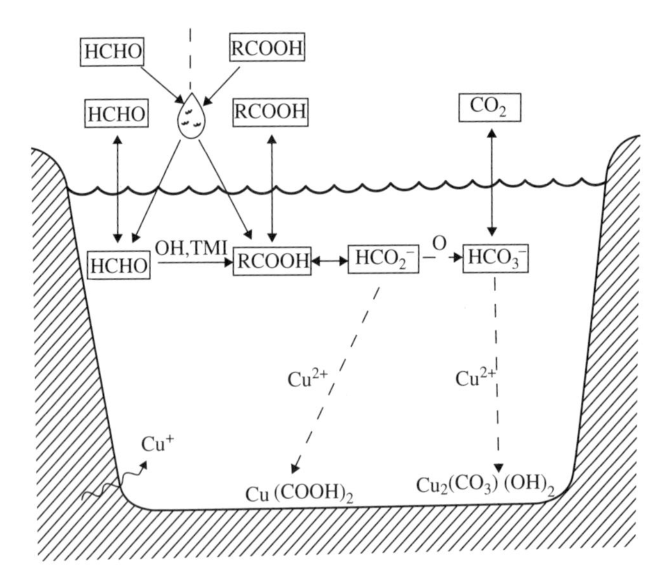

FIGURE E.5 A schematic representation of the processes involved in the formation of the carboxylic acid components of copper patinas (see Fig. E.2 legend for explanatory details).

particular compound that is formed being dependent on the precursor. Such processes are doubtlessly involved in the formation of the relatively substantial amounts of oxalate observed in many copper patina samples.

E.4.7 Nitrogen Chemistry

Stability data for the nitrate–copper–water system show that gerhardite, $Cu_2(NO_3)(OH)_3$, is stable only at very high nitrate concentrations. As a consequence, it would not be expected to be a patina component except possibly in an environment extremely close to sources of nitrate.

E.5 DISCUSSION

The information discussed in the previous sections not only reveals a substantial degree of knowledge concerning the patination of copper but also makes clear the complexity of the system. Gas, liquid, and solid phases are all present, and the diversity of species within each phase is substantial. It is not now possible to paint a complete and general picture of the patination process under the variety of conditions found in the atmosphere. One can, however, assemble the information into a form that represents the current level of understanding. As will be seen, this level is substantial.

In Figure 9.7, a schematic diagram of the patination process is outlined and discussed. That picture explains in a qualitative way important features of the patinas found on copper exposed to various atmospheres. However, a number of uncertainties remain. Perhaps the biggest of these is that the explanation is partly phenomenological. Secondly, the layered or onionskin nature of many patina films is not inherently predicted by the schematic description. It is possible, but unproven, that the layers are related to temperature or concentration variations during the vigorous phase of the initial patina formation. Thirdly, it is anticipated that solar photons must be involved in the patination process to some extent, probably as a source of reactive radicals; no provision is made herein for such involvement. Fourthly, the nature of the organic binder incorporated into the patina is not understood.

Despite the caveats previously mentioned, much useful perspective has been gained. First, the atmospheric origins of the different constituents of the copper patinas are identified with what seems to be reasonable certainty. Second, the dominance of brochantite in urban areas, the secondary nature of antlerite, and the trace existence of several other minerals are clearly seen to be consequences of the constraints imposed on the system by thermodynamics, by structural similarity, and by the atmospheric concentrations of reactive species. Third, in marine areas the high abundance of atacamite has a ready explanation. Finally, the information permits at least qualitative predictions of the environmental ruggedness of copper in different environments.

Details concerning the information in this appendix may be found in T.E. Graedel, K. Nassau, and J.P. Franey, Copper patinas formed in the atmosphere I. Introduction, *Corrosion Science*, 27, 639–657, 1987 and T.E. Graedel, Copper patinas formed in the atmosphere II. A qualitative assessment of mechanisms, *Corrosion Science, 27*, 721–740, 1987. A more comprehensive description of evolution of different corrosion products formed on copper as a result of atmospheric exposures is given in Chapters 9 and 13 together with suggestions of further reading.

APPENDIX F

THE ATMOSPHERIC CORROSION CHEMISTRY OF IRON AND LOW ALLOY STEELS

F.1 INTRODUCTION

The chemical complexities of the interaction of a surface with its environment are nowhere better illustrated that in the case of iron and its alloys. The corrosion of these metals has been studied for at least two centuries. Nonetheless, the chemical processes by which iron is degraded in use are fully understood only in simplified laboratory simulations and in the simplest and cleanest of real environments. This is particularly true for steel products put into field use within the past half century and thus subjected to environments containing a wider variety and greater concentrations of reactive trace constituents than in earlier times.

Although iron, the carbon steels (iron with $\leq 1\%$ carbon and small and varying amounts of Mn, S, P, and Si), and the weathering steels (iron with $<1\%$ Cu, Mn, Si, Ni, and Cr) differ in composition, their interactions with corrosive atmospheres demonstrate more similarities than differences. In particular, corrosion rates and surface morphologies vary, but many of the reactive atmospheric species and many of the corrosion products are the same. Such a picture is less appropriate for the stainless steels, since their passive surfaces and ability to repassivate after attack are unique within the steels. As a consequence, we concentrate on the low alloy steels, and the discussion of the applicability of our analysis to stainless steels is deferred until near the end of this appendix.

F.2 FORMATION RATES FOR RUST LAYERS

The formation of rust on iron and steels is a chemically complex process. For classification purposes, it is possible to identify three stages in the development of common corrosion layers. First is the formation of a thin oxide/hydroxide film

Atmospheric Corrosion, Second Edition. Christofer Leygraf, Inger Odnevall Wallinder, Johan Tidblad and Thomas Graedel.

(1–4 nm), which is stable and passivating in the absence of atmospheric impurities and/or high relative humidities (RH) or liquid water. Upon exposure to near-neutral aqueous environments, this oxide/hydroxide film changes into one of two types of green precipitate: "green rust I," $Fe^{II}_2Fe^{III}O_x(OH)_y$, and "green rust II," $Fe^{II}Fe^{III}O_x(OH)_y$. The green rusts are subsequently transformed into the third stage, a fragile brownish layer of iron oxides and hydroxides.

These stages consider only the incorporation of oxygen and water into the rust layer. In many rust-forming environments, however, other constituents, particularly sulfates and chlorides, are important. The time periods required for iron and steels to pass through the stages of rust development show wide variations in differing environments. In general, the first stage occurs on a timescale of milliseconds to seconds, and the second stage occurs in 2–3 h. The mineral lepidocrocite, γ-FeOOH, an indicator of the beginning of the third stage, has been observed after 2 weeks or less in the ambient atmosphere. The third stage is often followed in a few days by the formation of the mineral goethite, α-FeOOH, and magnetite, Fe_3O_4. The rate-limiting step, at least in cases in which reactive anions in solution are involved, is the diffusion of the oxidizing species in the corrosion product layer.

During stage 3, the corrosion rates for the carbon steels become quasi-linear, eventually encompassing the following ranges:

Rural	4–65 μm $year^{-1}$
Urban	23–71 μm $year^{-1}$
Industrial	26–175 μm $year^{-1}$
Marine	26–104 μm $year^{-1}$

On average, weathering steels corrode more slowly than carbon steels, typical urban rates being 4–10 μm $year^{-1}$.

F.3 THE MORPHOLOGY OF NATURAL RUST LAYERS

Rust layers on iron and carbon steel are porous and poorly adherent. Cracks in the layers are common and are concentrated in the outermost portions of the rust. Rust voids have typical diameters of a few nanometers; pore sizes as large as 15 nm in diameter have been found, with typical pore volume constituting 5–10% of the rust volume. Such a structure provides little barrier to the ingress of adsorbed water or precipitation, which penetrate to the underlying metal and perpetuate corrosion processes. Rust layers on the side of a specimen facing the ground are always looser, more porous, and rougher than those formed on the upper side. The rust films on low alloy steels are, in general, more compact and adherent than the powdery products formed on pure iron. The structure of rust on weathering steels is different from that on iron or carbon steels. It is characterized by a double-layer structure, with the inner phase providing a greater barrier to oxygen and water than does the outer phase. The outer phase is flaky and poorly adherent, while the inner phase adheres well.

Sulfate on the surface of corroded steel is reported to be localized into "nests." This sulfate may be initially present as bands at the metal–rust interface, becoming visible only after corrosion is far advanced. The "nest corrosion" may be attributable to an electrochemical process in concentrated electrolyte solutions. Possible aspects of this process are discussed in the following.

F.4 CHEMICAL MECHANISMS OF IRON AND STEEL CORROSION

F.4.1 Introduction and Overview

A knowledge of the ions and molecules present in the environment and of constituents in iron rust is a prerequisite to a survey of the mineralogical and chemical literature for possible reaction products. Some 20 different materials and compounds that have been identified as components of naturally formed rust layers are given in Table F.1. Also included in the table are index numbers (where the entries exist as cataloged minerals), crystal system, chemical formulas, and iron valence.

The rusting of iron and steel is known to require, or at least is rapidly accelerated by, water. Laboratory studies show that iron dissolution takes place immediately following brief periods of wetting, and field experiments have demonstrated good correlation between corrosion current and surface wetness during atmospheric exposure. For weathering steels, the beneficial effect of the minor alloying elements appears to be limited to samples that are wetted and dried repeatedly (see further discussion in Section 10.3).

F.4.2 Adsorbed and Absorbed Water

The amount of water present on iron and steel surfaces varies markedly with the composition and morphological properties of the surface. A number of studies give "order of magnitude" figures. At room temperature and approximately 60% RH, the equivalent of about two monolayers of water covers the surface. Much of this water is present in clusters rather than as a uniform thin layer. Over time, the water layer can thicken and, if salts are dissolved in it, assume the properties of an electrolyte solution. Similar behavior is observed for carbon steels and weathering steels.

The potential–pH diagram for the $Fe—H_2O$ system is shown in Figure F.1. The redox potential is established by the natural environment. For the atmosphere, as well as for ocean and freshwater systems, typical redox potentials are of order 0.3–0.5 V SHE, although those for highly acidic fogs are probably more positive. If the pH is less than about 4, the oxyhydroxide may be reduced to soluble Fe^{II} (Figure F.1). Such pH levels occur naturally in some rains and most fogs in urban areas. They can also occur as a consequence of the increased concentration of ionic species in the aqueous layer during its evaporation.

TABLE F.1 Minerals and Other Crystalline Substances Found on Corroded Iron[a]

Substance	Valence	Hey Index No.	Crystal System	Formula
Oxides and hydroxides				
Ferrous hydroxide	2		Hexagonal	$Fe(OH)_2$
Magnetite	2,3	7.20.2	Cubic	Fe_3O_4
Green rust I	2,3		Trigonal, hexagonal, rhombohedral	$Fe^{II}{}_2Fe^{III}O_x(OH)_y$
Green rust II	2,3		Hexagonal	$Fe^{II}Fe^{III}O_x(OH)_y$
Hematite	3	7.20.4	Trigonal	α-Fe_2O_3
Maghemite	3	7.20.3	Isometric	γ-Fe_2O_3
Ferric oxide (h)	3		Hexagonal	$Fe_2O_3 \cdot H_2O$
Ferric hydroxide	3		Cubic	$Fe(OH)_3$
Goethite	3	7.20.5	Orthorhombic	α-FeOOH
Akaganeite	3	7.20.6	Monoclinic	β-FeOOH
Lepidocrocite	3	7.20.8	Orthorhombic	γ-FeOOH
Feroxyhyte	3	7.20.7	Hexagonal	δ-FeOOH
Sulfates				
Szomolnokite	2	25.10.1	Monoclinic	$FeSO_4 \cdot H_2O$
Rozenite	2	25.10.2	Monoclinic	$FeSO_4 \cdot 4H_2O$
Melanterite	2	25.10.5	Monoclinic	$FeSO_4 \cdot 7H_2O$
Chlorides				
Lawrencite	2	8.11.10	Trigonal	$FeCl_2$
Ferrous oxychloride	2			$Fe_2Cl(OH)_3$
Ferrous ferric chloride (b)	2,3		—	$Fe_4Cl_2(OH)_7$
Ferric oxychloride	3		Orthorhombic	FeOCl
Nitrate				
Ferric nitrate (h)	3		Monoclinic	$Fe(NO_3)_3 \cdot 9H_2O$
Carbonates				
Siderite (chalybite)	2	11.13.1	Trigonal	$FeCO_3$
Ferrous ferric carbonate (b)	2,3		Hexagonal	$Fe^{II}{}_4Fe^{III}{}_2(OH)_{12}(CO_3) \cdot 3H_2O$

[a] See Table C1 footnotes for explanatory details.

F.4.3 Oxides and Hydroxides

The charge balance during oxidation of iron to Fe^{II} is maintained by the reduction of dissolved O_2 or water. The rates for these reactions are functions of the alloy and electrolyte composition, but typical times for maximum Fe^{II} generation under realistic atmospheric conditions are of order 1 h. Once Fe^{II} is present in solution, the formation and precipitation of a variety of corrosion products follow.

The generation of stable oxide and hydroxide products is preceded by formation of transitory green rust compounds identified as $[Fe^{II}{}_2Fe^{III}O_x(OH)_y]^{(7-2x-y)+}$ and

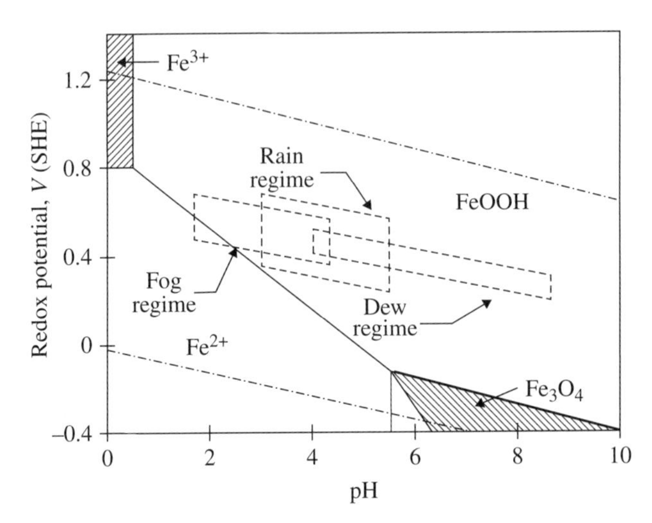

FIGURE F.1 Potential–pH diagram for the system Fe—H_2O at 25°C, for a concentration of iron ionic species of 0.1 M. The approximate regimes for fog and rain are indicated.

$[Fe^{II}Fe^{III}O_x(OH)_y]^{(5-2x-y)+}$. These compounds may include some carbonates or chlorides in environments favorable to the formation of mixed salts (see later).

About a dozen different oxides and hydroxides of iron occur in natural rust layers on iron and steels. FeOOH, in several crystal forms, is the predominant stable species except under conditions of high acidity. The simple thermodynamics of Figure F.1 provide only a starting point for a mechanistic understanding of the formation of rust layers, because the figure does not deal with the full chemical complexity of the system and because it applies only to the equilibrium state, which may not be attained under many circumstances. The rust layers generally include an inner region of dense, amorphous FeOOH and some crystalline Fe_3O_4 (magnetite) and an outer loose, crystalline assemblage of α-FeOOH (goethite), γ-FeOOH (lepidocrocite), and γ-Fe_2O_3 (maghemite). In some environments, other constituents are favored: for example, β-FeOOH (akaganeite) apparently is produced only in the presence of sufficient concentrations of chloride.

The corrosion products of weathering steels after some decades are distinctly different. The inner, more adherent layer is largely goethite, in which chromium atoms have been substituted for some of the iron atoms. The outer layer consists largely of lepidocrocite. The formation of protective corrosion products on weathering steels is more extensively discussed in Section 13.5.

In deriving a mechanistic understanding of iron corrosion, it is important to note that iron is initially oxidized to Fe^{II}, that the intermediate corrosion products include both Fe^{II} and Fe^{III}, and that the final corrosion products include only Fe^{III}. A central question is the identity of the principal oxidants and the reaction steps in the overall process. Molecular oxygen is the most likely oxidizer on a dry surface where solubility considerations are unimportant; this condition is usually not relevant for outdoor atmospheric corrosion. When surfaces are covered with a sufficiently thick layer of water, the relatively low aqueous solubility of O_2 can limit its supply to the iron surface so that the cathodic reaction (the reduction of O_2 to OH^-) becomes the rate-limiting step. An upper limit for the rate constant for O_2 oxidation of iron is perhaps $10^{-2}\,M^{-1}\,s^{-1}$, set by the O_2 diffusion coefficient in solution of $2.6 \cdot 10^{-5}\,cm\,s^{-1}$. With a moistened surface, aspects of both oxidation on a dry surface and in the presence of bulk water may be involved.

We next consider Fe^{II} oxidation by highly reactive species generated by photosensitive reactions in the atmosphere or within surface aqueous films. Such processes have been investigated for atmospheric droplets, freshwater, and seawater, in each of which hydrogen peroxide (H_2O_2) or related radical species such as the hydroxyl radical ($HO\cdot$), hydroperoxyl radical ($HO_2\cdot$), or superoxide radical ion ($O_2\cdot^-$) are involved. The three free radical oxidation reactions likely to be important are

$$Fe^{2+} + OH \rightarrow Fe^{3+} + OH^- \tag{F.1}$$

$$Fe^{2+} + HO_2\cdot + H_2O \rightarrow Fe^{3+} + H_2O_2 + OH^- \tag{F.2}$$

$$Fe^{2+} + O_2\cdot^- + 2H_2O \rightarrow Fe^{3+} + H_2O_2 + 2OH^- \tag{F.3}$$

as well as

$$Fe^{2+} + H_2O_2 \rightarrow Fe^{3+} + OH\cdot + OH^- \tag{F.4}$$

The dominant reaction depends on the initial conditions in the solution. Once the Fe^{III} ions are generated, some fraction will precipitate as oxyhydroxides. Once such corrosion products form, the decomposition rate of H_2O_2 becomes slow, as a relatively inert film appears to be formed by interaction with peroxy compounds. H_2O_2 is common in precipitation both by day and night, at least during the photochemically active summer months. It is also present in the gas phase, as are $OH\cdot$ and HO_2, and is an intermediate product in the reduction of O_2 in solution. Quantitatively, atmospheric droplet models show O_2^- and $OH\cdot$ to be the most important oxidizers of Fe^{II} in acidic raindrops during the day by the mechanisms of Equations F.1 and F.3. At night, the lack of solar photons inhibits production of the radical species, and the direct reaction (Eq. F.4) of H_2O_2 with Fe^{II} is more important.

F.4.4 Sulfides and Sulfates

After oxides and oxyhydroxides, sulfates are the most frequently reported constituents of rust layers produced by atmospheric exposure, although their relative abundance is much lower. Iron(II)sulfate tetrahydrate (rozenite) is the most abundant; heptahydrate (melanterite) and an anhydrous compound have also been identified. Sulfate concentrations are greatest on field samples exposed during the winter.

The corrosion of iron and its alloys is accelerated by the presence of sulfur dioxide, SO_2, and its progeny, particularly H_2SO_4. Little significant rusting appears to occur if sulfur dioxide is not present. An early effort to explain the involvement of atmospheric

TABLE F.2 Expected Upper Limits for Ratios of Metal Complex [*ML*] to Aquatic Metal Ions [M] in Atmospheric Droplets at Selected pH Values

		[*ML*]/[M][a]		
Metal Complex	K_1	pH 3	pH 4	pH 5
$[Fe^{III}(SO_3)]^+$	$1 \cdot 10^{10}$	1	100	10,000
$[Fe^{III}(SO_4)]^+$	$1.1 \cdot 10^4$	0.76	0.87	0.88
$[Fe^{III}(OH)]^{2+}$	$6.5 \cdot 10^{11}$	6.5	65	650
$[Fe^{III}(OH)_2]^{+}$[b]	$2.0 \cdot 10^{22}$	2.0	200	20,000
$[Fe^{III}(HO_2)]^{2+}$	$2 \cdot 10^9$	0.0002	0.002	0.02
$[Fe^{II}(SO_4)]$	158	0.011	0.013	0.013

[a] For upper limits of [*L*] expected in precipitation.
[b] $[ML^2]/[M]$.

sulfur envisioned a three-step process: (i) adsorption of SO_2 onto wetted rust; (ii) formation of $FeSO_4$ by oxidation of Fe involving SO_2 and O_2; and (iii) oxidation of $FeSO_4$ to rust, freeing SO_4^{2-}, which acidifies then corrodes more iron, giving fresh $FeSO_4$, hence more SO_4^{2-} and so on. The cycle continues until the SO_4^{2-} is removed as insoluble ferric hydroxysulfate. As will be seen, many aspects of this suggested process are consistent with the more detailed chemical picture now available from a consideration of the atmospheric and surface chemistry involved.

Aqueous solutions are acidified by the ionization of dissolved SO_2; since highly acidic solutions promote the dissolution of FeOOH (Fig. F.1), SO_2 adsorption can corrode iron without sulfur and iron species reacting directly. However, if HSO_3^- and Fe^{III} are present, S^{IV} is readily oxidized to S^{VI}, while Fe^{III} is reduced to Fe^{II} (the chemical details have not been determined). If sulfur enters the electrolyte as S^{VI} (most sulfur on particles and in precipitation is valence state S^{VI}), these redox processes would not occur.

Fe^{II} and Fe^{III} ions in aqueous solutions are coordinated with several water molecules and may also complex with any anions present. Expected abundances of complexes in atmospheric droplets are summarized in Table F.2. The most probable form of Fe^{II} is the hexaquo complex. For Fe^{III}, however, some of the water molecules are replaced by hydroxide, sulfite, and sulfate ions. Mono- and dihydroxy complexes and that of sulfite are dominant, while the complex with sulfate is relatively abundant at low pH levels. The rates of complex formation are rapid, and mixed complexes such as $Fe(OH)SO_4$ are anticipated if sulfate concentrations are relatively high. The transition of these complexes to sparingly soluble precipitates may be facile: $Fe(OH)SO_4$ precipitates in the pH range 2.3–3.7 and $Fe_2(OH)_4SO_4$ precipitates in the pH range 3.7–4.9. If Na^+ or K^+ is present in solution, as will be the case for iron or steel exposed to the atmosphere, $NaFe_3(OH)_6(SO_4)_2$ or $KFe^{III}_3(OH)_6(SO_4)^2$ (jarosite) will precipitate. Sulfate is present in solid corrosion films and is doubtless present in mixed salt compounds such as these.

Weathering steels are less sensitive to SO_2 than is carbon steel, especially for long exposure periods. The reason for this characteristic property remains in dispute.

The hydroxides formed by the alloy constituents may be important, with chromium and nickel being perhaps the most significant. Alternatively, the copper in the matrix might dissolve in the electrolyte and form the insoluble and protective mineral brochantite, $Cu_4SO_4(OH)_6$, the most common corrosion product of exposed copper or bronze. Brochantite or other copper-containing minerals have never been detected in the corrosion layers of copper-containing steel, but the amount of copper is small enough that they may have escaped detection. In any case, the alloy constituents promote the formation of amorphous iron hydroxides that are more protective than the crystalline hydroxides.

Sulfur chemistry provides a possible reason for the seasonal dependence of outdoor iron corrosion in that atmospheric SO_2 concentrations are about a factor of three higher in winter than in summer, largely a consequence of differences in meteorologically driven diffusion. This concentration difference outweighs rate changes due to seasonal temperature differences and operates in concert with variations in the concentrations of H_2O_2, since

$$HSO_3^- + H_2O_2 \rightarrow HSO_4^- + H_2O \tag{F.5}$$

The seasonal effect has been explained as follows: if upon the absorption of H_2O_2 and SO_2 into the aqueous surface layer their concentrations are such that $[H_2O_2] > [HSO_3^-]$, H_2O_2 will not be completely depleted by HSO_3^-, and some will be available to passivate the metal surface. This condition is generally satisfied during the summer. Conversely, if $[H_2O_2] < [HSO_3^-]$, as is usual in winter, H_2O_2 will react primarily with bisulfite and will be unavailable to form the protective film. The inversion of the inequality is probably more likely now than in the past, since atmospheric sulfur levels were much higher earlier in the century, at least in Europe and in the United States. In addition, the low levels of NO_x emission in the past would have produced less H_2O_2 than is the case at present.

To summarize, the sulfur chemistry thought to occur during the corrosion of iron in the atmosphere is shown in Figure F.2. The flux of iron from metal to solution is indicated by the wavy arrow. Three gas-phase species, carbonyl sulfide (COS), hydrogen sulfide (H_2S), and sulfur dioxide, are shown entering the liquid phase. In addition, SO_4^{2-} enters as a component of precipitation and of deposited atmospheric particles. The reduced sulfur species react with ferrous iron to form any of several insoluble sulfides. The small atmospheric concentrations of the reduced sulfur species suggest, however, that the rate of sulfide production in rust layers will be low, a prediction confirmed by analytical studies. More interesting chemically are the two processes that follow injection of oxidized sulfur into the system. The simpler is the formation of ferrous sulfates. The more complex is a multistep process involving Fe^{II}, Fe^{III}, and OH^- to produce hydroxysulfate mixed salts. (Iron hydroxysulfate salts have been found in laboratory experiments utilizing high concentrations of atmospheric sulfur.) If present, the mixed salts are likely to be relatively insoluble under normal atmospheric conditions. In contrast, the soluble ferrous sulfate species can be lost during solution runoff.

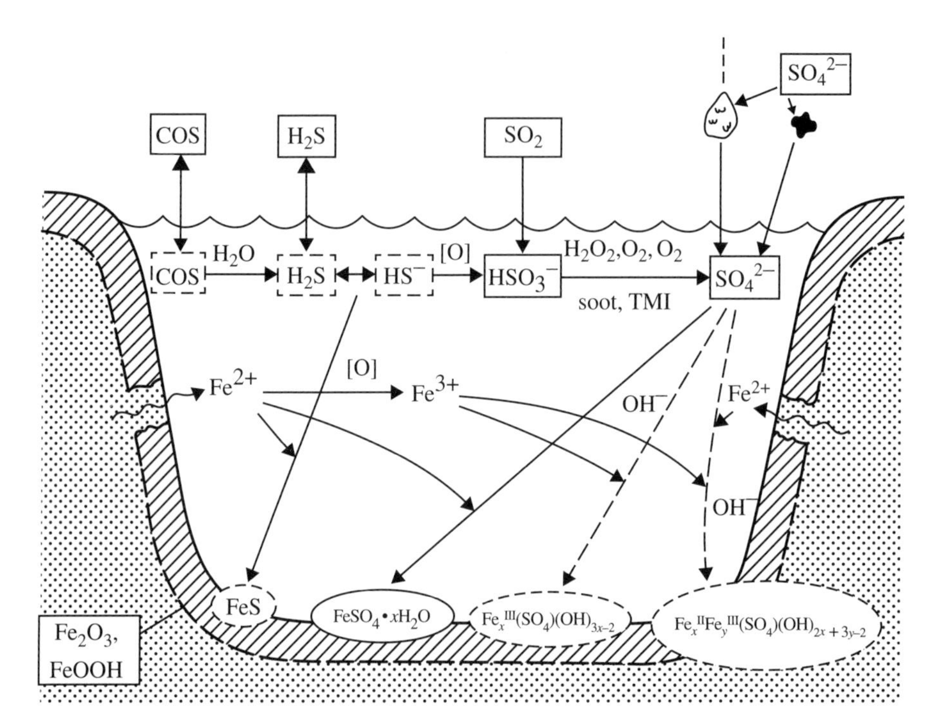

FIGURE F.2 A schematic representation of the sulfur chemistry involved in the corrosion of iron and steels. The atmosphere is at the top of this diagram, a surface water layer is in the center, and the corroding metal is at the bottom. For simplicity, the diagram pictures a depression in the metal surface, although deposited hygroscopic particles, heavy dew, or rain could produce similar conditions anywhere on the surface. The wavy arrows indicate iron going into solution as Fe^{2+}. In this and the following figure, species shown within rectangular boxes are present as solution constituents; those in ovals are present as precipitates. Dotted ovals or lines indicate constituents or reactions that are unconfirmed at present by laboratory or field studies. Solvation or reaction processes that have been confirmed by laboratory studies are shown as solid arrows; those for which the mechanism is uncertain are shown as dotted arrows. TMI indicates transition metal ions; they and soot catalyze the S^{IV} to S^{VI} oxidation. Rain is indicated by the atmospheric object. The diagram assumes that the HSO_3^- concentration exceeds that of H_2O_2 (see text).

F.4.5 Chlorides

Chloride accelerates rusting, although the influence of sulfur has received more analytical attention. Indeed, the dependence of the corrosion rate on distance from ocean beaches establishes that chloride ion is the most important factor under some circumstances. The transport of chloride ions through the rust film permits direct attack on the underlying bulk metal. While the vigorous corrosion of steels in marine locations is strong evidence for the effect of chloride, the solubility of chlorides may result in their removal from the corrosion products. Green rust I, often seen on iron and steel

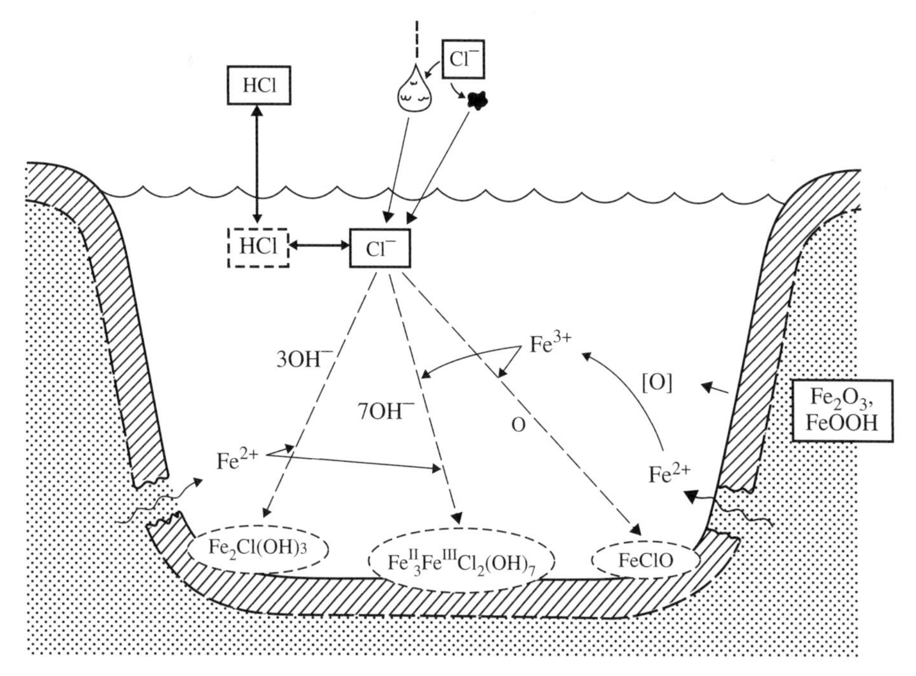

FIGURE F.3 A schematic representation of the processes involved in the formation of chlorine-containing components during the atmospheric corrosion of iron and steels. (See Figure F.2 legend for explanatory details.)

in marine environments, may not be solely a mixed valence iron oxyhydroxide but may also contain some chlorine in place of oxygen. In any case, corrosion films grown on carbon steel [but not on stainless steel] in chloride solutions contain chloride in some form or another. Chlorine may also be involved catalytically in processes such as the formation of β-FeOOH. Several basic chlorides and oxychlorides have been detected in rusts derived from exposures to bulk seawater, but only sparse evidence exists for those species upon indoor or outdoor exposure.

In solution, Fe^{III} forms complexes with Cl^- and mixed complexes with OH^- and Cl^-. The precipitated products will be hydroxychloride salts, as shown in Figure F.3. The solubility of these salts is not well established, but low solubility for the hydroxychlorides appears likely.

F.4.6 Nitrates

Laboratory experiments have clearly demonstrated a synergistic effect of SO_2 and NO_2 in the atmospheric corrosion of steel. The presence of nitrate in precipitation being established, the potential exists for nitrogen species to be involved in the corrosion of iron exposed to precipitation. The mechanism of the synergistic process is not understood, however, and the acceleration of the corrosion rate in the presence of NO_2 may be due simply to the increase in solution acidity produced by the formation of a strong acid from a gaseous precursor. Nitrate reduction in iron solutions can occur but apparently not under acidic conditions.

F.4.7 Carbonate Compounds

Although carbonates in rust films have seldom been studied, available evidence suggests that they are not uncommon. Table F.1 includes simple and mixed valence carbonate compounds that have been detected in samples exposed to indoor and outdoor air. The simple iron carbonate is soluble and readily washed from surfaces. The same is not true for the hydroxycarbonate mixed salt. More information is needed regarding its ubiquity, or lack thereof, in rust layers; it is of the pyroaurite group of minerals that is commonly found in nature. The mixed valence carbonate is closely related structurally to the oxyhydroxide and chlorine-containing green rusts, suggesting that formation of several mixed valence constituents can be controlled by local concentrations of trace species in the aqueous surface layer.

F.4.8 Processes Involving Particulate Matter

In addition to the oxidation products of iron, rust layers contain particles derived from soil dust and anthropogenic emissions, and it was demonstrated many years ago that the corrosion rate of iron is strongly inhibited by filtering particles from the air. This is explained either by the incorporation of atmospheric water vapor by the particles, the physical adsorption and concentration of reactive atmospheric species by the particles, or the direct reaction of particles containing chloride or sulfate with the iron. Crevice corrosion at the interface between an inert particle and the steel surface is another important process. It is probable that several processes occur simultaneously in most corrosion situations.

F.4.9 Photoprocesses

In addition to the oxidation of Fe^{II} to Fe^{III}, which is aided by photon-generated free radicals as described earlier, the corrosion of iron is enhanced by irradiation involving iron complexes themselves. Iron oxides absorb photons having wavelengths throughout the visible spectrum, with an absorption maximum near 400 nm. With α-Fe_2O_3, formation of an electron-hole pair may be followed by the formation of a surface-bound hydroxyl radical. Alternatively, reductive dissolution of the oxide may occur in aqueous solutions. Photoprocesses also occur in the aqueous films covering the metal surfaces. Although ferrous ions in water are generally present as hexaquo complexes, ferric ions have the mono- and dihydroxy complexes as the dominant forms. These complexes absorb visible and ultraviolet radiation over the 290–400 nm band that overlaps both the solar spectrum and the spectrum of fluorescent artificial illumination. Upon the absorption of a photon, the Fe^{III} complexes undergo charge transfer from the ligand to the metal center. The result is one electron reduction of the metal center, coupled with the oxidation of the ligand. The redox process is followed by the escape of the donor ligand radical into the bulk solution and solvation of the resulting Fe^{II} complex:

$$\left[Fe(H_2O)_5(OH)\right]^{2+} + H_2O \xrightarrow{h\nu} \left[Fe(H_2O)_6\right]^{2+} + OH\cdot \qquad (F.6)$$

$$\left[Fe(H_2O)_4(OH)_2\right]^{+} + H_2O \xrightarrow{h\nu} \left[Fe(H_2O)_5(OH)\right]^{+} + OH\cdot \qquad (F.7)$$

These reactions provide free hydroxyl radicals to enhance corrosion processes. The ensemble result is to transform the original oxide into a variety of products, including several soluble species that can be removed by runoff or precipitation.

F.5 STAINLESS STEELS IN THE ATMOSPHERE

It is now appropriate to consider how and if the stainless steels fit into the picture that has been drawn for the low alloy steels. The distinction between the stainless steels and the low alloy steels, besides composition, is that the former have a chromium-rich passive film of the order of 2 nm thickness that strongly inhibits atmospheric attack under most circumstances. If attack does occur, it is generally localized in pits that form beneath crevices at inclusions or flaws in the passive layer. Within pits the acidity and anion concentrations exceed those of the electrolyte on the metal surface, since diffusion and convection are inhibited. The attack can be accelerated by free radicals. Corrosion continues if the rate of metal dissolution exceeds the rate of repassivation but ceases if the reverse is true.

In the case of low alloy steels, the oxide does not passivate the metal but tends to be crystalline and porous, so that corrosion of the metal is aided by more efficient diffusion of the electrolyte within the rust layer.

Atmospheric corrosion rates of low alloy steels and stainless steels thus are functions of at least the following factors: composition of the electrolyte solution; quality of the passivating layer, if any; degree of mixing of the solution, which is influenced by the local humidity, wind, and temperature conditions; and pore and defect density of the corrosion products.

Thus, the tendency of stainless steels toward passivity renders them much less likely to attack. Only under extreme conditions, as in highly acid fogs or marine locations, will some chemical degradation processes of stainless steel begin to resemble those of the low alloy steels.

F.6 SUMMARY

The atmospheric corrosion of iron and low alloy steels is a multistep process. During the first few hours or days of exposure, a moderately protective surface layer of mixed oxides and hydroxides forms on the metal. If atmospheric hydrogen peroxide is available, the passivation properties of the layer are enhanced. Next, adsorption of water and deposition of corrosive gases and chemically complex particulate matter produces

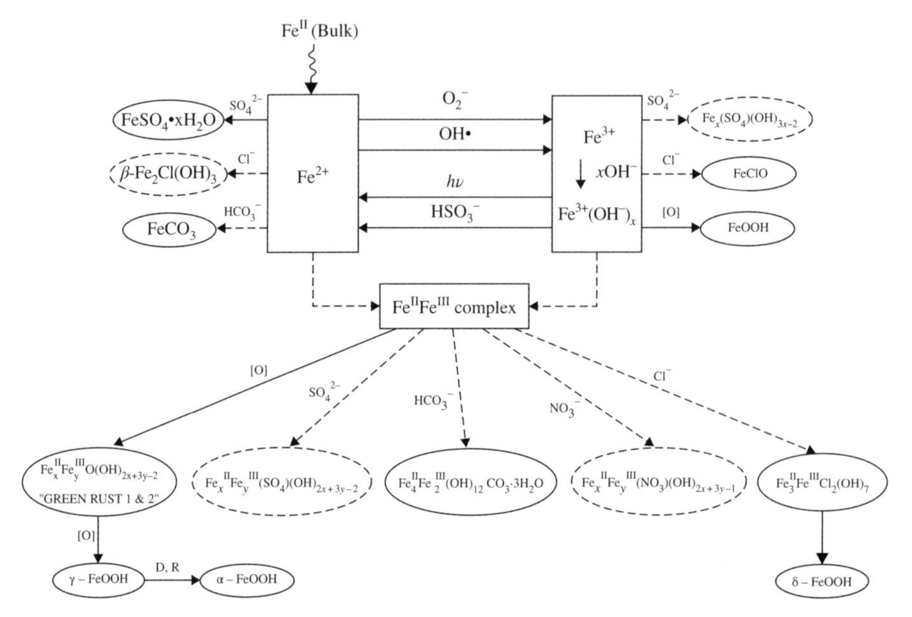

FIGURE F.4 A schematic diagram of the overall formation processes of intermediates and final products in the atmospheric corrosion of iron and low alloy steels. Because water layers and droplets exposed to the atmosphere are acidic, the diagram applies within the approximate pH range 3–6. Species shown within rectangular boxes are present as solution constituents; those in ovals are present as precipitates. Dotted ovals or lines indicate constituents or reactions that are unconfirmed at present by laboratory studies. D, R indicates dissolution and reprecipitation. This diagram is discussed in the text.

an electrolytic solution on the metal. Fe^{II} ions then enter the solution, either from oxidation of the bulk material or by reduction of Fe^{III}. The chemistry that follows is depicted schematically in Figure F.4, in which specific oxidation processes and effects of trace anions in solution are considered. The central transformations are the oxidation and reduction processes that link Fe^{II} and Fe^{III}. The available anions in the surface film compete for the iron cations, producing compounds that are either soluble or less protective than the oxyhydroxides that would form in their absence. The ability of iron to be readily transformed between the +2 and +3 valence states results in a tendency for mixed $Fe^{II}Fe^{III}$ complex formation. Hydroxysulfate and hydroxychloride complexes are the most common. Some portion of these products is retained on the surface, but a substantial fraction is lost to runoff. Indoors, the initial formation of oxyhydroxides follows the scenario outlined previously. If the humidity is greater than about 50%, multilayers of water vapor will be adsorbed, particularly if particles are present, and a less vigorous analog of the outdoor processes will occur. Indoor photon irradiation may be sufficient to liberate Fe(II) ions, although it is unlikely that the aqueous surface layer indoors will become highly acidic. Under some circumstances, sulfate or chloride aerosols in the indoor air may interact directly with the surface to produce rust components, which will not be affected by runoff.

Citations and additional information concerning this appendix can be found in T.E. Graedel and R.P. Frankenthal, Corrosion mechanisms for iron and low alloy steels exposed to the atmosphere, *Journal of the Electrochemical Society, 137*, 2385–2394, 1990. A more comprehensive description of evolution of different corrosion products formed on iron, low alloy steel, weathering steel, and stainless steel as a result of atmospheric exposures is given in Chapters 9 and 13 together with suggestions of further reading.

APPENDIX G

THE ATMOSPHERIC CORROSION CHEMISTRY OF LEAD

G.1 INTRODUCTION

Lead is one of the common engineering metals. Traditionally used in roofing, piping, and sheet lining, it has more recently been an essential component in storage batteries, radiation shields, and solders, though concerns about its toxicity are limiting many of the applications. Another property that occasionally limits lead's use is that it undergoes atmospheric corrosion fairly readily under some situations. Lead has been found to be resistant to sulfuric acid because of the formation of stable sulfate corrosion products but to be susceptible to nitric, hydrochloric, and especially carboxylic acids. Insight into the corrosion processes involved requires an understanding of the relationships between the chemical properties of the altered metal surface and the atmospheric constituents responsible for those changes.

G.2 ENVIRONMENTAL INTERACTIONS WITH LEAD SURFACES

The native oxide on lead is predominantly litharge, PbO, and is 3–6 nm thick. No studies of water adsorption and absorption on lead have apparently been performed. However, data on a variety of other metals make it likely that the oxidized surface of bulk lead is covered by at least 5 monolayers of water at room temperature and 70% relative humidity and by at least 10 monolayers of water at room temperature and 100% relative humidity. Lead corrosion is relatively humidity dependent, the crucial role of the water layer on the metal surface being to provide a medium for the deposition of atmospheric gases and the subsequent dissolution of the metal.

Since lead and its alloys are often used indoors, typical concentrations of chemical species present in indoor environments can be used to estimate the potential

---

Atmospheric Corrosion, Second Edition. Christofer Leygraf, Inger Odnevall Wallinder, Johan Tidblad and Thomas Graedel.

deposition to and concentrations of corrosive agents in aqueous layers on those surfaces. In the case of indoor exposures, the relative importance of the gas and particle fluxes to generic metal surfaces for particular chemical species has been determined under fixed sets of conditions; the general conclusions will apply also to lead surfaces. For chloride, the particle deposition rate is within an order of magnitude of the gas deposition rate, so both fluxes may need to be considered. In the case of sulfate, the sulfur dioxide, SO_2, deposition rate can be some three orders of magnitude higher than the particle sulfate deposition rate, so gaseous deposition will generally dominate. The adsorption of sulfur gases onto lead surfaces is consistent with facile transfer. Similarly, in the case of nitrate, the gaseous nitric acid flux dominates the nitrate flux from particles. For sulfide, the particle content is negligible, so the only source of significance is the gas phase. In the case of carbonate, ammonium, and the organics, the situation differs markedly from location to location, and no general assessment appears attainable.

For outdoor situations, sulfur and chlorine fluxes to the surface are generally comparable and are primarily from gases rather than from particles. The fluxes of nitrate and organic carbon are much lower, but those of the oxidizing species ozone (O_3) and hydrogen peroxide (H_2O_2) are higher than those of any of the anions. The implication is that the outdoor corrosion reactions are not oxidant limited.

G.3 PHYSICAL CHARACTERISTICS OF LEAD CORROSION

The limited corrosion rate data for lead upon atmospheric exposure in different environments are as follows:

Marine	0.1–2.2 μm year^{-1}
Rural	0.4–1.9 μm year^{-1}
Urban	0.5–0.7 μm year^{-1}

They are larger than those for copper and aluminum but smaller than those for zinc and steel. Lead corrosion in the presence of water occurs by successive processes of dissolution and crystallization.

G.4 CHEMICAL MECHANISMS OF LEAD CORROSION

G.4.1 Introduction

Ions or molecules that have been detected or suggested as potential participants in the corrosion chemistry of lead were used to direct a survey of the mineralogical and chemical literature for mineral constituents. The resulting list of species is given in Table G.1; it contains 13 entries, some of which have no mineralogical names and are therefore not known to be produced naturally by geochemical processes.

The oxide formed when lead is first exposed is apparently litharge, PbO. The most common minerals detected are anglesite, $PbSO_4$, and either cerussite, $PbCO_3$, or

TABLE G.1 Minerals and Other Crystalline Substances Found on Corroded Lead[a]

Substance	Hey Index No.	Crystal System	Formula	K_{sp}
Metal, oxides, and hydroxides				
Lead	1.21	Cubic	Pb	
Litharge	7.11.5	Tetragonal	PbO	$1.2 \cdot 10^{-15}$
Sulfides and sulfite				
Galena	3.6.3	Hexoctahedral	PbS	$1 \cdot 10^{-28}$
Scotlandite		Monoclinic	$PbSO_3$	
Sulfates				
Anglesite	25.7.1	Orthorhombic	$PbSO_4$	$1.6 \cdot 10^{-28}$
Basic lead sulfate	—	—	$Pb_2SO_4(OH)_2$	
Chlorides				
Cotunnite	8.8.2	Orthorhombic	$PbCl_2$	$1.6 \cdot 10^{-5}$
Laurionite	8.8.3	Orthorhombic	PbClOH	$2 \cdot 10^{-14}$
Carbonates				
Cerussite	11.9.1	Orthorhombic	$PbCO_3$	$3.3 \cdot 10^{-14}$
Hydrocerussite	11.9.2	Tetragonal	$Pb_3(CO_3)_2(OH)_2$	
Organics				
Lead formate	—	—	$Pb(HCO_2)_2$	
Lead acetate	—	—	$Pb(CH_3CO_2)_2$	
Lead oxalate	—	—	PbC_2O_4	$2.7 \cdot 10^{-11}$

[a] See Table C1 footnotes for explanatory details.

hydrocerussite, $Pb_3(CO_3)_2(OH)_2$. Litharge is also seen on samples exposed to either the atmosphere or seawater, and seawater exposures give, in addition, hydrocerussite, cotunnite ($PbCl_2$), laurionite (PbClOH), and a lead chloride carbonate, perhaps $Pb_2Cl_2(CO_3)$. Indoor exposures produce lead formate where formic acid, HCOOH, or formaldehyde, HCHO, sources are present.

G.4.2 Oxides

The potential–pH diagram for the system lead–water is shown in Figure G.1. Superimposed on the diagram are the approximate loci of typical ambient environments. Two features are immediately apparent. The first is that there is only a very small region of stability for the surface oxide. In fact, in the normally acidic environments in which atmospheric corrosion occurs, the surface oxide is prone to dissolution, and lead will continue to corrode unless overlain by a more stable corrosion deposit. The second is that virtually no stable region exists at any pH for lead in the +4 valence state except at high potentials and low acidities atypical of atmospheric or surface environments; one thus anticipates that $Pb^{2+} \leftrightarrow Pb^{4+}$ oxidation–reduction processes will be of negligible importance in the atmospheric corrosion of lead.

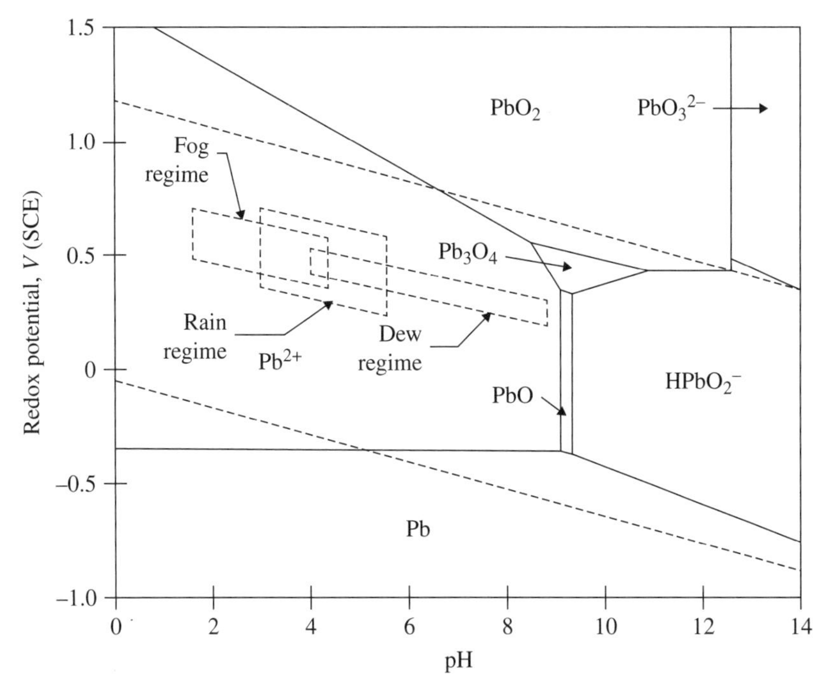

FIGURE G.1 Potential–pH equilibrium diagram for the system Pb—H_2O at 25°C, computed for a dissolved lead concentration of 1 μM. The dashed lines indicate the limiting conditions within which water is stable. (Adapted with permission from M. Pourbaix, C Si Ge Sn Pb, in *An Atlas of Electrochemical Equilibrium in Aqueous Solutions*, 2nd English edition, trans. J.A. Franklin, National Association of Corrosion Engineers, Houston, pp. 485–492, 1974.)

G.4.3 Sulfides and Sulfates

Figure G.2 is a potential–pH diagram for the system Pb—S—H_2O—CO_3^{2-}. In normal atmospheric regimes, $PbSO_4$ is the expected compound. It forms a protective passive layer, a characteristic that has been one of the more useful aspects of lead systems, since it permits the use of lead in the presence of sulfuric acid. Carbonate and hydrocarbonate salts are anticipated if the surface solution is basic. In highly reducing environments, as might occur in sediments, for example, the mineral galena, PbS, is expected. How might these stable minerals form, especially in a kinetically limited situation? A qualitative scenario is shown in Figure G.3. It is initiated by dissolution of the surface oxide as soon as depositing acidic gases lower the surface solution pH to about 6. (Note that carbon dioxide, CO_2, in the background atmosphere is sufficient to produce at equilibrium a pH of about 5.6, and other atmospheric acidic gases are stronger, so dissolution will be common.) Figure G.3 shows three gas-phase species, carbonyl sulfide, COS, hydrogen sulfide, H_2S, and sulfur dioxide, SO_2, entering the liquid phase by gaseous transport. In addition, SO_4^{2-} enters as a component of precipitation and of deposited atmospheric particles. The reduced sulfur species can potentially react with lead ions to form an

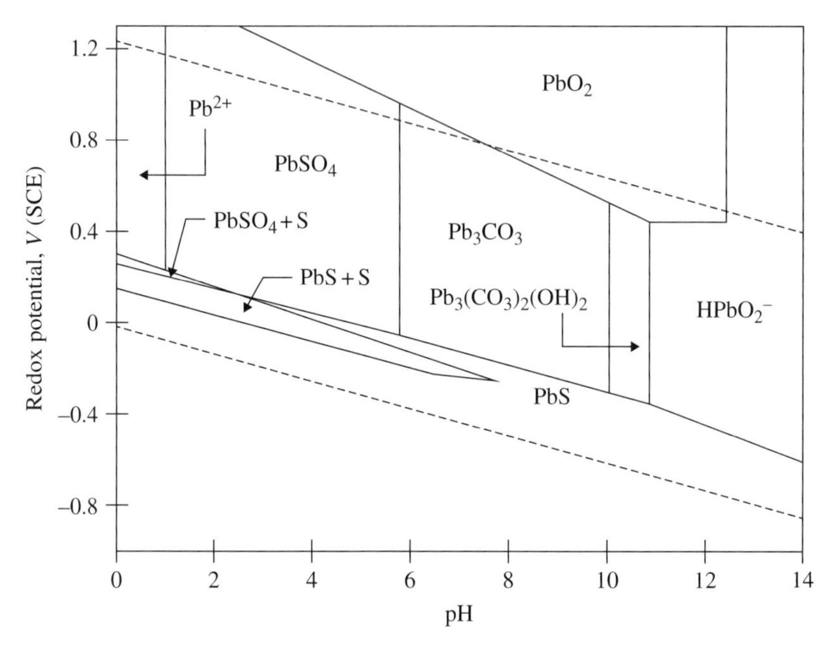

FIGURE G.2 Potential–pH equilibrium diagram for the system Pb—S—CO_2—H_2O at 25°C, computed for a dissolved lead concentration of 1 μM, a dissolved sulfur concentration of 10^{-1} M, and a dissolved carbonate concentration of 10^{-2} M. The dashed lines indicate the limiting conditions within which water is stable. (Adapted with permission from C. Klein, Group IVb Elements. Lead, in *Equilibrium Diagrams for Minerals at Low Temperature and Pressure*, ed. H.H. Schmitt, Geological Club of Harvard, Cambridge, MA, pp. 177–187, 1962.)

insoluble sulfide. The low atmospheric concentrations of the reduced sulfur species, together with the fact that the region of stability for galena is well separated from typical exposure conditions, suggest that the rate of sulfide production in lead surface layers will be very low, a prediction confirmed by analytical studies of the layer constituents. More interesting chemically are the processes that follow the injection of oxidized sulfur into the system. If the sulfur enters in the S^{IV} state, lead sulfite may result or the S^{IV} may be oxidized to S^{IV} in solution. Effective oxidizers of the tetravalent sulfur include H_2O_2 and O_3 dissolved from the atmosphere. In addition, it has been observed that lead produces H_2O_2 as it corrodes. In such a case, although the precise chemical mechanism of the dissolution step remains to be determined, one can envision a pathway to the production of the lead sulfate corrosion products that follows an overall chemical chain including the following steps:

$$PbO(s) + H_2O \rightarrow Pb^{2+} + H_2O_2 + 2e^- \quad (G.1)$$

$$H_2O_2 + HSO_3^- \rightarrow HSO_4^- + H_2O \quad (G.2)$$

$$HSO_4^- \leftrightarrow SO_4^{2-} + H^+ \quad (G.3)$$

$$Pb^{2+} + SO_4^{2-} \rightarrow PbSO_4 \downarrow \quad (G.4)$$

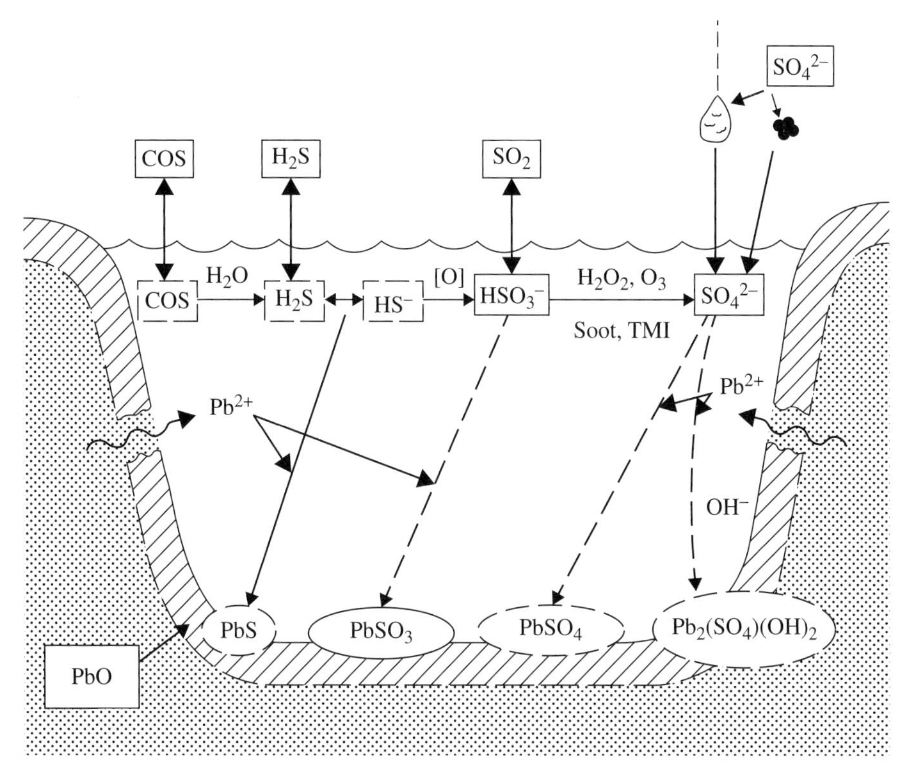

FIGURE G.3 A schematic representation of the processes potentially involved in the sulfur chemistry of the atmospheric corrosion of lead. The atmosphere is at the top of this diagram, a surface water layer is in the center, and the corroding metal is at the bottom. For simplicity, the diagram pictures a depression in the metal surface, although deposited hygroscopic particles or condensed water could produce similar conditions anywhere on the surface. The wavy line indicates Pb^{2+} ions diffusing into the aqueous solution from the bulk solid. Species shown within rectangular boxes are present as solution constituents; those in ovals are present as precipitates. Dashed rectangles or ovals indicate constituents that are unconfirmed by laboratory or field studies. Similarly, solvation or reaction processes that have been confirmed by laboratory studies are shown as solid arrows; those for which the mechanism is uncertain are shown as dashed arrows. Soot refers to elemental carbon on aerosol particles, TMI, to transition metal ions. Airborne particles are indicated by the irregular small atmospheric object. The rates of many of the reactions will be functions of the solution acidity.

G.4.4 Chlorides

Lead is susceptible to reacting with halogen ions in the presence of water. Lead chloride is obtained when lead is exposed to solutions of HCl. Since lead hydroxychlorides are moderately to poorly soluble, it seems likely that they are present as trace corrosion products also on lead samples exposed to the atmosphere, especially in marine atmospheres, as diagrammed schematically in Figure G.4. None have yet been reported for such samples, however.

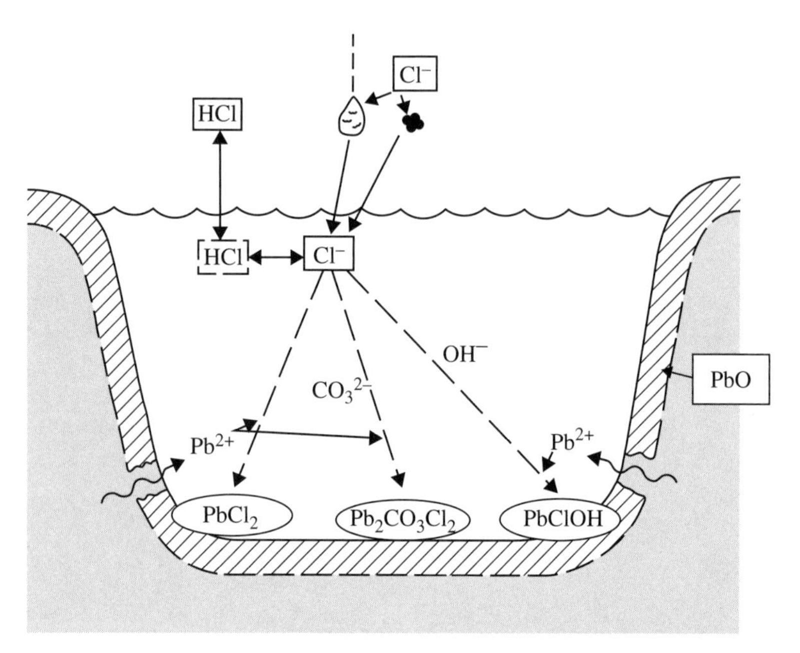

FIGURE G.4 A schematic representation of the processes potentially involved in the chlorine chemistry of the atmospheric corrosion of lead. In the marine atmosphere and at the edges of continents, sea-salt aerosol particles dominate the chlorine cycle. The display is described in more detail in the caption to Figure G.3.

G.4.5 Nitrogen Species

Lead is attacked readily by nitric acid, HNO_3, so it might be anticipated that atmospheric aerosol particles containing nitrate of the atmospheric gases NO_2 or HNO_3 would affect lead. No laboratory tests of these possibilities have been performed, and lead corrosion products containing nitrate have not been found on samples exposed at field sites. If nitrates of lead were formed, however, their high solubility would render them unlikely to be stable enough to be detected.

G.4.6 Carbonates and Organic Compounds

CO_2 is an abundant atmospheric gas, and its dissolution in aqueous surface films on lead produces anions for the formation of carbonate and hydroxycarbonate corrosion products. Such products are common on lead exposed outdoors and indoors where more aggressive competing anions are not present. The relatively straightforward chemical processes leading to carbonates are indicated in Figure G.5.

Organic acid vapors have long been known to be corrosive to lead, with acetic acid, CH_3COOH, being 5–10 times as aggressive as HCOOH. In fact, a traditional method for formulating the pigment known as "white lead," a basic lead carbonate, is to expose lead disks to vapors of CH_3COOH and CO_2, the CH_3COOH acting to promptly oxidize the lead surface.

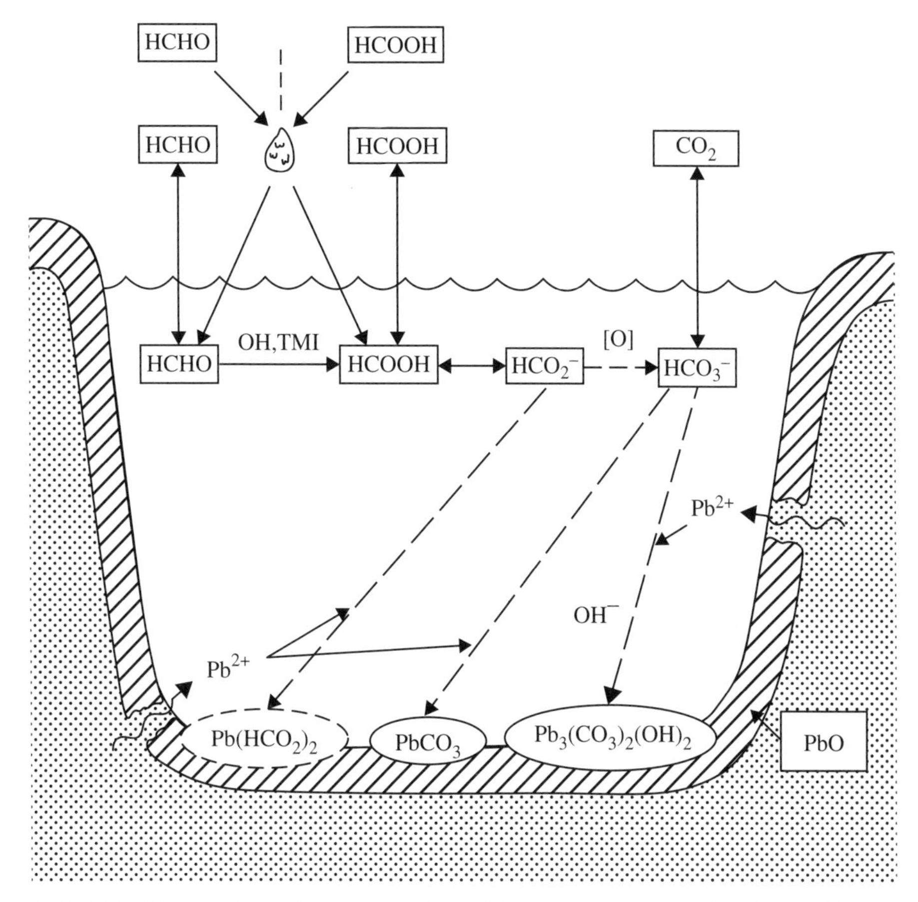

FIGURE G.5 A schematic representation of the processes potentially involved in the formation of carbon-containing compounds during the atmospheric corrosion of lead. The organic chemistry uses formaldehyde and formic acid for illustration but applies equally well to other aldehydes and carboxylic acids. TMI refers to transition metal ions. The display is described in more detail in the caption to Figure G.3.

In addition to being a potential reliability problem in technological applications involving lead, organic acid susceptibility is of substantial concern from a curatorial standpoint. Indeed, the corrosion of lead by organic acids has been seen in leaded church roofs as a consequence of proximity to uncured wood beams and in lead objects in museum display cases, as a consequence of formaldehyde emissions from sealing compounds. The frequent occurrence of metal carboxylates as constituents of indoor corrosion products is further discussed in Section 8.4.

Organic acids are fairly common in urban atmospheres, being products of smog chemistry, and are present in precipitation as well. Their participation in lead corrosion is diagrammed schematically in Figure G.5.

G.5 PHYSICAL AND CHEMICAL CHARACTERISTICS OF CORRODING LEAD ALLOYS

Lead alloys, especially the 60/40 eutectic with tin, find use in electronics as solders, components of solder pastes, and platings. These alloys produce lead corrosion products when they undergo atmospheric exposure. The products in the case of leaded brasses are lead sulfate and lead hydroxysulfate, and in the case of terne plate (steel coated with an 80Pb/20Zn alloy), the product is lead sulfate. For an alloy of 50Pb/50Sn exposed separately to O_2, SO_2, NO_2, H_2S, CO, Cl_2, and NH_3, no effects were seen for high or moderate concentrations of CO, NH_3, H_2S, or SO_2. Exposure to NO_2 produced $Pb(NO_3)_2$ as a corrosion product. Exposure to Cl_2 gave both $PbCl_2$ and $SnCl_2$. The results were relative humidity dependent. Thus, the laboratory results suggest that alloying of lead with tin, followed by oxidation, produces a surface in which the lead atoms react to anions in the overlying aqueous layer much as do the atoms in bulk lead.

Lead alloys have sufficient corrosion resistance in a variety of environmental conditions that they have sometimes been used as a plating on base metal electrical contacts. Field exposures for several months tend to produce increased contact resistances but generally no higher than a factor of two to three over the original levels. The performance of lead–tin alloys as electrical contacts is confirmed by laboratory studies of contact resistance levels following exposure to one or more corrosive gases. Nonetheless, the susceptibility of tin–lead alloys to fretting corrosion makes them generally unsatisfactory for modern high-reliability electrical and electronic products, as discussed further in Section 11.2.

G.6 SUMMARY

Lead is capable of forming crystalline compounds in the Pb^{IV} oxidation state but strongly favors the Pb^{II} state and probably does not exist as Pb^{IV} in solution. The aggressiveness and abundance of various potentially interacting anions produce the composite diagram in Figure G.6 for atmospheric laboratory exposures of lead involving the gases H_2S, SO_2, CO_2, and Cl_2. The principal corrosion products found are $PbSO_4$ and $PbCO_3$.

Lead is quite active in its response to corrosive gases, especially those generating sulfate, chloride, and perhaps nitrate. Laboratory experiments suggest that the sulfur gases are relatively inactive to lead if their solution oxidation state is S^{IV}, but if an oxidant for the sulfur is present, the corrosion proceeds readily.

Lead-rich alloys tend to dissolve faster than the pure metals under all aqueous conditions. Depending on the characteristics of the environment to which the lead is exposed, atmospheric corrosion products customarily include anglesite, cerussite, and lead formate. A careful analytical search would probably reveal galena, lead chlorides, and lead oxalate in appropriate environments.

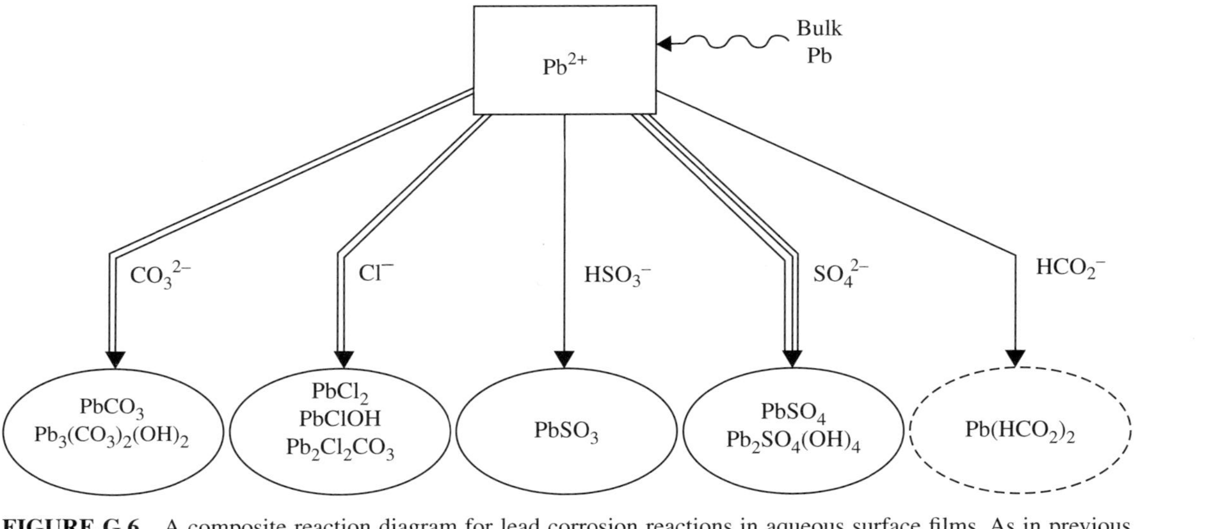

FIGURE G.6 A composite reaction diagram for lead corrosion reactions in aqueous surface films. As in previous figures, solid lines, rectangles, and ovals indicate processes or species confirmed by measurement, while dashes indicate merely the potential for reaction or detection. The width of the transformation lines reflects the estimated relative importance of the processes.

Details concerning the information in this appendix may be found in T.E. Graedel, Chemical mechanisms for the atmospheric corrosion of lead, *Journal of the Electrochemical Society, 141*, 922–927, 1994.

FURTHER READING

L. Black, G.D. Allen, and P.C. Frost, Quantification of Raman spectra for the primary atmospheric corrosion products of lead, *Applied Spectroscopy, 49*(9), 1299–1304, 1995.

S.A. Matthes, S.D. Cramer, B.S. Covino Jr., S.J. Bullard, and G.R. Holcomb, Precipitation runoff from lead, in *Outdoor Atmospheric Corrosion,* ASTM STP 1421, ed. H.E. Townsend, ASTM, West Conshohocken, pp. 265–274, 2002.

A. Niklasson, L.-G. Johansson, and J.-E. Svensson, Atmospheric corrosion of lead. The influence of formic acid and acetic acid vapors, *Journal of the Electrochemical Society, 154*(11), C618–C625, 2007.

A. Niklasson, L.-G. Johansson, and J.-E. Svensson, The influence of relative humidity and temperature on the acetic acid vapour-induced atmospheric corrosion of lead, *Corrosion Science, 50*, 3031–3037, 2008.

APPENDIX H

THE ATMOSPHERIC CORROSION CHEMISTRY OF NICKEL

H.1 INTRODUCTION

Nickel is a metal relatively resistant to atmospheric corrosion. Together with its favorable mechanical properties, nickel's corrosion resistance has resulted in the extensive use of nickel in the chemical, automotive, and electronics industries, among others. Electroplated coatings of nickel are common, the platings serving both corrosion protection and aesthetic functions.

The original corrosion experiments on nickel were performed by Vernon in the 1920s and 1930s. While considerable experience has been acquired over the years concerning nickel's environmental ruggedness, the understanding of the chemical processes involved in its degradation upon exposure to the atmosphere remains inadequate. Laboratory studies and detailed field studies have identified only a small number of corrosion products and have not related them comprehensively to the atmospheres in which the corrosion has occurred. This paper synthesizes the information concerning nickel surfaces and the atmosphere and uses that information to infer the interactions that commonly occur.

H.2 CORROSION LAYER FORMATION RATES

The corrosion rate of nickel exposed to the atmosphere is a function of the length of exposure and of the average outdoor concentration of sulfur dioxide. A UN/ECE study at 39 sites in Europe and North America produced the following equation:

$$R = 2.4 \cdot t \cdot [SO_2] \tag{H.1}$$

Atmospheric Corrosion, Second Edition. Christofer Leygraf, Inger Odnevall Wallinder, Johan Tidblad and Thomas Graedel.

where R is the corrosion rate of nickel in sheltered locations, expressed as mass loss ($\mu g\,cm^{-2}$), t is the time in years, and $[SO_2]$ is the average outdoor concentration in $\mu g\,m^{-3}$. Thus, in relatively unpolluted environments the rate of nickel corrosion can be quite low, whereas in polluted sites it can be as high as that of other commonly used metals such as copper and silver.

H.3 THE MORPHOLOGY OF ATMOSPHERIC CORROSION LAYERS ON NICKEL

The corrosion products that form on nickel during at least the first several weeks of exposure to the atmosphere are initially whitish in visual appearance. They are commonly dendritic in form, the dendrites being composed largely of nickel, oxygen, and sulfur. Over time, these coalesce to a thin, adherent film, as would be expected for a metal with good corrosion resistance. The film is highly insulating. The corrosion that eventually results may be manifested either uniformly or as pitting, but nickel is generally quite corrosion resistant in mild environments.

H.4 CHEMICAL MECHANISMS OF NICKEL CORROSION

H.4.1 Introduction

In initiating a study of the chemistry of atmospheric nickel corrosion, it is of interest to identify possible corrosion products. Accordingly, ions or molecules that have been detected or suggested as potential participants in corrosion chemistry were used to survey the mineralogical and chemical literature for potential mineral constituents. The resulting list of candidate species is given in Table H.1; it contains 21 entries, some of which have no mineralogical names and are therefore not known to be produced naturally by geochemical processes. Included in the table are mineral species index numbers (where available) and information on nickel valence, crystal habits, and chemical formulas. The appearance of an entry in this table presupposes that the nickel corrosion products are primarily crystalline rather than amorphous. As will be seen, this presupposition is only partly correct.

The Table H.1 list has been used as a guide to the analytic literature on nickel corrosion layers. In the following paragraphs, the chemical families are discussed individually.

H.4.2 Adsorbed Water

The atmospheric corrosion of nickel is strongly influenced by the presence or absence of moisture. The amount of moisture on a nickel surface varies slightly with temperature and greatly with relative humidity, as shown in Figure H.1. It can be seen that at temperatures common for exposed nickel, humidities of 70% or above result in the presence of several equivalent monolayers of water on the nickel surface. For nickel

TABLE H.1 Minerals Containing Nickel and Other Crystalline Substances with Possible Relevance to the Corrosion of Nickel[a]

Substance	Valence	Hey Index No.	Crystal System	Formula
Metal, oxides, and hydroxides				
Nickel	—	1.63a	Cubic	Ni
Bunsenite	2	7.22.1	Cubic	NiO
Theophrastite	2	7.22.2a	Trigonal	$Ni(OH)_2$
Nickel oxide (b)	2,3	7.22.3a	—	$Ni_2O(OH)_3$
Nickel hydroxide	3		—	$Ni(OH)_3$
Sulfates				
Dwornikite	2		Monoclinic	$NiSO_4 \cdot H_2O$
Retgersite	2	5.12.6a	Tetragonal	α-$NiSO_4 \cdot 6H_2O$
Nickel hexahydrite	2		Monoclinic	β-$NiSO_4 \cdot 6H_2O$
Morenosite	2	5.12.6a	Orthorhombic	$NiSO_4 \cdot 7H_2O$
Nickel sulfate (b)	2		—	$Ni_3(SO_4)_2(OH)_2$
Chlorides				
Nickel chloride	2		Hexagonal	$NiCl_2$
Nickel chloride (b)	2		Hexagonal	$NiClOH$
Nickel chloride (b)	2		Hexagonal	$Ni_2Cl(OH)_3$
Nitrates				
Nickel nitrate	2		Cubic	$Ni(NO_3)_2$
Nickel nitrate (h)	2		Triclinic	$Ni(NO_3)_2 \cdot 6H_2O$
Carbonates				
Gaspeite	2	11.14.9	Trigonal	$NiCO_3$
Hellyerite	2	11.14.7a	Triclinic	$NiCO_3 \cdot 6H_2O$
Nullaginite	2		Monoclinic	$Ni_2CO_3(OH)_2$
Otwayite	2		Orthorhombic	$Ni_2CO_3(OH)_2 \cdot H_2O$
Zaratite	2	11.14.7	Cubic	$Ni_3CO_3(OH)_4 \cdot 4H_2O$
Organics				
Nickel formate	2		—	$Ni(CHO_2)_2 \cdot 2H_2O$
Nickel oxalate	2		—	NiC_2O_4
Nickel acetate	2		—	$Ni(C_2H_3O_2)_2$

[a] See Table C1 footnotes for explanatory details.

with normal industrial surfaces, as opposed to the carefully polished surfaces used for the experiments that resulted in Figure H.1, additional layers can be expected. Because adsorbed water begins to exhibit bulk water behavior at about three layers of thickness, corrosion occurs in what is essentially the aqueous phase.

A crucial role for the water layer on the metal surface is to provide a medium for the mobilization of nickel ions. The oxidation step for this process can be rendered as

$$Ni(s) \rightarrow Ni^{2+} + 2e^- \qquad (H.2)$$

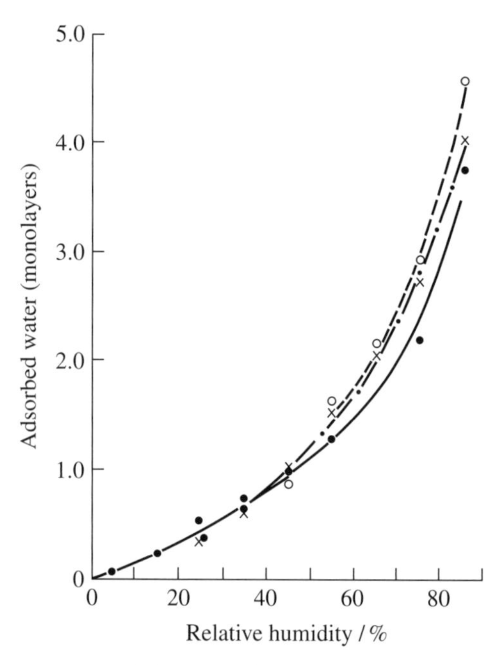

FIGURE H.1 The thickness of the water layer (in monolayers) adsorbed on nickel at 15° (×), 25° (○), and 35°C (●) for various relative humidities.

This reaction is balanced in acid solution by

$$O_2 + 4H_3O^+ + 4e^- \rightarrow 6H_2O \tag{H.3}$$

The details of the chemistry comprising these multistep reactions are not well understood and are, in any case, peripheral to the topic of this paper, which is to examine the processes that occur once nickel oxidation and ionic mobilization have taken place.

In the case of some materials that corrode on exposure to natural environments, the mechanisms and rates of oxidation of the mobilized ions from one valence state to another are items of interest as possible limiting steps in the degradation process. Such is not the case with nickel, which has only the +2 valence state normally available to it. As a consequence, the chemical degradation mechanisms need to consider only questions of reaction and dissolution of the Ni^{2+} ion and not of the ion's oxidation or reduction.

H.4.3 Oxides and Hydroxides

The initial corrosion film formed on nickel in the atmosphere is an inner layer of NiO (the mineral bunsenite) and an outer layer of $Ni(OH)_2$ (the mineral theophrastite). The hydroxide compound is perfectly understandable in view of the thermodynamic stability of nickel–water species. As shown in Figure H.2, nickel dihydroxide has a region of stability close to the region of the pH–eH diagram typical of dew.

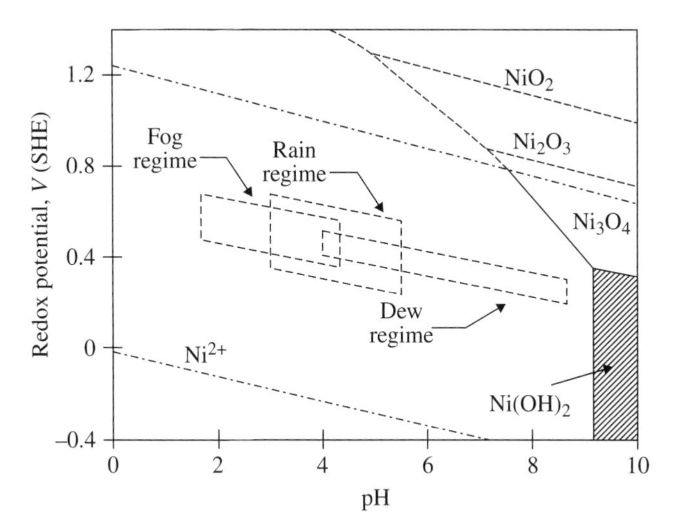

FIGURE H.2 Potential–pH diagram for the system Ni—H_2O at 25°C, for a concentration of nickel ionic species of 0.1 M. The dash-dotted lines indicate the limiting conditions within which water is stable. The approximate regimes for rain, fog, and dew are indicated.

H.4.4 Sulfur Compounds

Sulfur compounds are the most common products in the atmospheric corrosion of nickel. Of the several substances that have been identified or conjectured, all involve sulfur in the +6 oxidation state. Different researchers have reported the detection of different compounds, perhaps as a result of differing atmospheric conditions at the experimental field sites. Studies at 39 urban, rural, and industrial sites throughout Europe and North America have demonstrated that the corrosion product formed after 1 year in sheltered conditions is primarily an amorphous basic nickel sulfate, which subsequently metamorphoses into a crystalline basic nickel sulfate with slightly higher corrosion resistance. In a rural atmosphere, only a hydrated (nonbasic) nickel sulfate is generally seen on sheltered nickel samples. In all cases, the data indicate the presence of S^{VI}, the mobilization of nickel ions, and the importance of water.

The pattern observed in the field studies is consistent with the picture presented by crystalline thermodynamics, as shown in Figure H.3. At the sulfur concentrations and typical pH values of fog and rain, hydrated nickel sulfate is the stable species. In the case of dew, where low or negative redox potentials are present, a region of stability exists for nickel sulfide. Such conditions are not typical for surfaces exposed to the atmosphere, however, and no sulfides have been reported as components of the atmospheric corrosion of nickel.

While Figure H.3 indicates that nickel sulfates are anticipated thermodynamically, from a kinetic standpoint the limiting step will be the oxidation of S^{IV}, dissolved in the aqueous surface layer, to S^{VI}. A possible process diagram is given in Figure H.4. It shows that any sulfur dioxide gas dissolving into the surface film will be ionized to

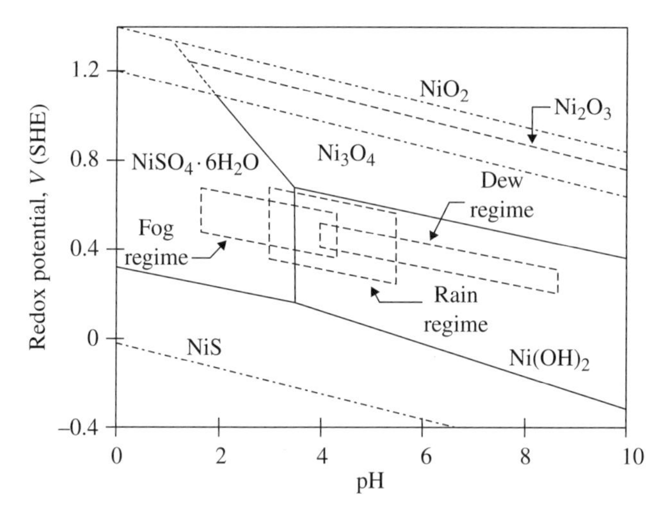

FIGURE H.3 Potential–pH diagram for the system Ni—S—H_2O at 25°C, for a concentration of sulfur species of 10^{-6} M. The dash-dotted lines indicate the limiting conditions within which water is stable. The approximate regimes for rain, fog, and dew are indicated. Thermodynamic data are not available for hydroxysulfates of nickel so those compounds do not appear on this graph.

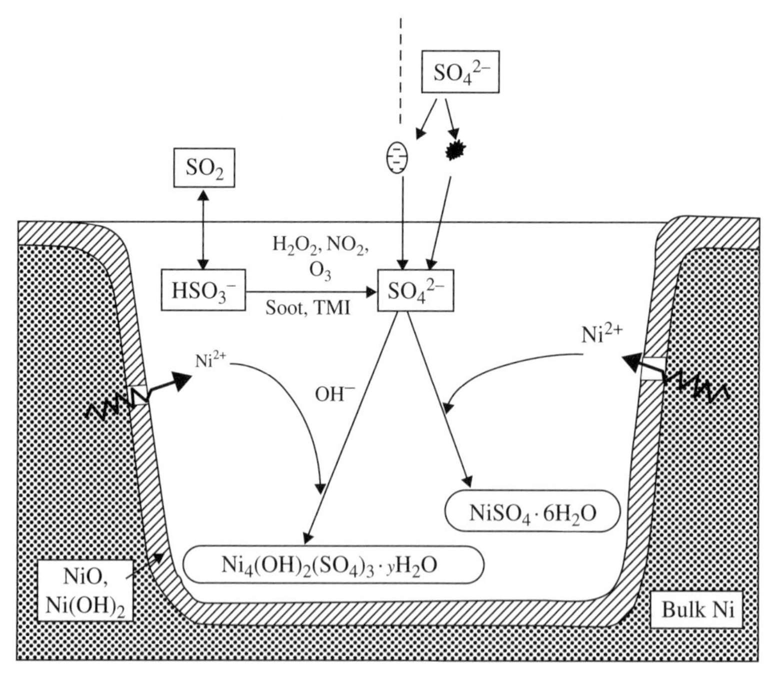

FIGURE H.4 A schematic representation of the processes potentially involved in the formation of sulfur-containing components during the atmospheric corrosion of nickel. The atmosphere is at the top of this diagram, a surface water layer is in the center, and the corroding metal is at the bottom. For simplicity, the diagram pictures a depression in the metal surface, although deposited hygroscopic particles, heavy dew, or rain could produce similar conditions anywhere on the surface. The wavy line indicates nickel ions diffusing into the aqueous solution from the bulk solid. Rain is indicated by the atmospheric water drop and airborne particles by the irregular small atmospheric object. TMI indicates transition metal ions.

bisulfite ion (HSO_3^-). The bisulfite ion is readily converted to sulfate ion by ozone, hydrogen peroxide, nitrogen dioxide, soot, or transition metal ions (largely iron or manganese). (In the absence of a catalyst, oxidation by dissolved O_2 is unimportant.) Once sulfate is formed by solution chemistry or supplied by particles or precipitation, it is available to form nickel sulfates or hydroxysulfates under the prevailing eH–pH conditions.

In the simplest of the mineral formation processes involving nickel and sulfur, the uncomplexed ions are mobilized in acidic aqueous solution, form ion pairs, and precipitate as the hydrated nickel sulfate mineral retgersite upon the evaporation of the surface water layer:

$$Ni^{2+} + SO_4^{2-} + 6H_2O \rightarrow NiSO_4 \cdot 6H_2O \downarrow \tag{H.4}$$

The process is enhanced to some degree in the presence of ozone.

This is a process that is reasonably facile, since all three of these species are abundant on nickel exposed to the outdoor atmosphere. Also appearing on exposed nickel, however, is a hydroxysulfate. Nucleation of the less soluble hydroxide mixed salt is probably initiated by seed crystals at the metal surface. Thus, one could write the formation reaction of the nickel hydroxysulfate as a combination of the simple salts

$$Ni(OH)_2 + 3NiSO_4 + yH_2O \rightarrow Ni_4(OH)_2(SO_4)_3 \cdot yH_2O \tag{H.5}$$

but it is almost certainly more realistic to assume that one of the ubiquitous nickel dihydroxide molecules functions as a seed crystal and subsequently adds the needed ions from the solution with which it is in contact:

$$Ni(OH)_2(s) + 3Ni^{2+} + 3SO_4^{2-} + yH_2O \rightarrow Ni_4(OH)_2(SO_4)_3 \cdot yH_2O \tag{H.6}$$

(Eqs. H.5 and H.6 are written here with the product stoichiometry being that identified in field studies but could obviously be adapted to any nickel hydroxysulfate species.)

Unlike some other metals, nickel is not known to form sulfides under atmospheric corrosion conditions. This is consistent with mixed gas laboratory experiments, in which the presence of H_2S was shown to have little influence on nickel corrosion.

H.4.5 Chlorides

Chloride is commonly detected on nickel exposed to the atmosphere, although the amounts are relatively small. The chloride ion (e.g., from sodium chloride on aerosol particles) can be incorporated into nickel oxide films and alter the composition of the corrosion products. Nickel corrodes rather rapidly in the presence of small amounts of Cl_2, but gaseous atmospheric chlorine is hardly ever this abundant and never in this highly reactive form. It mostly exists as HCl and CH_3Cl, toward which nickel is resistant.

H.4.6 Nitrates

Nitrates are found to a small degree on field samples of corroded nickel. In addition, the presence of NO_2 in laboratory corrosion chambers has been shown to enhance nickel corrosion. This latter effect probably occurs as a consequence of the ability of NO_2 to oxidize S^{IV} in aqueous solution. Overall the degree of response suggests that NO_2 has an accelerating effect on the SO_2- induced atmospheric corrosion, even though nitrates are rarely seen as corrosion products.

H.4.7 Carbonates and Organic Compounds

Although not commonly present, carbonates are potential constituents in atmospheric corrosion layers on nickel. In the thermodynamic equilibrium diagram for the Ni—CO_2—H_2O system, there is a region of stability for gaspeite, the simple nickel carbonate. This region does not overlap with those of common forms of atmospheric moisture but is close enough so that fluctuations in the chemical environment of the aqueous surface films should permit at least occasional $NiCO_3$ formation. Hydrated carbonates or hydroxycarbonates are likely to show some overlap with precipitation regimes once their thermodynamic properties are determined.

Other possible carbon-containing compounds in nickel corrosion layers include organic acids, which are fairly common in the atmosphere and have been detected on the surfaces of metals exposed to the atmosphere. Nickel is known to corrode in the presence of organic acids, and evidence for significant amounts of carboxylates has been found in corrosion films on nickel exposed indoors.

A speculative picture of nickel atmospheric corrosion involving carbon-containing compounds appears in Figure H.5. Most of the compounds shown have been identified in the atmosphere or on exposed nickel samples, but the exact processes and their rates remain to be determined.

H.4.8 Photocorrosion

It is clear that photons can play a role in the atmospheric corrosion of nickel, though the details remain uncertain. Vernon observed that surface films, presumably oxides and hydroxides, formed much more rapidly on nickel samples exposed to broadband light than on samples exposed in the dark. This characteristic is attributed to photon absorption in the nickel oxide lattice followed by increases in surface reaction rates. For nickel, the relevant band gap energy is exceeded by photons with wavelengths shortward of about 400 nm. It is thus reasonable to anticipate that orientation and illumination will need to be considered in order to gain a complete understanding of the atmospheric corrosion of nickel.

H.5 LABORATORY AND COMPUTATIONAL STUDIES OF NICKEL'S ATMOSPHERIC CORROSION

Two laboratory studies have usefully investigated the involvement of gaseous compounds in the atmospheric corrosion of nickel. In the first, nickel was exposed simultaneously to SO_2, NO_2, O_3, Cl_2, H_2S, and NH_3 in moist air. In different experiments,

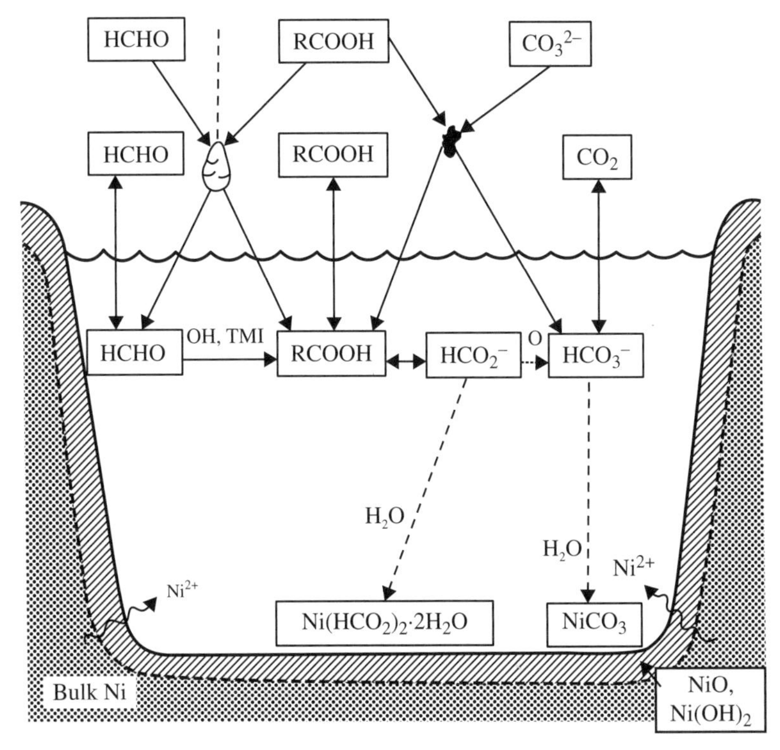

FIGURE H.5 A schematic representation of the processes potentially involved in the formation of carbonate components during the atmospheric corrosion of nickel. See Figure H.4 legend for explanatory details.

the concentration of one of the gases was varied while those of the others were held constant for periods of several days. In the cases of SO_2, NO_2, O_3, and Cl_2, the rate of nickel corrosion increased with increased gas concentration. H_2S showed no significant effect, and NH_3 had an inhibiting effect.

In the second study, nickel was exposed to SO_2 in moist air in the presence or absence of either NO_2 or O_3. Both NO_2 and O_3 promoted the atmospheric corrosion of nickel. The mechanism was not determined but was clearly more complex than a simple direct oxidation.

The only theoretical chemistry study of the atmospheric corrosion of nickel was devised to simulate corrosion in the presence of SO_2. Without a strong oxidizer present, the principal product was a surface complex of nickel sulfite (in agreement with experimental results).

H.6 CONCLUSIONS

Nickel is a metal much more corrosion resistant than most other industrial metals, and its atmospheric corrosion chemistry is simpler than most. A full quantitative analysis of the physical and chemical processes involved in the atmospheric corrosion of nickel would require much information not yet available: the solubilities in solutions of different acidities of the several hydroxy mixed salts of nickel, the kinetic parameters appropriate to their formation, more information on the temporal evolution of corrosion layer components, and so on. Lacking this information, it has nonetheless proved possible to produce a rather extensive qualitative picture of the corrosion process. The involvement of chemical reactions in the production of many species is clearly required. The species playing important roles are sulfates, chlorides, nitrates, and carbonates, and—indoors—formate and acetate should be considered as well.

Kinetic simulations of nickel corrosion are promising but clearly preliminary. Several processes that are probably important are thus far unaddressed. For example, although the enhancement of atmospheric nickel corrosion by photons has not been studied extensively, the limited literature suggests a definite effect. The same is true for corrosion enhancement due to ozone and nitrogen dioxide. In neither case have mechanisms been proposed. It is obvious that additional research is needed concerning the details of nickel's atmospheric corrosion chemistry.

Gaseous sulfur dioxide and sulfate ions in aerosol particles and precipitation are clearly the major factors in the atmospheric corrosion of nickel. With concentrations of these species decreasing in more developed countries as control technology is enhanced, nickel corrosion concerns in these areas are likely to decrease. In parts of the less developed world, however, rapidly increasing use of coal is raising the levels of atmospheric sulfur, and rates of nickel corrosion in those regions can be expected to increase rapidly.

Details concerning the information in this appendix may be found in T.E. Graedel and C.Leygraf, Corrosion mechanisms for nickel exposed to atmosphere, *Journal of the Electrochemical Society, 147*, 1010–1014, 2000.

FURTHER READING

S. Jouen, M. Jean, and B. Hannoyer, Atmospheric corrosion of nickel in various outdoor environments, *Corrosion Science, 46*(2), 499–514, 2004.

I. Odnevall and C. Leygraf, The atmospheric corrosion of nickel in a rural atmosphere, *Journal of the Electrochemical Society, 144*(10), 3518–3525, 1997.

D. Persson and C. Leygraf, Analysis of atmospheric corrosion products of field exposed nickel, *Journal of the Electrochemical Society, 139*(8), 2243–2249, 1992.

D.W. Rice, P.B.P. Phipps, and R. Tremoureux, Atmospheric corrosion of nickel, *Journal of the Electrochemical Society, 127*(3), 563–568, 1980.

APPENDIX I

THE ATMOSPHERIC CORROSION CHEMISTRY OF SILVER

I.1 INTRODUCTION

Silver is an attractive, lustrous metal whose electrical and thermal conductivities are the highest of any of the elements. Silver readily forms alloys, and many of its alloys possess attractive electrical and mechanical properties. As a consequence of these characteristics, silver and high-silver alloys are regularly exposed to the indoor atmosphere in uses such as electronics, solder, batteries, conductive pastes, dental alloys, silverware, jewelry, and decorative objects of various kinds. The use of silver in outdoor locations is rare, a fact particularly interesting in view of some evidence that the outdoor corrosion of silver proceeds more slowly than indoor corrosion (see Chapter 8).

Considerable experience has been acquired over the years concerning silver's behavior in the indoor and outdoor atmosphere, both in pure and alloy forms, but the understanding of the chemical processes involved in the degradation remains rudimentary. Laboratory studies with complex test atmospheres simulating real-world conditions have been few, as have detailed analytical studies of samples exposed at field sites. Insight into the processes requires a detailed understanding of the relationships between the chemical properties of the altered metal surface and the atmospheric constituents responsible for those changes.

I.2 ENVIRONMENTAL INTERACTIONS WITH SILVER SURFACES

The atmospheric corrosion of silver occurs only in the presence of moisture. The amount of moisture adsorbed onto a silver surface lightly covered by its native oxide (see in the following) is a function of relative humidity, as shown in the composite

Atmospheric Corrosion, Second Edition. Christofer Leygraf, Inger Odnevall Wallinder, Johan Tidblad and Thomas Graedel.

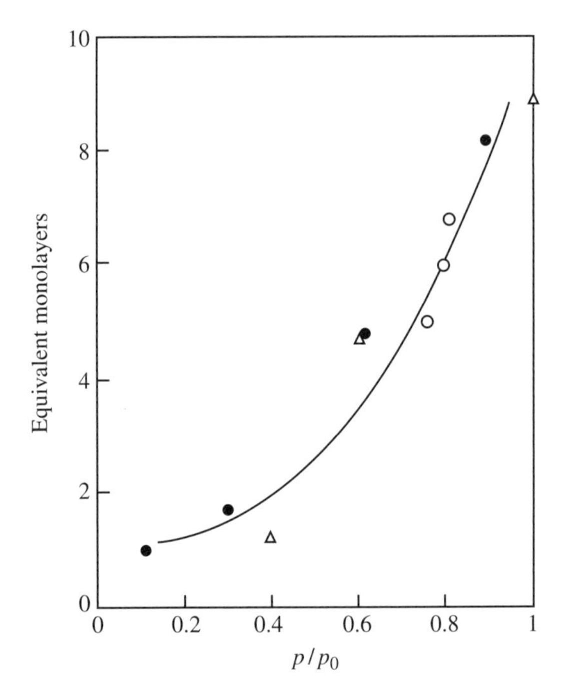

FIGURE I.1 Composite sorption isotherm at 25°C for water vapor on silver. The technique assumes the water to be uniformly spread over the surface, although surface tension may cause it to "pool"; the ordinate variable is thus "equivalent" monolayers. The line is a least-squares fit to the data, expressed as the linearized form of (ml) = $a \cdot \exp(b \cdot p/p_0)$. The different symbols refer to data from different experimental studies.

isotherm in Figure I.1. These data give a dependence of water monolayers on humidity that may be expressed as

$$\ln(\mathrm{ml}) = 2.73 \cdot p/p_0 - 0.366 \tag{I.1}$$

ml indicating the number of monolayers of water adsorbed on the silver surface. This monotonic behavior is typical of that seen for the adsorption of water to the surfaces of many different metals. About eight monolayers of water are present at 90% RH. At lower humidities the quantity of adsorbed water is decreased, but some will still be present. Extensive research makes it clear that silver corrosion increases with increasing relative humidity.

The crucial role of the water layer on the metal surface is to provide a medium for the absorption of atmospheric gases and the subsequent dissolution of solid silver. The oxidation step for the dissolution can be rendered as

$$\mathrm{Ag(s)} \rightarrow \mathrm{Ag^+} + \mathrm{e^-} \tag{I.2}$$

This reaction is thought to be balanced in acid solution by

$$\mathrm{O_2} + 4\mathrm{H_3O^+} + 4\mathrm{e^-} \rightarrow 6\mathrm{H_2O} \tag{I.3}$$

TABLE I.1 Minerals and Other Crystalline Substances Found on Corroded Silver[a]

Substance	Hey Index No.	Crystal System	Formula	K_{sp}
Metal/oxides				
Silver	1.3	Cubic	Ag	
Silver oxide		Cubic	Ag_2O	$2.6 \cdot 10^{-8}$
Sulfides, sulfite, and sulfates				
Acanthite	3.2.1	Monoclinic	Ag_2S	$6.0 \cdot 10^{-50}$
Argentite	3.2.1	Cubic	Ag_2S	
Silver sulfite		Monoclinic	Ag_2SO_3	$1.5 \cdot 10^{-14}$
Silver sulfate		Orthorhombic	Ag_2SO_4	$1.6 \cdot 10^{-5}$
Chloride and nitrate				
Chlorargyrite (cerargyrite)	8.3.1	Cubic	$AgCl$	$1.8 \cdot 10^{-10}$
Silver nitrate		Orthorhombic	$AgNO_3$	$1.6 \cdot 10^{-1}$

[a] See Table C1 footnotes for explanatory details.

and

$$2H_3O^+ + 2e^- \rightarrow 2H_2O + H_2 \uparrow \quad (I.4)$$

and in neutral solution (probably a relatively uncommon state for silver exposed to the atmosphere) by

$$2H_2O + 2e^- \rightarrow H_2 \uparrow + 2OH^- \quad (I.5)$$

and

$$O_2 + 2H_2O + 4e^- \rightarrow 4OH^- \quad (I.6)$$

I.3 CHEMICAL MECHANISMS OF SILVER CORROSION

I.3.1 Introduction

In initiating a study of the chemistry of atmospheric silver corrosion, it is of interest to identify possible corrosion products. Accordingly, ions or molecules that have been detected or suggested as potential participants in corrosion chemistry were used to survey the mineralogical and chemical literature for possible mineral constituents. The resulting list of candidate species is given in Table I.1. It contains eight entries, some of which have no mineralogical names and are therefore not known to be produced naturally by geochemical processes. Included in the table are mineral species index numbers (where available) and information on crystal habits, chemical formulas, and solubility products. The appearance of an entry in this table presupposes that the silver corrosion products are primarily crystalline rather than amorphous. It is potentially significant that the order of solubility for a unit structure of the substances (i.e., taking the number of ionic constituents into account) is as follows:

$$Ag_2S < AgCl < Ag_2SO_3 < Ag_2C_2O_4 < Ag_2O < Ag_2SO_4 < AgCO_2CH_3.$$

The final column in the table indicates whether a specific mineral has been found in corrosion layers on silver. (All of the relevant literature information refers to indoor exposures, where silver is invariably used.) By far the most common mineral detected is acanthite, Ag_2S. Much less abundant, but detected by several research groups, is chlorargyrite, AgCl. These constituents are also commonly seen as products in laboratory experiments involving mixed corrosive gases.

Alloys of silver produce corrosion products that reflect the most reactive of the alloyed metals. In the case of 90Ag/10Cu ("sterling silver"), the principal corrosion product in a reduced sulfur atmosphere is copper sulfide, reflecting the fact that the sulfidation rate of copper is an order of magnitude or more larger than that of silver. When silver is alloyed with gold, the principal corrosion product is Ag_2S. The corrosion product layer on Ag–Zn alloys contains Ag_2S, ZnO, and ZnS. Alloys of silver with palladium produce Ag_2S in sulfur-rich environments and $PdCl_2$ or AgCl in chloride-rich environments.

I.3.2 Oxides and Carbonates

The nobility of silver is shown in the potential–pH equilibrium diagram in Figure I.2. Unlike many metals, dry silver does not form a significant surface oxide at ambient temperature and pressure. Under wet conditions, the Ag_2O is stable only in a narrow region of Figure I.2 at high pH and in the presence of strong oxidizers. Atmospheric corrosion processes thus begin on a surface composed of a very thin to negligible oxide layer over the metal itself.

Atmospheric carbon dioxide, CO_2, is quite abundant and will dissolve in aqueous surface layers on silver to produce weakly acidic solutions. Silver carbonate is rather soluble, however (see Table I.1), and crystalline silver carbonate is expected only in strongly alkaline solutions.

I.3.3 Sulfides and Sulfates

Since the principal constituent of corrosion layers on silver is acanthite (Ag_2S), many laboratory studies have attempted to resolve the details of the rates and processes involved in the attack of the reduced sulfur species hydrogen sulfide (H_2S) on silver. Related studies have utilized other reduced or chemically neutral sulfur species: carbonyl sulfide (COS), organic sulfides, and flowers of sulfur. Of these species, H_2S and COS have occasionally been detected indoors, organic sulfides have not (but can be expected indoors near strong indoor or outdoor sources), and flowers of sulfur does not occur either indoors or out. The silver corrosion generated as a consequence of those experiments has been measured in different ways: as the weight gained during corrosion, as the thickness of the corrosion film induced, or as the contact resistance of the sulfided silver.

Some of the potential chemical reactions involving sulfur species are shown in Figure I.3. The formation of silver sulfide is clearly related to the presence of reduced sulfur in the indoor atmosphere. The HS^- ion is expected to be the principal reduced sulfur solution constituent at near-neutral pH from either H_2S or COS.

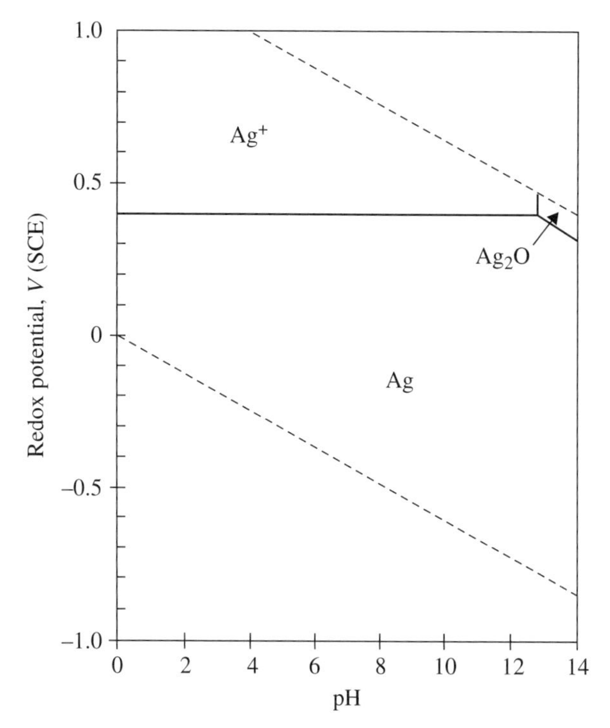

FIGURE I.2 Potential–pH equilibrium diagram for the system Ag—H_2O at 25°C, computed for a dissolved silver concentration of 1 μM. The dashed lines indicate the limiting conditions within which water is stable. (Adapted with permission from M. Pourbaix, *Atlas of Electrochemical Equilibrium in Aqueous Solutions*, 2nd English edition, trans. J.A. Franklin, National Association of Corrosion Engineers, Houston, pp. 393–398, 1974.)

HS^- can either react directly with silver ions or can sorb to the surface, subsequently reacting to form the sulfide salt.

No ready routes exist to transform reduced sulfur into higher oxidation states, but oxidized sulfur is readily supplied by gaseous sulfur dioxide, SO_2, and by sulfate in airborne particulate matter. In the former, oxidation to sulfate in solution is readily accomplished by reaction with either dissolved hydrogen peroxide, H_2O_2, or dissolved ozone, O_3. The sulfate or bisulfate ion (depending on the solution pH) may then be able to form solid silver sulfate, though little evidence exists for either the process or the product.

Laboratory studies provide qualitative support for this general picture. It is agreed that silver is very sensitive to the presence of H_2S and COS and about an order of magnitude less sensitive to SO_2. Silver sulfate can be formed by contact with SO_2 in moist air but only at SO_2 concentrations two to three orders of magnitude higher than typical of ambient environments. The presence of an oxidizing species enhances the rate of formation of silver sulfide, with O_3, NO_2, and gaseous chlorine (Cl_2) having been shown to be effective. The specific mechanism for the interaction of the

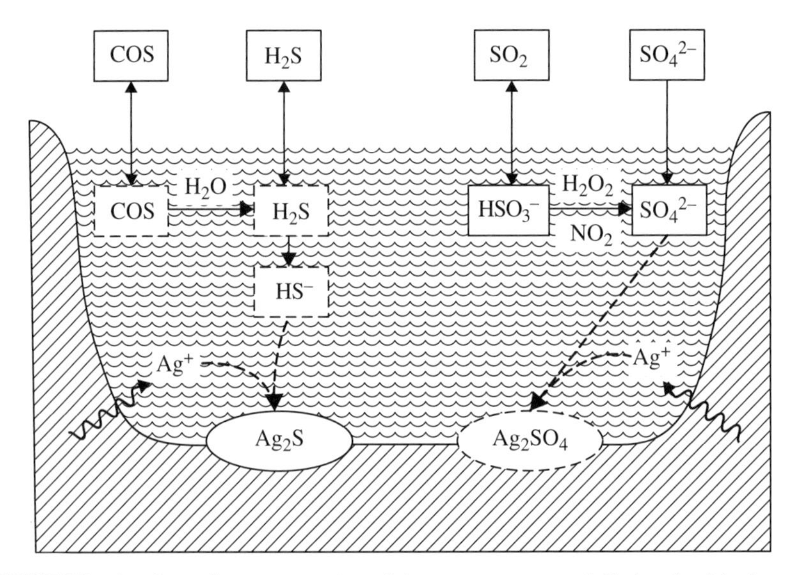

FIGURE I.3 A schematic representation of the processes potentially involved in the sulfur chemistry of the atmospheric corrosion of silver. The atmosphere is at the top of this diagram, a surface water layer is in the center, and the corroding metal is at the bottom. For simplicity, the diagram pictures a depression in the metal surface, although deposited hygroscopic particles or condensed water could produce similar conditions anywhere on the surface. The wavy line indicates Ag^+ ions diffusing into the aqueous solution from the bulk solid. Species shown within rectangular boxes are present as solution constituents; those in ovals are present as precipitates. Dashed rectangles or ovals indicate constituents that are unconfirmed by laboratory or field studies. Similarly, solvation or reaction processes that have been confirmed by laboratory studies are shown as solid arrows; those for which the mechanism is uncertain are shown as dotted arrows.

oxidizer has not been identified. It may be that oxidizers promote the S(IV) to S(VI) transition, increasing the acidity and perhaps the rate of bulk silver dissolution. Alternatively, the oxidizers may be involved in surface processes following interaction with the silver lattice.

I.3.4 Chlorides

Most of the laboratory studies of silver corrosion that have involved chlorine-containing gases have utilized Cl_2, despite the fact that this gas is virtually nonexistent in the atmosphere. All researchers agree that silver is quite sensitive to molecular chlorine. In the case of HCl, a common atmospheric gas but one much more difficult to deal with in the laboratory, there is a modest sensitivity of silver to gaseous HCl in moist air. Palladium–silver alloys show similar response to the two gases.

Once chlorine enters the aqueous surface layer on silver either through the incorporation of gaseous HCl or through the deposition of chloride-containing airborne particles, a solid product may form by precipitation of the aqueous ionic complex AgCl.

Alternatively, the chloride ion may be sorbed onto the silver surface and form silver chloride upon evaporation of the aqueous layer. AgCl is soluble in concentrated nitric or hydrochloric acid so may be subject to dissolution if an evaporating aqueous surface layer becomes highly concentrated:

$$AgCl(s) + Cl^- \rightarrow AgCl_2^-(aq) \tag{I.7}$$

I.3.5 Nitrogen Species

Only tentative evidence has been presented for the presence of silver nitrate in corrosion layers, and some uncertainty exists as to whether nitrite instead of nitrate might have been present. The only significant gas-phase atmospheric precursor for either is nitrogen dioxide, NO_2, which is poorly soluble in water and has been shown in laboratory experiments to be quite unreactive toward silver. Nitric acid is the principal gaseous form that dissolves in the aqueous surface layer; nitrate enters from particles as well. Deposition from gaseous nitric acid is expected to be larger than nitrate from particles if an aqueous surface layer is present.

A number of researchers have examined in the laboratory the corrosive effects of NO_2 on silver. Most of the work indicates that NO_2 by itself is ineffective. Only one study has used ammonia (NH_3) as a test gas; the resulting silver corrosion rate was found to be weakly sensitive to the gaseous ammonia concentration.

I.3.6 Multicomponent Equilibria

The potential–pH diagram for the conditions anticipated to be closest to those obtained during the atmospheric corrosion of silver is shown in Figure I.4. Two features of the diagram are particularly worth noting: large fields of stability exist for pure silver, for Ag_2S, and for AgCl. Since most water exposed to air and dust will have a redox potential of a few tenths of a volt and a pH in the range 2–6, silver and its sulfide are anticipated to be the predominant species found on bulk silver surfaces. Under extreme oxidizing conditions, AgCl is also anticipated. None of the following phases have significant fields of stability: Ag_2CO_3, Ag_2SO_4, or Ag_2O.

Comparison with Table I.1 demonstrates that, in the absence of kinetic constraints to the formation of the sulfide and chloride salts, the thermodynamic properties of the potential corrosion products explain precisely the occurrence of the species found.

I.4 PHYSICAL CHARACTERISTICS OF SILVER CORROSION

I.4.1 Corrosion Layer Formation Rates

The corrosion rates of silver upon atmospheric exposure are presented in Table I.2; they refer to the weight increase produced by the corrosion process. For urban, industrial, and marine conditions, these corrosion rates are comparable to those for

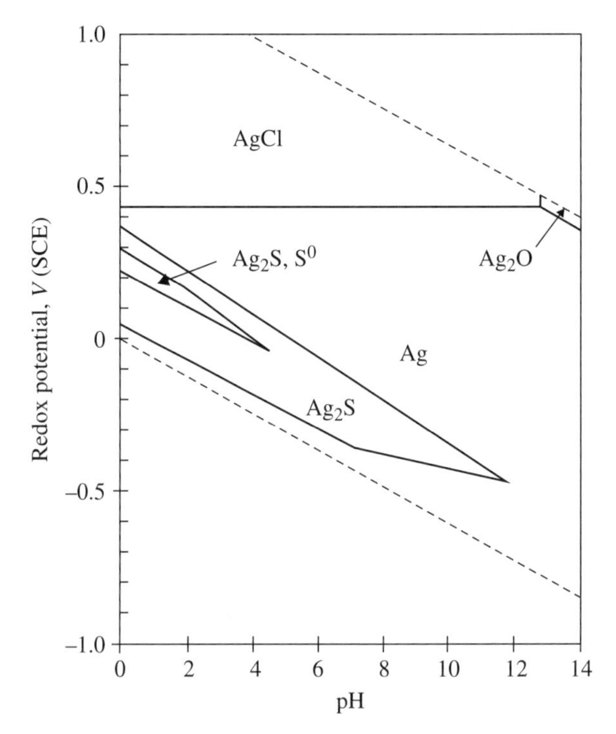

FIGURE I.4 Potential–pH equilibrium diagram for the system Ag—S—Cl—CO_2—H_2O at 25°C, for an activity of sulfur of 0.1 M, an activity of chlorine of 0.05 M, a total carbon concentration of 0.01 M, and a dissolved silver concentration of 1 μM. (These conditions are a reasonable representation of silver-phase stabilities in "natural freshwater.") The dashed lines indicate the limiting conditions within which water is stable.

TABLE I.2 Atmospheric Corrosion Rates for Silver

Locale	Number of Sites	Median Value ($ng\,cm^{-2}\,h^{-1}$)	Range ($ng\,cm^{-2}\,h^{-1}$)
Indoor	57	310	5–730
Marine	1	97	—
Urban	11	82	67–192
Industrial	4	410	80–1270

aluminum and iron, less than those for zinc, and much less than those for carbon steel. Lower corrosion rates occur indoors if relative humidity control and air filtration are present. Although the data are not extensive, they indicate that corrosion is frequently more rapid indoors than out. The exception is aggressive industrial environments. More rapid corrosion indoors is contrary to the normal behavior of all other metals for which rates have been determined (see Chapter 7).

I.4.2 The Morphology of Atmospheric Corrosion Layers on Silver

Silver does not tend to form a uniform film when it corrodes. Laboratory studies demonstrate the presence of "spikes" and "dendrites" in the corrosion layers. Silver sulfide films are sometimes adherent, sometimes not. Nonuniform growth also occurs outside the laboratory, with "clumps," "dendrites," "nodules," and "whiskers" being reported. Unlike the laboratory-grown films, corrosion films formed upon exposure to uncontrolled indoor environments often adhere tenaciously to the surface.

The behavior of silver is thus at variance with the natural inclination to think of corrosion films as uniform, an inclination encouraged by reports of average corrosion rates. In actuality, comprehensive analyses of silver corrosion must take account of the nucleating characteristics of surface sites that serve as corrosion initiation centers, perhaps where surface water collects in pools and drops.

I.4.3 Contact Resistance of Corroded Silver

Several researchers have measured contact resistance as a function of exposure time for silver exposed in field locations. Prior to exposure, typical contact resistances are satisfactorily low. At most sites, several years of indoor exposure results in a significant increase in contact resistance, and in many locations it increases over the course of a few years by five or six orders of magnitude, far too much for satisfactory electrical performance. Corrosion of electric contacts is further described in Chapter 11.

I.5 CHEMICAL TRANSFORMATION SEQUENCES

The discussions and diagrams in this appendix can be briefly summarized in Figure I.5, which shows possible chemical transformations in the silver corrosion process. Efficient reaction chains leading through the Ag^{2+} ion or free radical species are unlikely. Organic processes are similarly unimportant. In contrast, if H_2O_2 is present in the gas phase, it will be abundant in surface layers, where it will liberate silver ions from the bulk surface. H_2O_2, which may be generated outdoors by smog reactions and brought indoors by air handling systems or may be generated indoors by volatilization from industrial cleaning solutions, thus has the potential to be an important and unappreciated participant in silver corrosion.

Figure I.5 shows that the formation of the AgCl complex and its subsequent precipitation provide a straightforward route to chlorargyrite. In contrast, a compound such as acanthite, which contains two silver atoms, is more difficult to form in solution. If compounds do not form in the aqueous phase, the alternative is that they form by sorption of the appropriate anion onto the bulk silver surface followed by solid state formation of the corrosion product. The distinction between liquid- and solid-phase formation of AgCl and Ag_2S can perhaps be made on the basis of the observed characteristics of the corrosion layers; these tend to be strongly adherent, especially in the case of sulfide. Since solid particles that precipitate are fine-grained, amorphous materials that would be only loosely adherent, it seems likely that silver

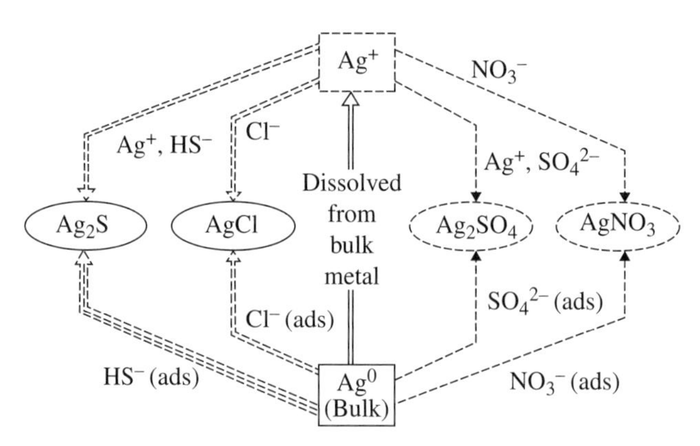

FIGURE I.5 A composite reaction diagram for silver corrosion reactions in aqueous surface films. Solid lines, rectangles, and ovals indicate processes or species confirmed by measurement, while dashes indicate merely the potential for reaction or detection. The width of the transformation lines reflects the estimated relative importance of the processes.

sulfide corrosion proceeds most commonly by an initial step involving sorption of a sulfur-containing moiety to the surface. In the case of the chloride, which is found much more rarely as a surface corrosion product and for which a much more straightforward solution formation pathway exists, the odds favor aqueous-phase reaction followed by precipitation.

A complication in Figure I.5 concerns the ability of oxidizers to increase the rate of formation of Ag_2S. If oxidizers are important, what is being oxidized? By the process of elimination, the oxidizer must in some way promote the dissolution of the bulk silver. Another complication that is inadequately understood is that of silver's demonstrated susceptibility to photocorrosion. While it is unclear exactly how photons enter into the corrosion process, it seems likely that they increase the reactivity of the surface toward adsorbed corrosive species.

Since the principal corrosion products have been identified, it is of interest to discuss the sources of the chemical species responsible for the corrosion. In the case of acanthite, Ag_2S, the associated atmospheric species are clearly H_2S and COS. Outdoor sources for both these species include pulp and paper mills, oil refineries, sewage treatment facilities, decaying vegetation, and volcanoes. Since outdoor air is brought indoors by building air handling systems or by infiltration, reduced sulfur from these sources is readily present in the indoor environment. However, the predominant sources of indoor-reduced sulfur gases are probably themselves indoors, such as the outgassing of H_2S from paper bags or cooking operations.

As discussed early in this appendix, both gaseous and particulate chlorines are deposited onto indoor surfaces. Outdoors, chlorine is generated by coal combustion, refuse incineration, volatilization from sea salt, airborne injection of sea-salt aerosol by wave action, and volcanoes. Indoor sources of chlorine include the use of industrial cleaning compounds and the slow degradation of PVC insulation. The particle chloride seems likely to be more important to silver corrosion than gaseous chlorine in many gases, but specific local conditions will determine which source dominates.

To summarize, when exposed to the ambient environment, silver forms corrosion films consisting largely of Ag_2S, with some AgCl in high chloride environments. Alloys of silver behave as does the metal itself unless the alloying element is more reactive than silver. Unlike many other metals, the corrosion layers on silver do not include carbonates, sulfates, or nitrates. The thermodynamic properties of the several compounds, together with their aqueous solubilities, readily explain the presence or absence of these products. Although the typical concentrations of corrosive species present in most environments are low, they are sufficient to initiate the formation of corrosion films. As a consequence, protection and/or regular maintenance of silver surfaces is necessary if surface degradation is to be minimized.

Details concerning the information in this appendix may be found in T.E. Graedel, Corrosion mechanisms for silver exposed to the atmosphere, *Journal of the Electrochemical Society, 139*, 1963–1970, 1992.

FURTHER READING

Z.Y. Chen, D. Liang, G. Ma, G.S. Frankel, H.C. Allen, and R.G. Kelly, Influence of UV irradiation and ozone on atmospheric corrosion of bare silver, *Corrosion Engineering, Science and Technology, 45*(2), 169–180, 2010.

H. Kim, Corrosion process of silver in environments containing 0.1 ppm H_2S and 1.2 ppm NO_2, *Materials and Corrosion, 54*, 243–250, 2003.

C. Kleber, R. Wiesinger, J. Schnöller, U. Hilfrich, H. Hutter, and M. Schreiner, Initial oxidation of silver surfaces by S^{2-} and S^{4+} species, *Corrosion Science, 50*(4), 1112–1121, 2008.

M. Watanabe, A. Hokazono, T. Handa, T. Ichino, and N. Kuwaki, Corrosion of copper and silver plates by volcanic gases, *Corrosion Science, 48*(11), 3759–3766, 2006.

APPENDIX J

THE ATMOSPHERIC CORROSION CHEMISTRY OF ZINC

J.1 INTRODUCTION

Zinc is one of the metals regularly exposed to the indoor and outdoor atmosphere. A large proportion of its use is in applications that take advantage of its favorable corrosion properties and its relatively inexpensive price, such as in zinc coating of carbon steel (galvanizing), which results in a product with much better corrosion resistance than the underlying carbon steel itself. Among the uses of galvanized steel is the fabrication of roofing, fencing, vehicle body panels, equipment frames, and housings. In electronics, zinc sees extensive use as an alloying element, particularly for copper, with subsequent utilization in contact, connector, and piece part applications.

J.2 CORROSION LAYER FORMATION RATES

The corrosion of zinc has historically been examined chiefly as a component of investigations of the corrosion of brass or galvanized steel. More recent studies have treated the pure metal itself. All of these studies agree that upon initial exposure zinc rapidly forms a thin film of zinc hydroxide. Under continued exposure, this film is transformed into various other atmospheric products.

The rate of zinc corrosion is generally greater than that for copper and less than that for many steels, at least under outdoor atmospheric exposure. The following ranges of corrosion rates are seen in different environments:

Rural	0.2–3 μm year^{-1}
Urban	2–16 μm year^{-1}
Industrial	2–16 μm year^{-1}
Marine	0.5–8 μm year^{-1}

Atmospheric Corrosion, Second Edition. Christofer Leygraf, Inger Odnevall Wallinder, Johan Tidblad and Thomas Graedel.

These rates are rapid enough that the original zinc surface is covered with corrosion products in hours or days.

Many metals form protective surface films upon exposure to the atmosphere, but zinc does not. Detailed field studies show that the rates of zinc corrosion do not decrease with time but rather appear to be responsive to changes in the local concentrations of corrosive species.

Indoors, zinc corrodes much more slowly than occurs in the outdoor environment, often beginning the process at points where dust particles have settled on the surface. Analyses demonstrate that indoor zinc surfaces have upon them adherent particles containing large concentrations of chloride and sulfate ions. The indoor zinc corrosion rate is affected by relative humidity, but corrosion only begins for $RH > 70\%$ or so. At of above that point, the deposited particles of the zinc corrosion products appear to absorb sufficient moisture to stimulate chemical degradation processes.

J.3 THE MORPHOLOGY OF ATMOSPHERIC CORROSION LAYERS ON ZINC

Under conditions of slow corrosion, the corrosion film formed on zinc exposed to the atmosphere is uniform, somewhat nodular, and fine grained. Where more rapid corrosion occurs, massive dissolution and precipitation features are superimposed on the underlying nodular material. It was proposed that these features are evidence for a process of sequential dissolution, concentration, redistribution, and precipitation that builds large structures on the metal surface and that these processes probably occur when the surface is drying.

The composition of the surface film produced by atmospheric corrosion is not constant with depth. It varies with environment and also with exposure time. The outer part is often enriched in sulfur (especially in rural, urban, and some industrial environments) or in chlorine (especially in marine environments) or in both. The inner part may be enriched in carbon, consistent with the formation of some zinc carbonate.

J.4 CHEMICAL MECHANISMS OF ZINC CORROSION

J.4.1 Oxides and Hydroxides

Table J.1 indicates in its rightmost column those compounds of zinc that have been detected in corrosion layers resulting from exposure to natural environments. Zincite, zinc(II)oxide, ZnO, is the initial surface component formed. In the presence of water, the oxide is promptly transformed to zinc hydroxide, probably in several different crystal structures. For exposures of a few months, zinc hydroxides are the principal constituents of the corrosion film produced by atmospheric exposure. Although several different forms of zinc hydroxide have been described in the scientific literature, the only one known to occur naturally as a mineral is wülfingite, ε-$Zn(OH)_2$.

TABLE J.1 Minerals and Other Crystalline Substances Found on Corroded Zinc[a]

Substance	Hey Index No.	Crystal System	Formula
Metal, oxides, and hydroxides			
Zincite	7.5.1	Hexagonal	ZnO
Wülfingite	7.5.4	Orthorhombic	$\varepsilon\text{-}Zn(OH)_2$
Sulfides			
Wurtzite	3.4.5	Hexagonal	$\beta\text{-}ZnS$
Sulfites			
Zinc sulfite (h)		Monoclinic	$ZnSO_3{\cdot}2H_2O$
Sulfates			
Zinkosite	25.5.1	Orthorhombic	$ZnSO_4$
Gunningite	25.5.11	Monoclinic	$ZnSO_4{\cdot}H_2O$
Boyleite	25.5.6	Monoclinic	$ZnSO_4{\cdot}4H_2O$
Bianchite	25.5.13	Monoclinic	$ZnSO_4{\cdot}6H_2O$
Goslarite	25.5.2	Orthorhombic	$ZnSO_4{\cdot}7H_2O$
Zinc hydroxysulfate			$Zn_4SO_4(OH)_6{\cdot}H_2O$
Zinc hydroxysulfate		Triclinic	$Zn_4SO_4(OH)_6{\cdot}4H_2O$
Gordaite		Hexagonal	$NaZn_4Cl(OH)_6SO_4{\cdot}6H_2O$
Chlorides			
Simonkolleite	8.5.1	Hexagonal	$Zn_5Cl_2(OH)_8{\cdot}H_2O$
Zinc oxychloride			$Zn_5Cl_2O_4{\cdot}H_2O$
Zinc chlorosulfate		Monoclinic	$Zn_4Cl_2(OH)_4SO_4{\cdot}5H_2O$
Carbonates			
Smithsonite	11.6.1	Hexagonal	$ZnCO_3$
Zinc carbonate			$ZnCO_4{\cdot}4H_2O$
Zinc hydroxycarbonate			$Zn_4CO_3(OH)_6$
Hydrozincite	11.6.2	Monoclinic	$Zn_5(CO_3)_2(OH)_6$
Zinc carbonate oxychloride			$Zn_a(CO_3)_b(OH)_cOCl$
Nitrates			
Zinc nitrate			$Zn(NO_3)_2$

[a] See Table C1 footnotes for explanatory details.

In all its contacts with the environment, zinc is exposed to carbon dioxide as well as water, and the potential–pH diagram for this three-component system is shown in Figure J.1. At the slightly positive potential, which is anticipated in active precipitation, the stable form of zinc in mildly or strongly acidic solution is the divalent ion. In the near-neutral or basic solutions anticipated from dew or adsorbed water vapor, however, zinc hydroxide is expected to predominate. Between pH6 and 7, a region of stability exists for zinc carbonate.

In addition to molecular oxygen, an important reactant to consider is hydrogen peroxide, H_2O_2, which is always present in precipitation, is a strong oxidizer and is a species identified as a participant in the atmospheric corrosion mechanisms of iron. In support of the concept of exploring the chemistry of strong oxidizers on the zinc

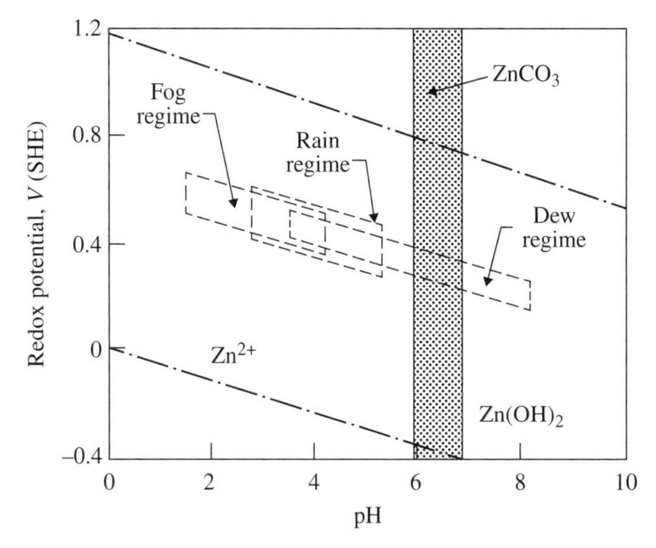

FIGURE J.1 Potential–pH diagram for the system Zn—CO_2—H_2O at 25°C, for a concentration of zinc ionic species of 0.1 M, and a concentration of H_2CO_3 of 1×10^{-5} M (this is the concentration in equilibrium with the current atmospheric CO_2 level of about 340 ppmv). $ZnCO_3$ is stable within the stippled region. At more acid pHs, Zn^{2+} is the stable form of zinc; at more basic pHs, $Zn(OH)_2$ is the stable form. The dash-dotted lines indicate the limiting conditions within which water is stable. The approximate regimes for fog, rain, and dew are indicated.

surface, some laboratory and field data suggest that the corrosion of zinc in at least some environments involves a reactive oxygen intermediate such as one might expect from H_2O_2.

J.4.2 Carbonates and Organic Compounds

Carbonates are second only to hydroxides in abundance in most natural corrosion layers on zinc. Although several compounds have been detected, the most abundant are smithsonite, $ZnCO_3$, and the hydroxycarbonate mixed salt hydrozincite, $Zn_5(CO_3)_2(OH)_6$. This latter mineral is found on corroded zinc exposed not only to the atmosphere but also to freshwater and seawater.

As shown in Figure J.1, zinc dihydroxide, $Zn(OH)_2$, the common initial surface constituent, is readily dissolved in acidic solutions, even the weakly acidic ones characteristic of dew and some rains. As a consequence, it appears likely that the corrosion products that form on zinc do so largely in the surface aqueous layer and not in the solid phase. Nucleation of the low solubility hydroxide mixed salts is probably initiated by seed crystals at the solid–liquid interface. Thus, one could write the formation reaction of hydrozincite as a combination of the simple salts:

$$3Zn(OH)_2 + 2ZnCO_3 \rightarrow Zn_5(CO_3)_2(OH)_6 \downarrow \tag{J.1}$$

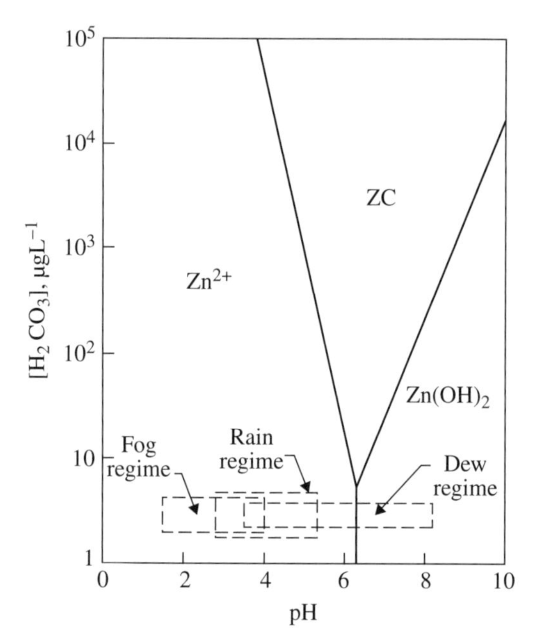

FIGURE J.2 Stability domains for zinc carbonate (ZC) in aerated aqueous solutions with varying H_2CO_3 content and pH values at 25°C and with concentrations of zinc ionic species of 0.1 M.

but it is almost certainly more realistic to assume that one of the ubiquitous zinc dihydroxide molecules function as a seed crystal and sequentially adds the needed ions from the solution with which it is in contact:

$$Zn(OH)_2(s) + 4Zn^{2+} + 4OH^- + 2CO_3^{2-} \rightarrow Zn_5(CO_3)_2(OH)_6 \downarrow \qquad (J.2)$$

The importance of carbonate salts suggests that it may be useful to express the information in Figure J.1 as a function of carbonic acid concentrations rather than the redox potential: such a diagram is presented in Figure J.2. Diagrams of this sort have substantial utility for defining and constraining corrosion chemistry provided two limitations are recognized. The first is that the diagram represents chemical equilibrium conditions, which may not be met in the exposure of materials to the atmosphere. The second is that the diagram applies in a strict sense only to pure systems, that is, those without any additional constituents. In reality, a variety of additional ions and molecules is always present, often at fairly high concentrations. The potential clearly exists for metastable compounds, complex species, and synergistic effects. As a result, the stability domain diagrams should be regarded as only crude guides to the chemical products expected as a result of atmospheric exposure.

As seen in Figure J.2, the stability domain for $ZnCO_3$ varies with the H_2CO_3 content of the solution and thus with the atmospheric or aqueous concentration of dissolved or dissolvable carbon dioxide (CO_2) as well as with the acidity of the aqueous surface layer. Figure J.2 also indicates the regimes appropriate to rain or fog in typical urban atmospheric situations as well as that for dew. It is evident that zinc carbonate salts might not be expected on zinc exposed to rain or fog and allowed to reach thermodynamic equilibrium but are reasonable products if formed in the presence of dew at slightly elevated total carbonate concentrations. Since hydrozincite does not appear in the thermodynamic equilibrium diagram of Figure J.2, it is probably merely a metastable state in the presence of even mildly acidic precipitation.

J.4.3 Sulfides and Sulfates

The presence of sulfur in zinc corrosion layers is a usual occurrence, particularly for extended exposures. Different compounds have been reported by different investigators; it is not clear whether all can exist of whether the same compound has been identified differently by different experimentalists. In the case of sulfur, SO_2 has been shown in laboratory experiments to enhance zinc corrosion and is known to be incorporated from the atmosphere into zinc surface films. The possible process diagram is given in Figure J.3. It shows that any sulfur gas dissolving into the surface film will eventually be oxidized to bisulfite ion, HSO_3^-. The bisulfite ion is readily converted to sulfate ion by O_3, H_2O_2, or transition metal ions (largely iron or manganese). (In the absence of a catalyst, bisulfite oxidation by dissolved O_2 is unimportant). Once sulfate is formed by solution chemistry or supplied by particles or precipitation, it is available to form the zinc hydroxysulfates; the presence of the sulfate ion has been confirmed in the corrosion layers of both outdoor and indoor zinc surfaces.

In the simplest of the mineral formation processes involving zinc and sulfur, the simple ions are mobilized in acidic aqueous solutions, form ion pairs, and precipitate as hydrated zinc sulfates upon the evaporation of the surface water layer:

$$Zn^{2+} + SO_4^{2-} + xH_2O \rightarrow ZnSO_4 \cdot xH_2O \downarrow \quad (x = 0,1,4,6,\text{ or } 7) \tag{J.3}$$

This is a process which is reasonably facile, since all three of these species are abundant on zinc exposed to the outdoor atmosphere. Also among the common corrosion products, however, are two hydroxysulfates. Nucleation of the less soluble hydroxide mixed salts is probably initiated, as with hydroxycarbonates, by seed crystals at the metal surface. Thus, one could write the formation reactions of zinc hydroxysulfates as combinations of the simple salts:

$$3Zn(OH)_2 + ZnSO_4 + xH_2O \rightarrow Zn_4SO_4(OH)_6 \cdot xH_2O \downarrow \quad (x = 0,1,4,5,\text{ or } 5) \tag{J.4}$$

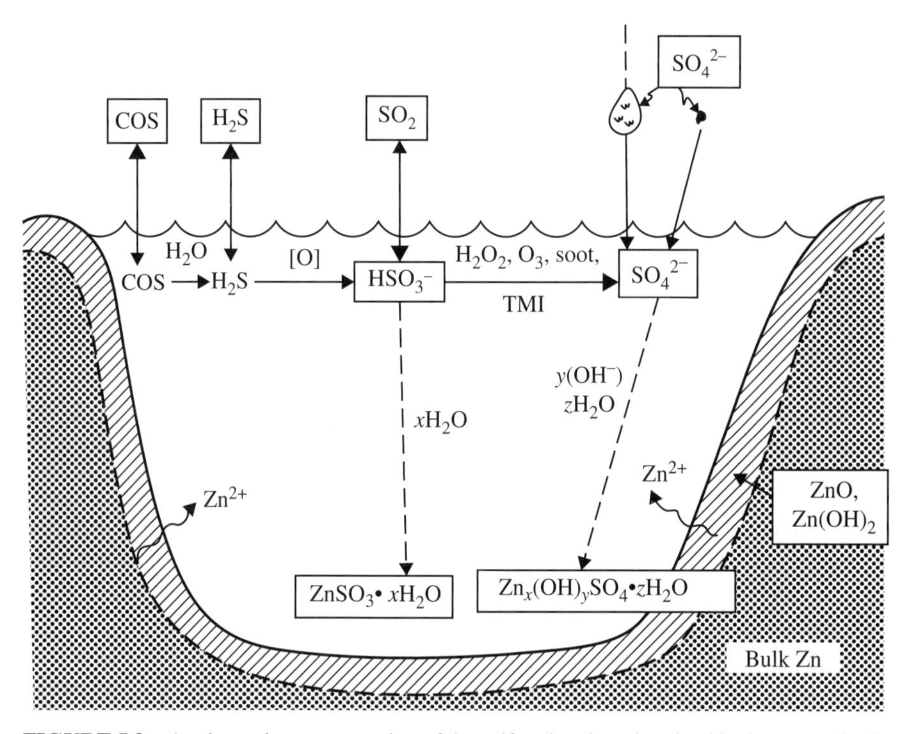

FIGURE J.3 A schematic representation of the sulfur chemistry involved in the atmospheric corrosion of zinc. The atmosphere is at the top of this diagram, a surface water layer is in the center, and the corroding metal is at the bottom. For simplicity, the diagram pictures a depression in the metal surface, although deposited hygroscopic particles, heavy dew, or rain could produce similar conditions anywhere on the surface. The wavy line indicates zinc ions diffusing into the aqueous solution from the bulk solid. A species whose presence in the indicated phase has been confirmed by field measurement are enclosed in a rectangular box; the presence of species not so enclosed is unconfirmed but is thought to be reasonable. Solvation or reaction processes that have been confirmed by laboratory studies are shown as solid arrows; those for which the mechanism is uncertain are shown as dotted arrows. Rain is indicated by the atmospheric water drop and airborne particles by the irregular small atmospheric object.

but it is almost certainly more realistic to assume that one of the ubiquitous zinc dihydroxide molecules functions as a seed crystal and sequentially adds the needed ions from the solution with which it is in contact, for example,

$$Zn(OH)_2(s) + 3Zn^{2+} + 4OH^- + SO_4^{2-} \rightarrow Zn_4SO_4(OH)_6 \cdot 4H_2O \downarrow \quad (J.5)$$

The formation of the zinc hydroxysulfates in acid solution (i.e., in rain or fog) is rather uncertain since one, at least, is not thermodynamically stable (Figure J.4). The near-neutral conditions present in dew might be suitable for their formation, however. Zinc hydroxysulfates are frequently observed in highly polluted urban and industrial environments.

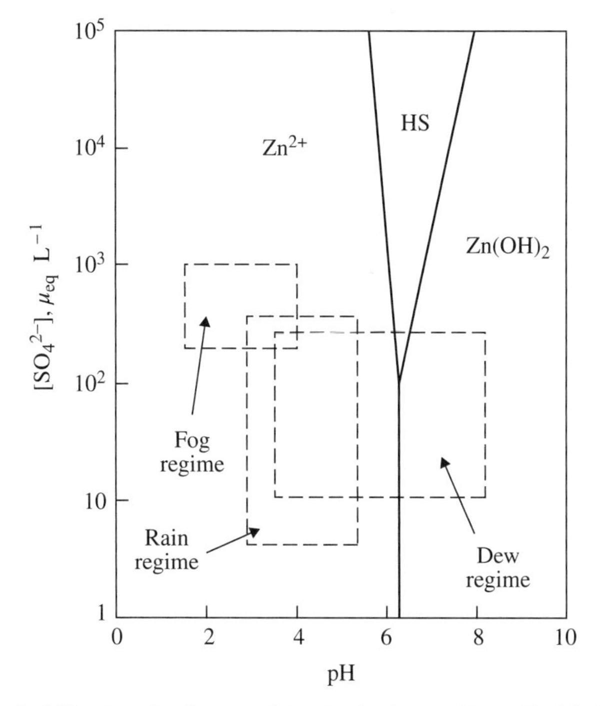

FIGURE J.4 Stability domains for one of the zinc hydroxysulfates, $Zn_4SO_4(OH)_6$, abbreviated HS, in aerated aqueous solutions with varying sulfate content and pH values at 25°C and with concentrations of zinc ionic species of 0.1 M.

J.4.4 Chlorides

Compounds containing chlorine are identified on zinc under some, but not all, conditions. Extractable chloride ions are abundant on indoor zinc, less so on outdoor zinc. The mixed salt $Zn_5(OH)_8Cl_2 \cdot H_2O$ is found in samples exposed to industrial atmospheres and marine atmospheres. When one water of hydration is added to that mixed salt, the resulting compound is simonkolleite, a natural mineral formed by weathering processes. A zinc oxychloride, $Zn_5Cl_2O_4 \cdot H_2O$, is also occasionally found in samples exposed to marine atmospheres. In none of these cases are the chlorine compounds very abundant, a result which presumably reflects the tendency of the simple zinc chloride to be washed away rather than to be available for incorporation into more complex minerals.

At high chloride concentrations and near-neutral acidities, a stability regime exists for zinc hydroxychloride, as shown by the thermodynamic diagram for the chloride–zinc–water system, Figure J.5. As seen in the figure, the normal atmospheric characteristics of rain and fog are well outside the stability region. A different circumstance holds for dew, since it is considerably more basic than rain and may have high chloride ion concentration as it evaporates; here the formation of simonkolleite or other zinc hydroxychloride minerals is thermodynamically possible. In support of

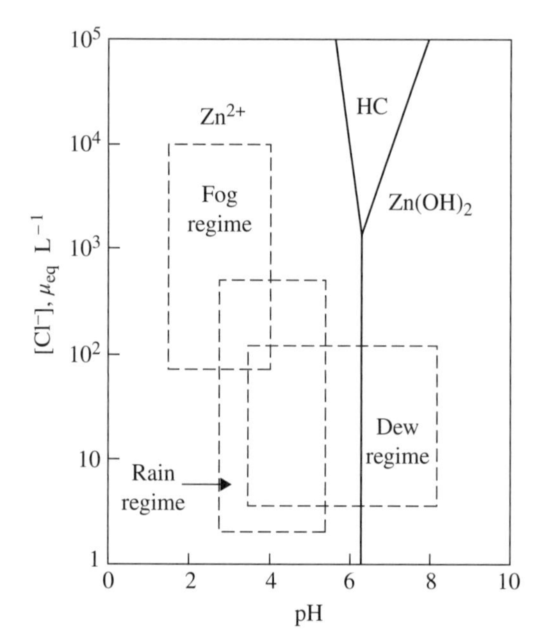

FIGURE J.5 Stability domains for zinc hydroxychloride (HC) in aerated aqueous solutions with varying chloride content and pH values at 25°C and with concentrations of zinc ionic species of 0.1 M.

this idea, it is worth noting that hydroxychlorides are seen in field-exposed zinc samples, Table J.1, as well as in zinc corrosion layers produced by laboratory exposures to hydrochloric acid (the latter being a ubiquitous atmospheric constituent). The formation processes would be expected to follow the same general paths as for hydroxycarbonate and hydroxysulfate mixed salts. Thus, one could write the formation reaction of simonkolleite as a combination of the simple salts:

$$4Zn(OH)_2 + ZnCl_2 \rightarrow Zn_5(OH)_8Cl_2 \cdot H_2O \downarrow \tag{J.6}$$

But it is again likely that one of the ubiquitous zinc dihydroxide molecules functions as a seed crystal and sequentially adds the needed ions from the solution with which it is in contact:

$$Zn(OH)_2(s) + 4Zn^{2+} + 6OH^- + 2Cl^- \rightarrow Zn_5(OH)_8Cl_2 \cdot H_2O \downarrow \tag{J.7}$$

In Figure J.6 a schematic diagram for chloride deposition and reactions is shown. Chloride enters the aqueous surface layer through the incorporation of gaseous HCl or through the deposition of sea salt particles or precipitation. Alternatively, chlorine ions may be supplied by the degradation of atmospheric methyl chloride or freons on the oxidized zinc surface. Once present, sequential combination reactions lead to the formation of zinc hydroxychlorides. The latter are soluble in weak acid solutions, so would be expected to be present only if they can be isolated from the often acidic

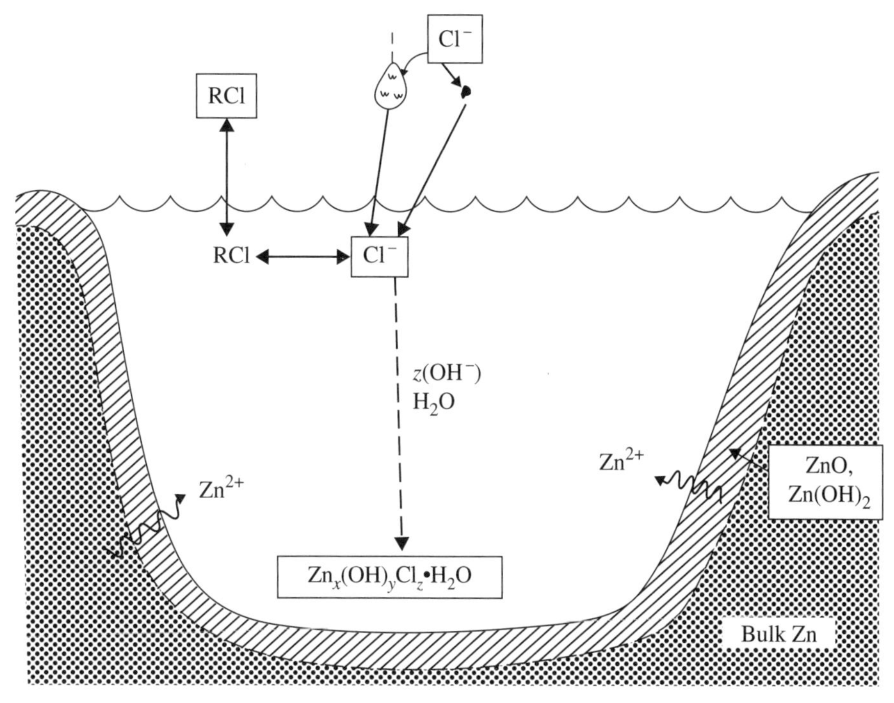

FIGURE J.6 A schematic representation of the processes involved in the formation of components containing chlorine during the atmospheric corrosion of zinc (see Fig. J.3 legend for explanatory details).

aqueous surface layer. This solubility property seems a reasonable explanation for the tendency of chlorine to be found deep within the corrosion layer, if at all. An alternative explanation is that of Biestek and coworkers who suggest that hydroxychlorides are formed rapidly upon exposure but are metastable, gradually being transformed into the less soluble hydroxysulfates.

J.5 TRANSFORMATION PROCESSES

J.5.1 Chemical Processes

The general picture of the interaction of zinc with the atmosphere is quite consistent in an overall sense with the picture anticipated from the equilibrium information. The first step is the formation of oxides and hydroxides, primarily $Zn(OH)_2$. Equilibrium with atmospheric CO_2 then leads to the formation of zinc carbonates or zinc hydroxycarbonates in neutral or near-neutral surface solutions. Zinc is lost from the corrosion layer by one of two mechanisms, the most simple of which is the dissolution of the zinc salts in rain of pH 5–6 (the natural level in uncontaminated atmospheres) followed by washing from the surface. An enhanced process occurs in urban or industrial areas where atmospheric concentrations of nitrogen and sulfur compounds are high.

TABLE J.2 Rates of Accumulation of Ions from the Atmosphere on Indoor Zinc Surfaces

Ion	Accumulation Rate ($\mu g\ cm^{-2}\ year^{-1}$)	Reactive Constituent Ratio Rate (Ions Per Surface $Zn(OH)_2$ Molecule Per Year (Central Value))
SO_4^{2-}	0.1–0.8	0.2
NO_3^-	0.02–0.23	0.2
Cl^-	0.06–0.6	0.4

In those regions, prospects for the formation of sulfuric and nitric acids are greater, particularly if oxidizing species are generated by smog chemistry. At the high acidities typical of urban fog, clouds, or rain, the zinc salts dissolve readily as do any potential products such as $ZnSO_4$, $Zn(NO_3)_2$, or $ZnCl_2$. Only if the relatively insoluble hydroxide mixed salts are formed are signatures of the corrosion chemistry retained by the zinc surface. The solubility of even those species increases as the covering solutions become more highly acid, however, and the high rates of zinc corrosion in urban areas are probably due to the relatively high acidity now common in precipitation rather than to the chemical properties of some specific anion.

J.5.2 Anion Balances

A significant amount of information is available concerning the rates of deposition of ions to zinc surfaces inside electronic equipment buildings. The ranges of measured rates are given in Table J.2 for three anions: sulfate, chloride, and nitrate. The last column of Table J.2 presents ion deposition densities. As can be seen, there is approximately one-half anion supplied per square Angstrom of surface per year. It follows that the anion supply is ample for reactions to occur and that the indoor zinc corrosion rate will not be constrained by a shortfall of reactants. Outdoors, where the greater concentrations of reactive gases and aerosol particles and the presence of precipitation provide sources of ions of much higher magnitude than those indoors, anion supply will never be a limitation to the rate of zinc corrosion. Rather, the rate will be controlled by the pH of the surface water and by the typically slow crystalline rate of the minerals constituting the corrosion layer and could be substantially affected by the washing action of precipitation.

J.6 SUMMARY

A full quantitative analysis of the physical and chemical processes involved in the atmospheric corrosion of zinc would require much information not yet available: the solubilities in solutions of different acidities of the several hydroxy mixed salts of zinc, the kinetic parameters appropriate to their formation, more information on the temporal evolution of corrosion layer components, and so on. Lacking this information, it has nonetheless proved possible to produce a rather extensive qualitative picture of the process. The involvement of chemical reactions in the production of many species is clearly required; among the species playing important roles are carbonate, sulfate, chloride, and nitrate; formate, acetate, and perhaps several cations should be considered as well. It is of interest that zinc is the only one of the common engineering metals for which a carbonate plays an important role in atmospheric corrosion. This is probably a function of the aqueous solubilities of the hydroxy mixed salts. The supply rates of the major anions to both indoor and outdoor zinc surfaces are ample to sustain the observed corrosion rates, and zinc corrosion thus appears constrained by chemical and not physical bottlenecks.

Details concerning the information in this appendix may be found in T.E. Graedel, Corrosion mechanisms for zinc exposed to the atmosphere, *Journal of the Electrochemical Society, 136*, 193C–203C, 1989. A more comprehensive description of evolution of different corrosion products formed on zinc as a result of atmospheric exposures is given in Chapters 9 and 13 together with suggestions of further reading.

APPENDIX K

INDEX OF MINERALS RELATED TO ATMOSPHERIC CORROSION

Substance	Hey Index Number	Chemical Formula
Acanthite	3.2.1	Ag_2S
Akaganeite	7.20.6	β-FeOOH
Akdalaite	7.6.8	$Al_2O_3{\cdot}\frac{1}{4}H_2O$
Aluminite	25.6.5	$Al_2SO_4(OH)_4{\cdot}7H_2O$
Anglesite	25.7.1	$PbSO_4$
Anhydrite	25.4.1	$CaSO_4$
Antlerite	25.2.7	$Cu_3SO_4(OH)_4$
Argentite	3.2.1	Ag_2S
Atacamite	8.2.4	$Cu_2Cl(OH)_3$
Azurite	11.2.2	$Cu_3(CO_3)_2(OH)_2$
Bassanite	25.4.2	$CaSO_4{\cdot}\frac{1}{2}H_2O$
Bayerite	7.6.5	$Al(OH)_3$
Bianchite	25.5.13	$ZnSO_4{\cdot}6H_2O$
Boehmite	7.6.3	γ-AlOOH
Botallackite	8.2.6	$Cu_2Cl(OH)_3$
Brochantite	25.2.10	$Cu_4SO_4(OH)_6$
Bunsenite	7.22.1	NiO
Cadwaladerite	8.6.17	$AlCl(OH)_2{\cdot}4H_2O$
Calcite	11.4.1	$CaCO_3$
Cerussite	11.9.1	$PbCO_3$
Chalcanthite		$CuSO_4{\cdot}5H_2O$
Chalcocite	3.1.1	Cu_2S
Chalcocyanite	25.2.1	$CuSO_4$
Chlorargyrite	8.3.1	AgCl
Cotunnite	8.8.2	$PbCl_2$
Cuprite	7.3.1	Cu_2O

Atmospheric Corrosion, Second Edition. Christofer Leygraf, Inger Odnevall Wallinder, Johan Tidblad and Thomas Graedel.

Substance	Hey Index Number	Chemical Formula
Dawsonite	11.7.3	$NaAlCO_3(OH)_2$
Dwornickite		$NiSO_4 \cdot H_2O$
Feroxyhyte	7.20.7	δ-FeOOH
Galena	3.6.3	PbS
Gaspeite	11.14.9	$NiCO_3$
Gerhardite	13.4	$Cu_2(NO_3)(OH)_3$
Gibbsite	7.6.4	$Al(OH)_3$
Goethite	7.20.5	α-FeOOH
Gordaite		$NaZn_4Cl(OH)_6SO_4 \cdot 6H_2O$
Goslarite	25.5.2	$ZnSO_4 \cdot 7H_2O$
Gunningite	25.5.1a	$ZnSO_4 \cdot H_2O$
Gypsum	25.4.3	$CaSO_4 \cdot 2H_2O$
Hannebachite	27.1.1	$CaSO_3 \cdot H_2O$
Heazlewoodite	3.11.2a	Ni_3S_2
Hellyerite	11.14.7a	$NiCO_3 \cdot 6H_2O$
Hematite	7.20.4	α-Fe_2O_3
Hydrocerrusite	11.9.2	$Pb_3(CO_3)_2(OH)_2$
Hydrozincite	11.6.2	$Zn_5(CO_3)_2(OH)_6$
Jurbanite		$Al(SO_4)(OH) \cdot 5H_2O$
Langite	25.2.9	$Cu_4SO_4(OH)_6 \cdot 2H_2O$
Laurionite	8.8.3	PbClOH
Lawrencite	8.11.10	$FeCl_2$
Lepidocrocite	7.20.8	γ-FeOOH
Lesukite		$Al_2Cl(OH)_5 \cdot 2H_2O$
Litharge	7.11.5	PbO
Maghemite	7.20.4	γ-Fe_2O_3
Magnetite	7.20.3	Fe_3O_4
Malachite	11.2.1	$Cu_2CO_3(OH)_2$
Melanterite	25.10.5	$FeSO_4 \cdot 7H_2O$
Millerite	3.11.1	γ-NiS
Morenosite	5.12.6a	$NiSO_4 \cdot 7H_2O$
Nantokite	8.2.1	CuCl
Nitrocalcite	13.5	$Ca(NO_3)_2 \cdot 4H_2O$
Nullaginite		$Ni_2CO_3(OH)_2$
Otwayite		$Ni_2CO_3(OH)_2 \cdot H_2O$
Paratacamite	8.2.5	$Cu_2Cl(OH)_3$
Posnjakite	25.2.11a	$Cu_4SO_4(OH)_6 \cdot H_2O$
Retgersite	5.12.6a	α-$NiSO_4 \cdot 6H_2O$
Rozenite	25.10.2	$FeSO_4 \cdot 4H_2O$
Scotlandite		$PbSO_3$
Siderite	11.13.1	$FeCO_3$
Simonkolleite	8.5.1	$Zn_5Cl_2(OH)_8 \cdot H_2O$
Smithsonite	11.6.1	$ZnCO_3$
Spertiniite	7.3.4	$Cu(OH)_2$
Strandbergite		$Cu_{2.5}SO_4(OH)_3 \cdot 2H_2O$
Szomolnokite	25.10.1	$FeSO_4 \cdot H_2O$
Tenorite	7.3.2	CuO

Substance	Hey Index Number	Chemical Formula
Theophrastite	7.22.2a	$Ni(OH)_2$
Tucanite		$Al(OH)_3 \cdot ½H_2O$
Weddellite	31.1.6	$Ca(C_2O_4) \cdot 2H_2O$
Whewellite	31.1.5	$Ca(C_2O_4) \cdot H_2O$
Wülfingite	Orthorhombic	ε-Zn $(OH)_2$
Wurtzite	3.4.5	β-ZnS
Zaratite	11.14.7	$Ni_3(CO_3)(OH)_4 \cdot 4H_2O$
Zincite	7.5.1	ZnO
Zincosite	25.5.1	$ZnSO_4$

GLOSSARY

Basic corrosion terms in the book not listed below can be found in the Standard EN ISO 8044:2015 Corrosion of metals and alloys-Basic terms and definitions.

Absorption	the incorporation of a gas or dissolved material into a liquid or solid
Acid deposition	the deposition of acidic constituents to a surface; this occurs not only by precipitation but also by the deposition of atmospheric particulate matter and the incorporation of soluble gases
Adsorption	the adherence of a gas or dissolved material to the surface of a solid
Amorphous phase	A solid phase without crystalline structure
Anodic reaction	electrode reaction equivalent to a transfer of positive charge from the metal to the electrolyte
Atmospheric corrosion	corrosion in which Earth's atmosphere at ambient temperature is the corrosive environment
Bimetallic cell	corrosion cell, where the two electrodes are formed by dissimilar metals
Bioaccessibility	the fraction of a substance that is released/dissolved into an aqueous setting and potentially may become available for an organism
Bioavailability	the extent to which a substance is taken up by an organism and is available for interaction
Biosphere	that spherical shell encompassing all forms of life on Earth; the biosphere extends from the ocean depths to a few thousand meters of altitude in the atmosphere and includes the surface of land masses. Alternatively, the life forms within that shell

Atmospheric Corrosion, Second Edition. Christofer Leygraf, Inger Odnevall Wallinder, Johan Tidblad and Thomas Graedel.

Brass an alloy of copper and zinc and possibly with other metals in lesser amounts

Bronze an alloy with copper as main constituent and with other metals, including tin, lead, aluminum, or zinc in lesser amounts

Brown rust (see *Green rust*)

Calcareous stone a stone composed primarily of calcium carbonate; common examples are limestone and marble

Cathodic reaction electrode reaction equivalent to a transfer of negative charge from the metal to the electrolyte

Conservation the care and treatment of cultural artifacts

Corrodent substance that when in contact with a given metal will cause corrosion

Corrosion cell short-circuited galvanic cell in a corrosion system, the corroding metal forming one of its electrodes

Corrosion product substance formed as a result of corrosion

Corrosion rate the corrosion effect on a material per time unit; the corrosion rate may be expressed as an increase in corrosion depth per time unit or the mass of material transformed into corrosion products per unit time and surface area

Corrosivity ability of an environment to cause corrosion of a metal in a given corrosion system

Crystalline phase a solid phase with crystalline structure, that is, a three-dimensional atomic, ionic, or molecular structure consisting of periodically repeated, identically constituted unit cells

Deposition rate the rate at which a gas or particle in the atmosphere is transferred from the atmosphere to a surface

Deposition velocity the coefficient of proportionality in the expression relating the vertical flux of a gas or particle to a surface to the atmospheric concentration of that gas or particle

Dissolution a process involving the displacement of ions from the solid into the liquid phase

Dose–response function a mathematical relationship that expresses the rate of deterioration of a material when exposed to stresses induced by one or more corrodents

Dry deposition the transfer of trace species (gases or particles) from the atmosphere to a surface as a consequence of molecular diffusion, Brownian diffusion, or gravitational settling, in the absence of active precipitation; the term refers to the transfer process and not to the surfaces themselves, which may be moist

Ecotoxicity the damaging action of a chemical species upon an environmental system in the biosphere

Galvanized steel steel coated with a zinc layer to hinder corrosion

Green rust greenish corrosion layer formed on iron and steel; the layer is subsequently transformed into a more brownish layer (brown rust), both layers consisting of iron oxyhydroxides

Hydrophilic a substance having a strong tendency to combine with water

Hydrophobic a substance having a strong tendency to reject association with water

Mass transport control (see *Transport-controlled process*)

Microenvironment the local ambient environment to which a corroding object is exposed

Mixed control limitation of the corrosion rate by the simultaneous action of two or more controlling factors

Passive layer a thin, adherent, protective layer of corrosion products formed on a metal surface through reaction between the metal and the environment

Patina a corrosion layer, usually green or brown, on copper or copper alloys

Pewter an alloy with tin as main constituent and with other metals, including antimony, copper, or lead in lesser amounts

Photochemical smog classically, smog is a mixture of smoke plus fog; today the term has the more general meaning of any anthropogenic haze. Photochemical smog involves the production, in stagnant sunlit atmospheres, of oxidants such as by the photolysis of NO_2 and other substances, generally in combination with haze-causing particles

Rate constant the coefficient of proportionality in the expression relating the rate of a reaction to the concentration of reactants and/or products; also called rate coefficient

Recession rate the rate of loss of thickness of, for example, a stone material as a result of corrosion or some other deterioration process

Redox potential the electrode potential of an inert metal in contact with a solution; the redox potential is a measure of the oxidizing power of the solution

Restoration careful cleaning and subsequent conservation of cultural artifacts; although the term implies a return to the original appearance of the object, such a result is seldom achievable in practice

Runoff rate the rate of dissolution of a metal from its surface corrosion products into the ambient environment; may be expressed as the total mass of dissolved metal per time unit and surface area unit

Rust visible corrosion products consisting of mainly hydrated iron oxides

Scenario a detailed, carefully constructed story that describes a plausible alternative future

Selective membrane a membrane that responds selectively to specific chemical species, for example, by allowing the species in question to transfer through the membrane

Solder an alloy, usually of tin and lead, used to join metallic parts when applied in a molten state to the solid metal surfaces

Surface-controlled processes a process that is limited by reactions on the surface or interface involved in the process

Tarnishing dulling, staining, or discoloration of a metal surface, due to the formation of a thin layer of corrosion products

Technosphere that portion of Earth's surface influenced by human action (urban areas, agriculture, roads, etc.)

Time of wetness the time during which a surface, exposed to the atmospheric environment, is covered by a film of water; the water film thickness may range from very thin (1 nm) to very thick (1 mm) and originate from adsorption of water vapor or from rain, wet snow, fog, or dew

Transport-controlled processes a process whose rate is limited by the transport of species involved in the process

Washoff (runoff/release) the removal of species from a surface by the action of precipitation or dew

Weathering steel a low alloy steel that can resist many atmospheric environments due to alloying with copper, chromium, nickel, and/or silicon in amounts up to about 3 mass percent

Wet deposition the transfer of trace species (gases or particles) from the atmosphere to surfaces as a consequence of their absorption into atmospheric condensed water particles

INDEX

Atmospheric Corrosion, Second Edition. Christofer Leygraf, Inger Odnevall Wallinder, Johan Tidblad and Thomas Graedel.

THE ELECTROCHEMICAL SOCIETY SERIES

Corrosion Handbook
Edited by Herbert H. Uhlig

Modern Electroplating, Third Edition
Edited by Frederick A. Lowenheim

Modern Electroplating, Fifth Edition
Edited by Mordechay Schlesinger and Milan Paunovic

The Electron Microprobe
Edited by T. D. McKinley, K. F. J. Heinrich, and D. B. Wittry

Chemical Physics of Ionic Solutions
Edited by B. E. Conway and R. G. Barradas

High-Temperature Materials and Technology
Edited by Ivor E. Campbell and Edwin M. Sherwood

Alkaline Storage Batteries
S. Uno Falk and Alvin J. Salkind

The Primary Battery (in Two Volumes)
Volume I
Edited by George W. Heise and N. Corey Cahoon

Volume II
Edited by N. Corey Cahoon and George W. Heise

Zinc-Silver Oxide Batteries
Edited by Arthur Fleischer and J. J. Lander

Lead-Acid Batteries
Hans Bode
Translated by R. J. Brodd and Karl V. Kordesch

Thin Films-Interdiffusion and Reactions
Edited by J. M. Poate, M. N. Tu, and J. W. Mayer

Lithium Battery Technology
Edited by H. V. Venkatasetty

Quality and Reliability Methods for Primary Batteries
P. Bro and S. C. Levy

Techniques for Characterization of Electrodes and Electrochemical Processes
Edited by Ravi Varma and J. R. Selman

Electrochemical Oxygen Technology
Kim Kinoshita

Synthetic Diamond: Emerging CVD Science and Technology
Edited by Karl E. Spear and John P. Dismukes

Corrosion of Stainless Steels
A. John Sedriks

Semiconductor Wafer Bonding: Science and Technology
Q.-Y. Tong and U. Göscle

Uhlig's Corrosion Handbook, Second Edition
Edited by R. Winston Revie

Atmospheric Corrosion
Christofer Leygraf and Thomas Graedel

Electrochemical Systems, Third Edition
John Newman and Karen E. Thomas-Alyea

Fundamentals of Electrochemistry, Second Edition
V. S. Bagotsky

Fundamentals of Electrochemical Deposition, Second Edition
Milan Paunovic and Mordechay Schlesinger

Electrochemical Impedance Spectroscopy
Mark E. Orazem and Bernard Tribollet

Fuel Cells: Problems and Solutions
Vladimir S. Bagotsky

Atmospheric Corrosion, Second Edition
Christofer Leygraf, Inger Odnevall Wallinder, Johan Tidblad, and Thomas Graedel